Agricultural Fertiliser Production

Principles, Processes, and Practice

AF820904

Richard Skiba

AFTER MIDNIGHT
PUBLISHING

Skiba, Richard (Author)

Agricultural Fertiliser Production: Principles, Processes, and Practice

ISBN 978-1-7644453-6-8 (Paperback) 978-1-7644453-7-5 (eBook)

978-1-7644453-8-2 (Hardcover)

Non-fiction

Contents

Introduction 1
Chapter 1 – Introduction to Fertilisers and Global Agriculture 3
Role of Fertilisers in Modern Agriculture 5
Historical Development of Fertiliser Use 10
Overview of Macro-, Secondary-, and Micronutrients 15
Global Fertiliser Demand and Food Security 26
Sustainability Challenges in Fertiliser Production 29
Chapter 2 – Soil Science and Plant Nutrition 32
Soil Composition and Structure 34
Nutrient Cycles (Nitrogen, Phosphorus, Potassium) 41
Soil Fertility and Deficiency Symptoms 48
Soil Profile Layers (O, A, B, C, R Horizons) 58
pH, Salinity, and Nutrient Availability 61
Interaction between Soil Biology and Fertilisers 67
Chapter 3 – Classification of Fertilisers 71
Classification Overview 73
Organic vs Inorganic Fertilisers 75
Straight, Compound, and Blended Fertilisers 77
Controlled-Release and Slow-Release Fertilisers 81
Liquid vs Solid Fertilisers 82
Biofertilisers and Emerging Products 87
Chapter 4 – Raw Materials and Feedstocks 95
Natural Gas (Ammonia Production) 97
Phosphate Rock Sources and Beneficiation 101
Potash Mining and Processing 111
Sulphur and Other Industrial Inputs 113
Supply Chains and Geopolitical Considerations 116
Chapter 5 – Ammonia Production (Haber–Bosch Process) 119
Overview of Ammonia Production (Haber–Bosch Process) 120

Chemistry of Ammonia Synthesis 122
Steam Methane Reforming (SMR) 128
Hydrogen Production and Purification 132
Reactor Design and Catalysts 137
Energy Consumption and Efficiency Improvements 143
Small Scale Implementation 145
Chapter 6 – Nitrogen Fertiliser Production 151
Urea Production Process 152
Ammonium Nitrate and Ammonium Sulphate 166
Nitric Acid Production (Ostwald Process) 173
Granulation and Prilling Techniques 178
Storage, Handling, and Safety Risks 182
Structuring a Urea Manufacturing Operation 191
Chapter 7 – Phosphate Fertiliser Production 200
Mining and Beneficiation of Phosphate Rock 202
Types of Fertilisers Made from Phosphate Rock 208
Wet-Process Phosphoric Acid Production 209
Superphosphate (SSP, TSP) Manufacturing 225
DAP and MAP Production 236
Impurities and Environmental Issues (e.g., Phosphogypsum) 247
Chapter 8 – Potash Fertiliser Production 250
Potash Fertiliser 252
Potash Ore Types (Sylvinite, Carnallite) 253
Mining Methods (Conventional, Solution mining) 259
Processing and Refining Potassium Salts 262
Production of MOP (KCl) and SOP (K_2SO_4) 263
Market Applications and Crop Requirements 275
Chapter 9 – Fertiliser Blending and Formulation 279
Bulk Blending vs Chemical Granulation 281
NPK Formulation Strategies 283
Additives and Coating Technologies 286
Quality Control and Uniformity 288

Custom Fertiliser Solutions for Crops 289
Chapter 10 – Fertiliser Application Technologies 295
Broadcasting, Banding, Fertigation 296
Precision Agriculture and Variable Rate Technology (VRT) 310
Timing and Efficiency of Nutrient Uptake 313
Minimising Losses (Leaching, Volatilisation) 317
Chapter 11 – Environmental and Sustainability Considerations 319
Greenhouse Gas Emissions (e.g., N_2O, CO_2) 321
Water Pollution and Eutrophication 322
Waste Management and By-products 324
Green Ammonia and Low-Carbon Fertilisers 326
Circular Economy Approaches 331
Chapter 12 – Industry Trends, Regulation, and Future Innovations 335
Global Fertiliser Regulations and Standards 337
Supply Chain and Market Dynamics 345
Digital Agriculture Integration 351
Biotechnology and Enhanced-Efficiency Fertilisers 353
Future Outlook: Sustainable Intensification 360
Appendices 363
Appendix A - Chemical Equations, Process Flow Diagrams and Fertiliser Calculations 364
Appendix B - Fertiliser Nutrient Conversion Tables 374
Appendix C - Safety Data and Handling Guidelines 378
Appendix D – Sample Business Plan for Small Scale Production 399
References 406
Index 423

Introduction

As populations grow, diets evolve, and pressures on land and natural resources intensify, the ability to produce sufficient, high-quality food has become one of the defining challenges of our time. At the centre of this challenge lies fertiliser production, a discipline that combines chemistry, biology, engineering, and environmental management to sustain and enhance plant productivity on a global scale.

This book, *Agricultural Fertiliser Production: Principles, Processes, and Practice*, has been developed to provide a comprehensive and technically grounded exploration of fertilisers, from their scientific foundations through to their industrial manufacture and real-world application. It is designed to bridge the gap between theory and practice, enabling readers to understand not only what fertilisers are, but how and why they function within complex agricultural systems.

Fertilisers are far more than simple nutrient inputs. They are integral to the biological processes that govern plant growth, the engineering systems that produce essential compounds at industrial scale, and the economic and geopolitical frameworks that influence global food security. From the molecular interactions within plant cells to the large-scale synthesis of ammonia and the global movement of fertiliser commodities, this field operates across multiple interconnected levels. Understanding these connections is essential for anyone involved in agriculture, agribusiness, engineering, or environmental management.

The structure of this book reflects this integrated perspective. It begins with the fundamentals of plant nutrition, soil science, and the historical development of fertiliser use, establishing the biological and agronomic context. It then progresses into the chemical and industrial processes that underpin fertiliser production, including key reactions, process systems, and manufacturing technologies. Subsequent chapters explore operational considerations such as handling, storage, safety, and environmental management, as well as emerging innovations in areas such as enhanced-efficiency fertilisers, biotechnology, and digital agriculture.

A central theme throughout the text is the concept of nutrient use efficiency and sustainability. While fertilisers have enabled unprecedented increases in agricultural productivity, they also present environmental and resource challenges when mismanaged. This book therefore emphasises the importance of precision, balance, and responsible use, aligning fertiliser practices with broader goals of environmental stewardship, climate resilience, and long-term soil health.

In addition to technical knowledge, the book aims to develop practical understanding. You will encounter real-world examples, process explanations, calculations, and applied scenarios that reflect how fertiliser systems operate in practice. This approach ensures that the content is not only rigorous but also directly relevant to industry applications, training environments, and professional development contexts.

Ultimately, this book is intended to serve as both a foundational reference and a practical guide. Whether the you are a student, educator, industry practitioner, or policymaker, the goal is to provide the knowledge and insight required to understand fertiliser production as a critical component of modern agriculture. By connecting science, engineering, and application, this text highlights the essential role fertilisers play in sustaining global food systems while navigating the challenges of a rapidly changing world.

Chapter 1

Introduction to Fertilisers and Global Agriculture

Modern agriculture is built upon the ability to understand, manage, and optimise plant nutrition. At the heart of this capability lies fertiliser production, a complex intersection of chemistry, biology, engineering, and global resource management. This chapter introduces the foundational principles that underpin fertiliser science, establishing the critical link between plant physiology, nutrient cycles, and the industrial systems that sustain food production worldwide.

Plants, unlike animals, must synthesise their own food through photosynthesis, relying on sunlight, water, carbon dioxide, and a precise balance of essential nutrients. These nutrients, ranging from primary macronutrients such as nitrogen, phosphorus, and potassium, to secondary and trace elements, govern every aspect of plant growth, from root development and energy transfer to enzyme function and reproductive success. Understanding how these nutrients interact within plant systems is essential for designing effective fertilisers and ensuring optimal crop performance.

Equally important is an appreciation of plant anatomy and internal transport systems. The root and shoot systems work in coordination to absorb, distribute, and utilise nutrients, while cellular structures such as chloroplasts and vacuoles support critical biochemical processes. These biological mechanisms directly influence how fertilisers are formulated, applied, and absorbed, highlighting the need for an integrated approach that connects plant science with fertiliser engineering.

This chapter also explores the biochemical processes that drive plant growth, including photosynthesis and the role of ATP as the energy currency of the cell. These processes are not only central to plant survival but also determine how efficiently nutrients are converted into biomass and yield. By examining these relationships, readers gain insight into why nutrient availability is so closely tied to agricultural productivity.

In addition, the chapter introduces the concept of nutrient classification, macronutrients, secondary nutrients, and micronutrients, and emphasises the importance of nutrient balance. It highlights how deficiencies, excesses, and interactions between nutrients can significantly impact plant health, crop yield, and soil sustainability. This understanding forms the basis for modern fertiliser formulation and application strategies.

Finally, the chapter situates fertiliser use within a broader global context, addressing the role of fertilisers in food security, the increasing demand driven by population growth, and the sustainability challenges associated with their production and use. By linking plant biology with industrial processes and global systems, this chapter provides a comprehensive foundation for the detailed exploration of fertiliser production presented in subsequent chapters.

Together, these concepts establish the essential knowledge required to understand how fertilisers are developed, produced, and applied, setting the stage for a deeper examination of the technologies, processes, and innovations that define the modern fertiliser industry.

Role of Fertilisers in Modern Agriculture

Fertilisers are foundational to modern agriculture, underpinning global food production, crop quality, and farm productivity by enabling high-yield, intensive cropping systems and serving as a core component of precision agriculture and sustainable soil resource management [1-3]. Across diverse farming contexts, fertilisers fulfill more than a mere nutrient supply; they sustain the scale and efficiency of operations necessary to meet rising global food demand while shaping soil fertility trajectories over multiple cropping cycles [1, 3, 4]. In plant nutrition, the primary macronutrients nitrogen (N), phosphorus (P), and potassium (K) are central to key physiological processes: nitrogen drives leaf area, chlorophyll synthesis, and protein formation; phosphorus supports root development and energy transfer via ATP and seed/fruit formation; potassium regulates water relations, enzyme activation, and disease resistance. Together, these nutrients, delivered through fertilisers, correct soil nutrient deficiencies and help crops reach their genetic yield potential, especially under high-yield production systems where nutrient supply can become limiting [1-3].

In simple terms, plants need certain key nutrients to grow properly, just like humans need food to stay healthy. The three most important nutrients are nitrogen (N), phosphorus (P), and potassium (K), often referred to as NPK.

- Nitrogen helps plants grow big, green leaves. It is important for making chlorophyll, which allows plants to use sunlight to produce energy. It also helps plants build proteins, which are essential for growth.
- Phosphorus helps plants develop strong roots and supports energy movement within the plant. This energy is needed for important processes like growing flowers, seeds, and fruit.
- Potassium helps plants manage water, making them more resistant to drought. It also helps activate important plant processes (enzymes) and improves the plant's ability to fight diseases.

In many soils, these nutrients are not available in the right amounts, especially when crops are grown repeatedly. This is where fertilisers come in. Fertilisers add these nutrients back into the soil so plants can grow properly.

By providing the right balance of nutrients, fertilisers help crops grow stronger and produce higher yields. This allows farmers to get the best possible results from their crops, especially in modern farming systems where high production is required.

The contribution of fertilisers to global yields is substantial and widely acknowledged. Modern agricultural systems rely heavily on fertiliser inputs to sustain high productivity levels, and many analyses attribute a large share of observed yields to carefully managed nutrient inputs; estimates in the literature suggest that fertilisers account for a significant portion, often framed as a substantial fraction, of total food production globally, reflecting the central role of nutrient availability in realizing crop yield potential [1-3]. Incremental yield gains from fertiliser use are particularly critical in the context of expanding populations and finite arable land, where intensification and yield per hectare must be maximised to meet demand; thus, fertilisers are integral to maintaining global food security through enhanced productivity and crop quality [1-3].

The yield relationship concept explains the direct link between crop productivity and the availability of essential nutrients in the soil. In agricultural systems, plant growth is highly dependent on the supply of key macronutrients—particularly nitrogen (N), phosphorus (P), and potassium (K). These nutrients support fundamental biological processes such as photosynthesis, energy transfer, root development, and overall plant health. When these nutrients are available in adequate amounts, crops are able to grow more vigorously and produce higher yields.

The expression *yield is proportional to nutrient supply* (Yield $\propto$ Nutrient Supply (N, P, K)) reflects this relationship. As the availability of nutrients increases, crop yield typically increases as well, because plants are no longer limited by deficiencies. However, this relationship is not unlimited; yield increases up to an optimal point, beyond which additional nutrients may have little effect or even cause negative impacts such as nutrient toxicity or environmental loss. This concept aligns with Liebig's Law of the Minimum, which states that plant growth is controlled by the most limiting nutrient.

In practical terms, soils often do not contain sufficient quantities of available nutrients to support high-yield agriculture, especially under continuous cropping systems. Over time, nutrients are removed from the soil through harvest, and without replenishment, crop yields decline. Fertilisers are therefore used to restore and optimise nutrient levels, ensuring that plants have access to the nutrients they require throughout the growing cycle.

This relationship highlights the critical role of fertilisers in modern agriculture. By increasing nutrient availability, fertilisers enable farmers to significantly increase the amount of food produced per hectare of land. This is particularly important in the context of a growing global population, where demand for food continues to rise while the amount of arable land remains limited. Through effective nutrient management, fertilisers help maximise productivity, improve efficiency, and support global food security.

Figure 1: "AN 33,5 N" ammonium nitrate big bags EC-fertilizer produced by Borealis. Cjp24, CC BY-SA 4.0, via Wikimedia Commons.

Soil fertility is not a one-off input but a dynamic parameter that necessitates nutrient replenishment across cropping cycles. Harvest removal of nutrients in biomass lowers soil fertility unless those nutrients are replenished; continuous cropping without adequate fertilisation leads to nutrient depletion and declining productivity, underscoring the long-term soil health rationale for fertiliser use and the need for integrated nutrient management to sustain soil productive capacity [1-3]. In intensive systems, such as monocropping, irrigated agriculture, and high-density planting, nutrient reserves are rapidly depleted, making precise and timely fertiliser inputs essential to maintain soil fertility and crop performance over time [1-3].

Beyond yield, fertilisers influence crop quality through nutrient-mediated modulation of key quality traits. Nitrogen availability, for instance, can elevate grain protein content in cereals and influence quality attributes such as fruit size, colour, and post-harvest shelf life, outcomes that matter for processing industries and consumer standards; potassium and phosphorus likewise impact fruit quality and seed development. The capacity of fertilisers to enhance specific quality parameters reinforces their role in matching agronomic practices with end-use quality requirements, an important consideration in markets with stringent quality standards [1-3].

The emergence of precision agriculture has amplified the importance of fertilisers through data-driven, site-specific management. Technologies such as GPS-guided machinery, variable-rate application, soil and crop sensors, and decision-support tools enable fertilisers to be applied where and when nutrients are needed, reducing waste and environmental losses while improving nutrient use efficiency (NUE). This precision paradigm seeks to align nutrient supply with crop demand dynamically, thereby enhancing NUE and diminishing environmental footprint; the literature consistently emphasizes the synergy between advanced sensing, modelling, and targeted fertilisation as a cornerstone of modern fertiliser practice [1-3].

A central objective in fertiliser science is improving nutrient use efficiency (NUE), defined as the fraction of applied nutrients actually taken up by crops. Innovations in fertiliser technology, such as slow-release formulations, stabilised nitrogen products with urease or nitrification inhibitors, and polymer-coated or coated granules, aim to synchronize N release with crop demand, thereby reducing losses to volatilisation, leaching, and denitrification. Enhanced-efficiency fertilisers (EEFs), encompassing chemically modified slow-release products, barrier-coated controlled-release fertilizers, and inhibitors, are widely studied as means to boost NUE and minimize environmental losses; meta-analyses and reviews show that coatings and inhibitors can reduce N losses and, in many cases, improve crop yields or NUE, albeit with context-dependent outcomes related to soil type, climate, and crop species [1-3]. Coatings, inhibitors, and stabilization strategies are therefore central to maintaining productive, environmentally responsible fertiliser use, particularly as climate variability and resource constraints intensify pressure on nutrient management [1-3].

Efficiency Concept

The concept of nutrient use efficiency (NUE) is central to modern fertiliser management and sustainable agriculture. NUE describes how effectively plants utilise the nutrients that are applied through fertilisers. It is commonly expressed as the ratio of the amount of nutrient absorbed by the plant to the total amount of nutrient applied. In simple terms, it measures how much of the fertiliser actually contributes to plant growth rather than being lost to the environment.

$$\text{NUE} = \frac{\text{Nutrient absorbed by plant}}{\text{Nutrient applied}}$$

A high NUE indicates that a large proportion of the applied nutrients are taken up by the crop and used for growth, resulting in efficient fertiliser use and better economic returns for farmers. Conversely, a low NUE means that a significant portion of the applied nutrients is lost through processes such as leaching, runoff, volatilisation, or denitrification. These losses not only reduce fertiliser effectiveness but can also contribute to environmental problems, including water pollution and greenhouse gas emissions.

Improving NUE is therefore a key objective in both agricultural productivity and environmental management. Advances in fertiliser technology have led to the development of products specifically designed to enhance nutrient availability and uptake by plants while minimising losses. One such approach is the use of slow-release fertilisers, which gradually release nutrients over time in alignment with plant demand. This reduces the risk of excess nutrients being present in the soil at any one time, thereby limiting losses.

Another important innovation is the use of stabilised nitrogen fertilisers, which include additives such as urease and nitrification inhibitors. These compounds slow down the chemical processes that convert nitrogen into forms that are easily lost from the soil. By delaying these transformations, stabilised fertilisers help retain nitrogen in forms that are more available to plants for longer periods.

Additionally, coated fertiliser granules are used to physically control the rate at which nutrients are released. These coatings, often made from polymers or sulphur, act as a barrier that regulates the movement of nutrients into the soil. This controlled release ensures a more consistent supply of nutrients to the plant, improving uptake efficiency and reducing waste.

Coatings and inhibitors also offer a practical path to reduced environmental impacts associated with fertiliser N. By delaying hydrolysis of urea and slowing nitrification, urease and nitrification inhibitors can substantially lower ammonia volatilisation and nitrate leaching, contributing to lower nitrous oxide (N_2O) emissions in some systems; however, the magnitude of these benefits is not universal and depends on soil texture, moisture, seasonal rainfall, and management practices. Reviews and empirical studies consistently report that the effectiveness of inhibitors and coated products varies with soil type and management, with notable reductions in certain settings but mixed or modest effects in others; these nuances highlight the importance of local site conditions when selecting EEFs and the need for integrated approaches that combine inhibitors with controlled-release coatings or other strategies to optimise NUE and environmental outcomes [1-3].

The environmental dimension of fertiliser use is intertwined with food security and economic considerations. While fertilisers enable higher yields, improved crop quality, and greater profitability for farms, they also carry the risk of environmental pollution if mismanaged, including nutrient runoff leading to water body eutrophication, soil acidification, and greenhouse gas emissions. Consequently, sustainable fertiliser use emphasises precision application, integrated nutrient management, and the blending of synthetic fertilisers with organic amendments and soil-conservation practices. This integrated framework seeks to reconcile productivity with environmental stewardship and long-term soil health, aligning fertiliser practices with broader sustainability goals and climate-smart agriculture objectives [1-3].

In addition to input efficiency and environmental concerns, economic considerations shape fertiliser use. Farmers balance the cost of fertilisers against expected yield gains and quality improvements, making NUE a key economic as well as agronomic objective. Improved NUE through advanced fertiliser technologies can translate into higher net returns by increasing yields and crop quality while reducing losses; however, the cost of EEFs, variability in response, and the need for precise management mean that uptake benefits from farmer education, extension services, and farmer-friendly decision support. This pragmatic view integrates agronomic potential with real-world constraints, acknowledging that the profitability and sustainability of fertiliser strategies hinge on context-specific optimization of rate, timing, placement, and product choice [1-4].

Integrated and multi-pronged approaches are increasingly promoted to sustain soil resources and productivity in the long term. These include combining synthetic fertilisers with organic amendments, adopting crop rotations, and implementing soil-conservation practices to maintain soil organic matter and nutrient cycling. Such integrated nutrient management aims to preserve soil fertility while minimising negative environmental impacts, supporting resilient production systems capable of withstanding climatic variability and land-use pressures. The general position across the cited literature is that fertilisers remain essential inputs for modern agriculture, with sustainability achievable through improved NUE, precision management, and integration with broader soil and ecosystem management practices [1-4].

Historical Development of Fertiliser Use

Overview

The historical development of fertiliser use reflects humanity's ongoing effort to improve soil fertility and increase agricultural productivity. From early organic practices to modern industrial chemistry, fertiliser use has evolved alongside advances in science, technology, and farming systems.

In ancient agricultural societies, farmers relied entirely on natural and organic sources of nutrients to maintain soil fertility. Early civilisations such as those in Mesopotamia, Egypt, China, and the Indus Valley used animal manure, composted plant material, and crop residues to replenish nutrients removed by harvest. In some regions, farmers also practised crop rotation and fallowing, allowing land to recover naturally. The use of materials such as wood ash and bone meal provided additional nutrients, particularly phosphorus and potassium. These practices were based on observation and tradition rather than scientific understanding, but they were effective in maintaining productivity in low-intensity farming systems.

During the medieval period, fertiliser use remained largely organic, but agricultural practices became more structured. The three-field crop rotation system in Europe helped maintain soil fertility by alternating crops and including legumes, which naturally fix nitrogen in the soil.

Farmers also increasingly used livestock manure as a key fertiliser input. Despite these improvements, agricultural productivity remained relatively low, and soil exhaustion was a common problem, particularly in areas with continuous cultivation.

The 18th and early 19th centuries marked the beginning of a more scientific approach to fertiliser use, driven by the Agricultural Revolution in Europe. Advances in soil science and plant physiology began to emerge, and farmers experimented with new fertilising materials such as lime (to correct soil acidity) and imported nutrients like guano (bird droppings) and Chilean nitrate. Guano, in particular, became highly valued due to its high nitrogen and phosphorus content, leading to large-scale extraction and international trade. This period highlighted the growing recognition that soil fertility could be actively managed to increase crop yields.

A major turning point came in the mid-19th century with the work of German chemist Justus von Liebig, who established the scientific basis of plant nutrition. Liebig demonstrated that plants require specific chemical elements for growth and proposed the Law of the Minimum, which states that plant growth is limited by the most deficient nutrient. This breakthrough shifted agriculture from empirical practices to a science-based understanding of fertilisation. It also led to the development of the first manufactured (inorganic) fertilisers, including superphosphate, produced by treating phosphate rock with sulfuric acid.

The late 19th and early 20th centuries saw rapid industrialisation of fertiliser production. One of the most significant developments was the invention of the Haber–Bosch process in the early 1900s, which allowed for the large-scale synthesis of ammonia from atmospheric nitrogen and hydrogen. This process provided a virtually unlimited supply of nitrogen fertiliser and revolutionised agriculture. Ammonia became the foundation for a wide range of nitrogen fertilisers, including urea and ammonium nitrate. At the same time, advances in mining and processing enabled large-scale production of phosphate and potash fertilisers.

The mid-20th century, particularly during the Green Revolution, marked a period of unprecedented growth in fertiliser use. High-yield crop varieties were developed that responded strongly to fertiliser inputs, especially nitrogen. Combined with improved irrigation, mechanisation, and pest control, fertilisers played a critical role in dramatically increasing global food production. This period helped prevent widespread famine and supported rapid population growth, particularly in developing countries.

In recent decades, the focus of fertiliser use has shifted towards efficiency, environmental sustainability, and precision agriculture. Concerns about nutrient runoff, water pollution, greenhouse gas emissions, and soil health have led to improved fertiliser technologies and management practices. Innovations such as slow-release fertilisers, stabilised nitrogen products, and precision application techniques aim to maximise nutrient use efficiency while minimising environmental impact. There is also growing interest in integrating organic and inorganic fertilisers, as well as developing alternative approaches such as biofertilisers and green ammonia.

Overall, the history of fertiliser use reflects a progression from natural, experience-based practices to highly advanced, science-driven systems. Each stage of development has contributed to increasing agricultural productivity, with modern fertilisers now playing a vital role in sustaining global food security while also presenting new challenges related to sustainability and environmental management.

Key Developers

The transformation of European agriculture from a traditional, practice-based system to a science-informed, industrialised regime was the cumulative result of interlinked theoretical insights, experimental demonstrations, and commercial ventures. Across the nineteenth and early twentieth centuries, a sequence of chemists, agronomists, metallurgists, and industrial entrepreneurs introduced key concepts and technologies that redefined plant nutrition, enabled large-scale fertiliser production, and laid the groundwork for a global fertiliser industry. Synthesising the historical threads from the sources, we see a progression from early soil chemistry and crop nutrition experiments, through foundational theories of mineral nutrients and nutrient limitation, to industrial innovations in fertiliser manufacture and by-product utilisation, culminating in the modern nitrogen fertiliser era that underpins contemporary agriculture.

Early empirical inquiries into soil amendments and crop nutrition set the stage for reformulation of fertiliser science Johann Friedrich Mayer, a German clergyman and agricultural reformer, emerges as one of the earliest European figures investigating mineral amendments to soils, notably gypsum (calcium sulphate). His experimental demonstrations that gypsum could improve crop growth stimulated wide interest and seeded subsequent inquiry into soil amendment mechanisms, a debate that persisted into the early nineteenth century about whether gypsum acted via sulfuric acid, atmospheric nutrition, or soil decomposition processes. Although Mayer's mechanistic understanding was incomplete, his empirical work helped catalyse a broader European research ethos on soil amendments and crop productivity [5]. Later, the English chemist Humphry Davy contributed to this discourse by characterising gypsum as a direct nutrient source for plants, reinforcing the view that mineral inputs could directly affect plant nutrition beyond organic matter [5]. In addition to gypsum, Mayer promoted improved crop rotation systems, an agronomic practice that maintained soil fertility and supported the broader shift toward scientifically informed husbandry [5].

The foundational shift toward chemistry-based fertilisers is crystallised in the work of Justus von Liebig, often hailed as the founder of modern agricultural chemistry. Liebig challenged humus-centric nutrient theories and argued that inorganic mineral nutrients supplied by the soil were fundamental to plant growth, highlighting nitrogen, phosphorus, and potassium as key elements and articulating the Law of the Minimum, the idea that plant growth is constrained by the scarcest essential nutrient. His theoretical contributions provided a rigorous scientific framework for fertiliser use, even though some early formulations were simplified; his emphasis on mineral nutrition anchored subsequent fertiliser development and agronomic optimisation [6]. Liebig also attempted commercialisation with bone meal-derived fertilisers, an endeavour that underscored the link between scientific insight and industrial production, though the initiative faced limitations in nutrient availability for plants at the time [6].

Sir John Bennet Lawes, an English entrepreneur, translated laboratory insights into industrial practice. In the 1830s he conducted systematic fertiliser experiments and, crucially, patented a process in 1842 for producing superphosphate by treating phosphate rock with sulfuric acid, marking the first commercially viable artificial fertiliser. Lawes's partnership with Joseph Henry Gilbert and the establishment of Rothamsted Experimental Station produced long-running, rigorous datasets on fertiliser use and crop responses, creating a durable knowledge base that shaped agricultural practice globally [6, 7]. The Rothamsted results provided empirical validation for inorganic phosphate fertilisers and established a model for integrating scientific inquiry with industrial-scale production and agronomic testing [6, 7].

Nitrogen fertilisation and nitrogen cycle understanding in France In France, Jean Baptiste Boussingault contributed to plant nutrition understanding, particularly concerning nitrogen content in fertilisers and its impact on growth. His experiments helped clarify nutrient cycles and fed into the growing scientific consensus on fertiliser application, complementing the Liebig-Lawes-Gilbert line of research and reinforcing the idea that precise nutrient management supports yield improvements [8].

Industrial by-products and phosphorus fertilisers Industrial developments also supplied fertiliser feedstocks through by-products of metal and mineral processing. The Gilchrist–Thomas process, developed by Percy Gilchrist and Sidney Gilchrist Thomas in the late nineteenth century, enabled the removal of phosphorus from iron ore during steelmaking. The phosphorus-rich slag by-product, Thomas phosphate, emerged as a fertiliser input, expanding phosphorus supply for agriculture and linking steel production with fertiliser markets. This industrial-phosphorus recycling pipeline is documented in contemporary discussions of the Thomas–Gilchrist process and its fertiliser by-products, illustrating a key bridge between metallurgy and agronomy in Europe [7, 9]. The value of the Thomas slag by-product as a plant nutrient was demonstrated in agricultural and laboratory experiments in the 1880s, further illustrating how industrial by-products fed the fertiliser economy [9].

A major turning point occurred with early twentieth-century nitrogen fixation technologies. The Birkeland–Eyde process (1903) used electric arcs to fix atmospheric nitrogen into nitric acid, representing an innovative but energy-intensive approach to nitrogen fertiliser precursors. Although it demonstrated the feasibility of atmospheric nitrogen utilisation, its energy demands limited long-term viability compared with later technologies [7].

Figure 2: Birkeland–Eyde electric arc furnace outside the Norwegian Museum of Science and Technology at Kjelsås. An electric arc furnace used for the production of artificial fertiliser by extracting nitrogen. Developed by Sam Eyde and Kristian Birkeland. The furnace shown is among the largest, with a capacity of 3300 kW. Used at Rjukan from 1916 to 1940. Erik den yngre, CC BY-SA 3.0, via Wikimedia Commons.

The Haber–Bosch process, developed by Fritz Haber and Carl Bosch in the early twentieth century, revolutionised fertiliser production by enabling large-scale synthesis of ammonia from atmospheric nitrogen and hydrogen. This breakthrough established ammonia as the foundational feedstock for nitrogen fertilisers and remains central to global fertiliser production [7, 9]. The Ostwald process, which oxidised ammonia to nitric acid, complemented Haber–Bosch by providing nitric acid for nitrate-based fertilisers, thereby forming the backbone of modern nitrogen fertiliser manufacture [7]. In parallel, Erling Johnson's Odda process (1927) advanced nitrophosphate fertilisers, combining phosphorus and nitrogen production routes to improve efficiency and diversify fertiliser options [9]. Together, these processes underpin the chemical supply chains that sustain large-scale European and global fertiliser industries [7, 9].

The late nineteenth and early twentieth centuries witnessed rapid industrialisation of fertiliser production, driven by chemistry, raw materials such as coprolites (phosphate-rich fossil materials), and new fertiliser varieties. In Britain and continental Europe, these advances supported large-scale production of superphosphate and related fertilisers, with an expanding global market. The broader expansion of fertiliser use accelerated after World War II, during the Green Revolution, when high-yield crop varieties increased nutrient demand. Large chemical companies and international markets spurred production and trade, and the fertiliser industry eventually consolidated through mergers and acquisitions, leading to the multinational firms dominant today. The historical record shows a trajectory from experimental trials and agronomic reform to an integrated, globally linked fertiliser sector, with Europe playing a central role in each phase of development [5-9].

Across the sources, a coherent narrative emerges in which early empirical agronomy and soil science laid the groundwork for later chemical fertiliser theory. Mayer's gypsum work, Liebig's inorganic nutrient emphasis, and Lawes and Gilbert's commercial production of superphosphate collectively established a framework linking plant nutrition to tangible fertiliser products. Boussingault's nitrogen-focused insights enriched the understanding of nutrient cycles, while the Gilchrist–Thomas process connected steelmaking by-products to phosphate fertilisers, expanding phosphorus supply chains. The shift from organic-humus theories to mineral-nutrient paradigms culminated in the Haber–Bosch and Ostwald processes, which provided scalable routes to nitrogen fertilisers and nitric acid, thereby enabling industrial fertiliser production on a global scale. The post-war Green Revolution era further intensified fertiliser use, reinforcing Europe's role as both a producer and innovator in fertiliser chemistry and industry. Where there is nuance, some early theories, such as Liebig's Law of the Minimum, are recognised as foundational but simplified in retrospect, and later work clarified mechanisms and context. The literature consistently emphasizes a continuum from foundational agronomy and chemistry to industrial chemistry and global fertiliser systems, with Europe at the forefront of both scientific discovery and industrial deployment [5-7, 9].

Overview of Macro-, Secondary-, and Micronutrients

Plant growth and productivity depend on a balanced supply of essential nutrients. These nutrients are classified into macronutrients, secondary nutrients, and micronutrients based on the quantities required by plants. While all are essential, they differ in function, mobility, and the amounts needed for healthy plant development.

Plant Anatomy

Plants are made up of two main systems: the root system and the shoot system, which work together to support the plant's structure, growth, and overall function. The root system is located below the ground and is primarily responsible for anchoring the plant and absorbing water and nutrients from the soil. It contains several important structures, including root hairs, which increase the surface area for absorption; the epidermis, which forms a protective outer layer; and the cortex, which stores nutrients and helps transport water inward. At the centre of the root are the vascular tissues, xylem and phloem, which are responsible for transporting water, nutrients, and sugars throughout the plant. In addition to absorption, roots can also store energy, as seen in crops like carrots and beets, and they interact with soil microorganisms that assist in nutrient uptake.

The shoot system includes all above-ground parts of the plant, such as stems, leaves, flowers, and fruits. The stem provides structural support and acts as a transport pathway between the roots and the leaves. Within the stem, the xylem carries water and dissolved minerals from the roots to the leaves, while the phloem transports sugars produced during photosynthesis to other parts of the plant. The stem also plays an important role in positioning leaves so they can capture maximum sunlight.

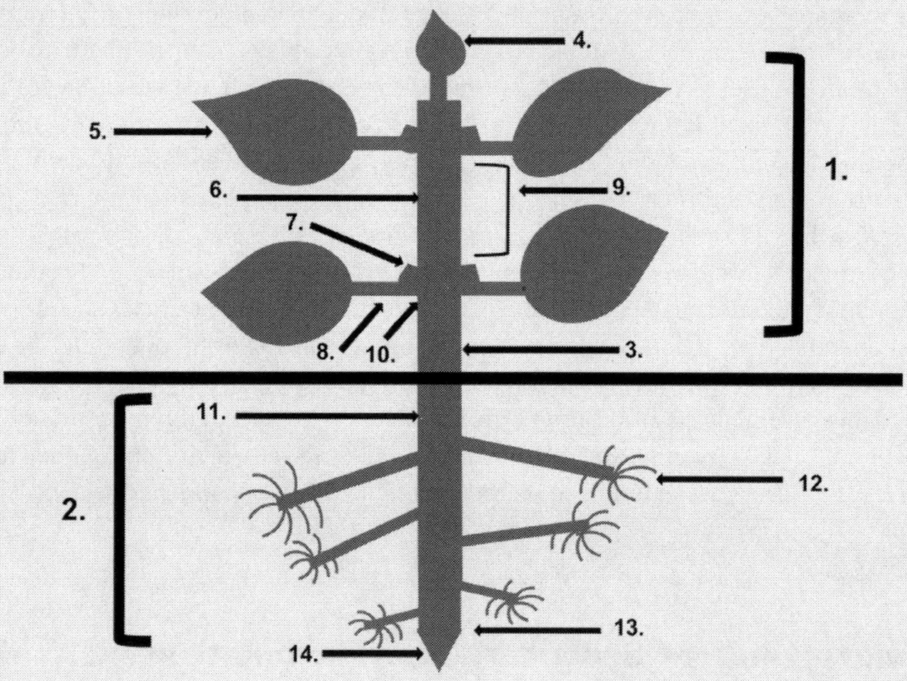

Figure 3: Diagram of the anatomy of a plant with labels of structural parts of the plants and the roots. 1. The Shoot System. 2. The Root System. 3. Hypocotyl. 4. Terminal Bud. 5. Leaf Blade. 6. The Internode. 7. Axillary Bud. 8. Node. 9. Stem. 10. Petiole. 11. Tap Root. 12. Root Hairs. 13. Root Tip. 14. Root Cap. Bigman24225, CC BY-SA 4.0, via Wikimedia Commons.

Leaves are the primary site of photosynthesis and are specially adapted to capture light efficiently while regulating water loss. They have a waxy cuticle that reduces water evaporation, and an epidermis that serves as a protective outer layer. Inside the leaf, the mesophyll contains chloroplasts where photosynthesis occurs. Leaves also contain stomata, which are small pores that allow carbon dioxide to enter and oxygen to exit, facilitating gas exchange essential for photosynthesis.

Flowers and other reproductive structures are responsible for plant reproduction. Flowers contain stamens, which produce pollen (the male reproductive cells), and the pistil or carpel, which contains the ovules (female reproductive structures). After fertilisation occurs, the ovules develop into seeds, and the ovary forms the fruit. These structures ensure the continuation of plant species and the production of crops.

Plants are composed of specialised tissues that perform different functions. Dermal tissue forms the outer protective layer and helps prevent water loss and physical damage. Vascular tissue includes the xylem and phloem, which transport water, minerals, and sugars throughout the plant. Ground tissue makes up much of the plant body and is involved in photosynthesis, storage, and structural support.

At the cellular level, plant cells have unique features that support their function. The cell wall provides strength and rigidity, allowing plants to maintain their structure. Chloroplasts enable photosynthesis by capturing light energy, while the large central vacuole stores water and helps maintain internal pressure, keeping the plant upright and stable.

Plants also rely on internal transport systems to move substances efficiently. Water and minerals absorbed by the roots travel upward through the xylem to the leaves, while sugars produced in the leaves are transported through the phloem to growing tissues or storage areas. This continuous movement of materials connects all parts of the plant and supports growth and development.

Understanding plant anatomy is essential in agriculture, as it helps farmers and agronomists make informed decisions about crop management. Knowledge of plant structure allows for more effective fertiliser application, whether through the soil or directly onto leaves. It also assists in diagnosing nutrient deficiencies, which often appear as visible symptoms in leaves or stems, and in improving irrigation practices. For example, damaged roots can reduce a plant's ability to absorb nutrients, while poor leaf health can limit photosynthesis and reduce crop yields.

Macronutrients (Primary Nutrients)

Macronutrients are required in the largest amounts and are fundamental to plant growth and metabolic processes. The three primary macronutrients are nitrogen (N), phosphorus (P), and potassium (K).

<u>Nitrogen (N)</u>

Nitrogen is the most critical nutrient for plant growth. It is a major component of chlorophyll, which enables photosynthesis, and is also essential for the formation of amino acids, proteins, and nucleic acids. Nitrogen promotes vigorous vegetative growth, particularly leaf and stem development. Deficiency typically results in pale or yellowing leaves (chlorosis) and reduced growth.

Photosynthesis

Photosynthesis is the process by which green plants convert sunlight into chemical energy, enabling them to produce their own food. This process takes place primarily in the leaves, within specialised structures called chloroplasts that contain the green pigment chlorophyll. Chlorophyll absorbs sunlight and uses this energy to drive a series of chemical reactions that sustain plant life.

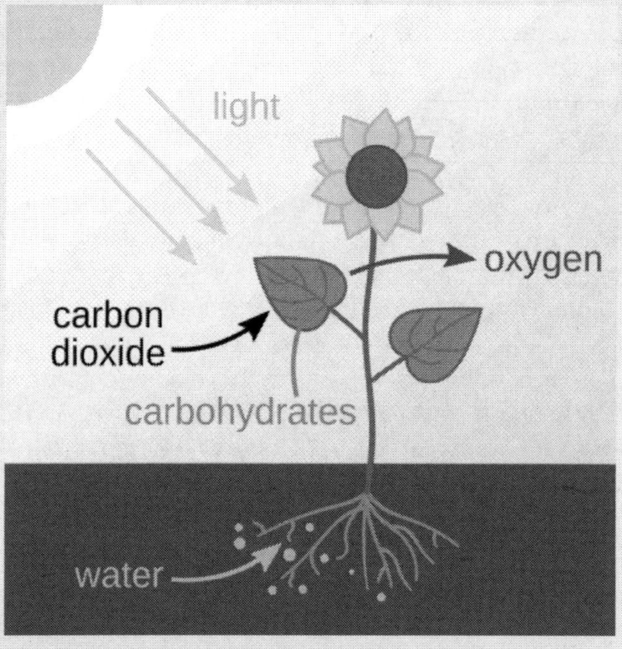

Figure 4: The overall process of photosynthesis. At09kg, Wattcle, NefronusAt09kg: originalWattcle: vector graphicsNefronus: redoing the vector graphics, CC BY-SA 4.0, via Wikimedia Commons.

At its core, photosynthesis relies on three main inputs: sunlight as the energy source, carbon dioxide (CO_2) from the air, and water (H_2O) absorbed from the soil. These inputs are transformed into glucose ($C_6H_{12}O_6$), a sugar that serves as a source of energy and a building block for growth, and oxygen (O_2), which is released into the atmosphere. The overall process can be summarised by the equation: $6CO_2 + 6H_2O + \text{light} \rightarrow C_6H_{12}O_6 + 6O_2$.

Photosynthesis occurs within the chloroplasts of plant cells. Inside the chloroplast, there are two main regions with distinct roles. The thylakoids are responsible for capturing light energy, while the stroma is where the production of sugars takes place. Leaves are the primary site of photosynthesis because they are well exposed to sunlight and contain stomata—tiny pores that allow carbon dioxide to enter the plant.

The process of photosynthesis takes place in two interconnected stages. The first stage, known as the light-dependent reactions, requires sunlight and occurs in the thylakoids. During this stage, chlorophyll absorbs light energy, water molecules are split into hydrogen and oxygen, oxygen is released into the atmosphere, and energy-rich molecules such as ATP and NADPH are produced. The second stage, called the light-independent reactions or Calvin Cycle, occurs in the stroma and does not require direct sunlight. In this stage, carbon dioxide is converted into organic molecules using the energy stored in ATP and NADPH, ultimately producing glucose.

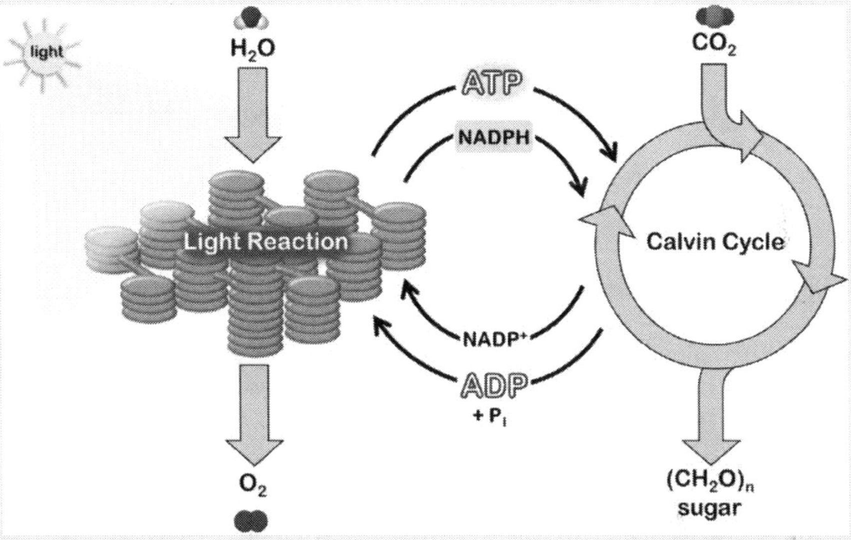

Figure 5: This illustration summarizes photosynthesis and the connection between the light reaction and the Calvin Cycle. ELaurent, CC BY-SA 4.0, via Wikimedia Commons.

Photosynthesis is essential for life on Earth. It produces the food that plants need for growth, energy, and storage, often in the form of starch. It also releases oxygen, which is vital for respiration in animals and humans. Furthermore, photosynthesis forms the foundation of the food chain, as all living organisms depend either directly or indirectly on plants for energy.

The efficiency of photosynthesis is closely linked to plant nutrition. Key nutrients play important roles in supporting this process. Nitrogen is required for the production of chlorophyll, magnesium is a central component of the chlorophyll molecule, phosphorus is necessary for energy transfer through ATP, and potassium helps regulate water balance and enzyme activity. If any of these nutrients are lacking, photosynthesis becomes less efficient, leading to reduced plant growth and lower crop yields.

Several environmental factors influence the rate of photosynthesis. Light intensity affects how much energy is available for the process, with higher light levels generally increasing the rate up to a certain limit. Carbon dioxide concentration also plays a role, as higher levels can enhance the rate of photosynthesis. Temperature affects the activity of enzymes involved in the process, while water availability is essential because water is a key input. When any of these factors are limited, the rate of photosynthesis decreases, which in turn affects plant growth and productivity.

Phosphorus (P)

Phosphorus plays a key role in energy transfer within the plant, particularly through molecules such as ATP (adenosine triphosphate). It supports root development, early plant establishment, and reproductive processes such as flowering and seed formation. Phosphorus deficiency often leads to stunted growth and dark green or purplish foliage.

Role of ATP in Plants (Detailed Explanation)

Adenosine triphosphate (ATP) is often described as the “energy currency” of the cell, and it plays a central role in virtually all plant processes. In plants, ATP stores and transfers energy that is required for growth, development, and survival. Without ATP, plants would not be able to carry out the essential biochemical reactions needed to sustain life.

ATP is produced primarily during photosynthesis and cellular respiration. During the light-dependent reactions of photosynthesis, energy from sunlight is captured and used to generate ATP within the chloroplasts. This ATP is then used in the next stage of photosynthesis, the Calvin Cycle, where carbon dioxide is converted into glucose. In addition, ATP is continuously produced in plant mitochondria through respiration, where stored sugars are broken down to release energy for cellular activities.

One of the most important roles of ATP in plants is its involvement in energy transfer. ATP acts as a temporary storage molecule for energy, which can be released when needed. When ATP is broken down into ADP (adenosine diphosphate) and a phosphate group, energy is released and used to drive chemical reactions. This energy is essential for processes such as synthesising sugars, proteins, and other vital compounds.

ATP is also critical for biosynthesis, the process by which plants build complex molecules. For example, during photosynthesis, ATP provides the energy needed to convert carbon dioxide into glucose. It is also involved in the synthesis of amino acids, nucleic acids, and lipids. These compounds are necessary for plant structure, metabolism, and reproduction.

Another key role of ATP is in active transport. Plants must move nutrients, ions, and water across cell membranes, often against concentration gradients. This movement requires energy, which is supplied by ATP. For instance, ATP powers the uptake of mineral nutrients from the soil through root cells, ensuring that plants receive essential elements such as nitrogen, phosphorus, and potassium.

ATP also supports growth and cell division. Processes such as cell elongation, DNA replication, and mitosis all require energy. ATP provides this energy, enabling plants to grow, develop new tissues, and reproduce. In rapidly growing regions of the plant, such as root tips and young leaves, ATP demand is particularly high.

In addition, ATP plays a role in enzyme activation and metabolic regulation. Many enzymes require ATP to function effectively, and ATP helps regulate metabolic pathways within the plant. This ensures that energy is used efficiently and that processes occur in a coordinated manner.

ATP is also essential for stress responses and adaptation. When plants experience environmental stresses such as drought, heat, or nutrient deficiency, ATP is required to activate defence mechanisms, repair damaged tissues, and maintain cellular balance.

Potassium (K)

Potassium is involved in water regulation, enzyme activation, and stress tolerance. It helps plants regulate stomatal opening, improving water-use efficiency and drought resistance. Potassium also enhances disease resistance and improves the quality of fruits and grains. Deficiency symptoms include leaf edge scorching and weak plant structure.

Figure 6: Avocado with Potassium deficiency: Symptom - Scorching of leaf margins, especially on older leaves. Scot Nelson, Public Domain, via Flickr.

Secondary Nutrients

Secondary nutrients are required in moderate amounts but are still essential for proper plant function. These include calcium (Ca), magnesium (Mg), and sulphur (S).

Calcium (Ca)

Calcium is important for cell wall structure and stability, helping to maintain plant integrity and strength. It also plays a role in cell division and growth, particularly in root and shoot tips. Calcium deficiency often affects new growth, leading to distorted or necrotic tissues (e.g., blossom end rot in tomatoes).

Magnesium (Mg)

Magnesium is the central component of the chlorophyll molecule, making it essential for photosynthesis. It also plays a role in enzyme activation and energy transfer. Deficiency typically appears as interveinal chlorosis (yellowing between leaf veins), especially in older leaves.

Figure 7: Raspberry vein chlorosis. Jerzy Opioła, CC BY-SA 4.0, via Wikimedia Commons.

Sulphur (S)

Sulphur is a component of certain amino acids and proteins, as well as vitamins and enzymes. It contributes to overall plant metabolism and improves crop quality, particularly in crops like oilseeds. Sulphur deficiency can resemble nitrogen deficiency but usually appears first in younger leaves.

Micronutrients (Trace Elements)

Micronutrients are required in very small amounts but are critical for specific physiological and biochemical processes. Even slight deficiencies can significantly impact plant health and yield.

Iron (Fe)

Iron is essential for chlorophyll synthesis and plays a role in electron transport during photosynthesis. Deficiency results in interveinal chlorosis in young leaves.

Zinc (Zn)

Zinc is involved in enzyme function, protein synthesis, and growth regulation. It is important for hormone production, particularly auxins. Deficiency can cause stunted growth and small, distorted leaves.

Copper (Cu)

Copper contributes to photosynthesis, respiration, and lignin formation (important for plant structure). Deficiency may lead to wilting and poor reproductive development.

Manganese (Mn)

Manganese is involved in photosynthesis, nitrogen metabolism, and enzyme activation. Deficiency causes interveinal chlorosis and reduced growth.

Boron (B)

Boron is critical for cell wall formation, flowering, and pollination. It also aids in sugar transport within the plant. Deficiency can result in poor fruit and seed development.

Molybdenum (Mo)

Molybdenum is essential for nitrogen metabolism, particularly in the conversion of nitrate to ammonia within plants. Deficiency can impair nitrogen utilisation.

Chlorine (Cl)

Chlorine plays a role in osmotic regulation and photosynthesis. It is usually available in sufficient quantities in most soils.

Nickel (Ni)

Nickel is required in very small amounts for enzyme activity, particularly in nitrogen metabolism. Although needed in trace amounts, it is essential for plant growth.

Nutrient Balance and Interaction

Plant nutrition is not just about supplying individual nutrients but ensuring a balanced supply. Nutrients interact with each other in the soil and within the plant. For example:

- Excess nitrogen can reduce potassium uptake
- High phosphorus levels can limit micronutrient availability (e.g., zinc, iron)
- Soil pH strongly influences nutrient availability

An imbalance—whether deficiency or excess—can limit plant growth, even if other nutrients are present in adequate amounts.

Macronutrients, secondary nutrients, and micronutrients each play distinct but interconnected roles in plant growth. Macronutrients drive major growth processes, secondary nutrients support structural and metabolic functions, and micronutrients regulate critical biochemical reactions. Together, they form a complete nutrient system that must be carefully managed through fertiliser application and soil management practices to ensure optimal crop performance and sustainable agricultural production.

Fertiliser Production

Fertiliser production is fundamentally driven by plant biology; manufacturers design products based on how plants absorb, transport, and utilise nutrients. Without a detailed understanding of plant anatomy and nutrient requirements, fertiliser products would be inefficient, unbalanced, or ineffective.

At the most basic level, the classification of nutrients into macronutrients, secondary nutrients, and micronutrients directly informs fertiliser composition. Manufacturers must produce fertilisers that supply nitrogen, phosphorus, and potassium in larger quantities because these are required in the greatest amounts by plants. At the same time, they must also include or supplement secondary nutrients such as calcium, magnesium, and sulphur, as well as micronutrients like iron and zinc, even though these are needed in smaller amounts. This classification determines formulation design, blending ratios, and product specifications in fertiliser manufacturing.

Plant anatomy, particularly the structure and function of the root system, is critical to fertiliser design. Since roots are responsible for nutrient uptake, fertilisers must be produced in forms that are readily available in the soil solution and accessible to root hairs. This influences decisions such as whether nutrients are supplied in soluble forms (e.g., nitrates, phosphates, potassium salts) or in controlled-release forms. Understanding how roots absorb nutrients also drives the development of fertilisers that minimise losses and maximise uptake efficiency.

The plant's vascular system (xylem and phloem) is also highly relevant. Once nutrients are absorbed, they must be transported throughout the plant. Fertiliser manufacturers must consider nutrient mobility within the plant, for example, nitrogen is highly mobile, while calcium is not. This affects how fertilisers are applied and formulated, particularly in the development of products designed for foliar application or targeted delivery to specific plant parts.

The role of leaves and photosynthesis is another key factor influencing fertiliser production. Since photosynthesis drives plant growth, fertilisers must support this process by supplying nutrients that are directly involved in it. For example, nitrogen is required for chlorophyll production, magnesium is central to the chlorophyll molecule, and phosphorus is essential for energy transfer through ATP. Fertiliser manufacturers therefore design products that enhance photosynthetic efficiency, which in turn improves crop yield and productivity.

The inclusion of processes such as photosynthesis and ATP production highlights the importance of energy transfer within plants. Fertilisers must provide nutrients that support these biochemical processes. For instance, phosphorus is critical for ATP formation, which powers nearly all plant metabolic activities, including nutrient uptake, growth, and reproduction. This understanding influences the production of phosphorus-based fertilisers and their integration into balanced fertiliser products.

The discussion of nutrient use efficiency (NUE) and plant metabolic processes is directly linked to modern fertiliser manufacturing innovations. Manufacturers aim to produce fertilisers that are not only effective but also efficient, reducing losses to the environment. This has led to the development of technologies such as slow-release fertilisers, coated granules, and stabilised nitrogen products. These innovations are designed based on how plants absorb and utilise nutrients over time.

The concept of nutrient balance and interaction is also critical in fertiliser production. Nutrients do not act independently; they interact within the soil and plant system. For example, excessive nitrogen can interfere with potassium uptake, while high phosphorus levels can reduce the availability of micronutrients such as zinc and iron. Fertiliser manufacturers must therefore carefully design products to ensure balanced nutrient supply, avoiding antagonistic effects and ensuring optimal plant growth.

Understanding plant structure and nutrient function is essential for diagnosing deficiencies and tailoring fertiliser products to specific crops and conditions. Different crops have different nutrient requirements, and fertiliser products must be formulated accordingly. This has led to the development of customised fertilisers, crop-specific blends, and precision agriculture solutions, all of which are grounded in plant science.

Global Fertiliser Demand and Food Security

Global fertiliser demand is tightly linked to the imperative of producing enough food to feed a growing world population. As populations expand, demand rises for staple and specialty crops such as grains, fruits, and vegetables, while the available arable land remains finite and is increasingly constrained by urbanisation, land degradation, and climate change. This combination creates a scenario in which higher land-use efficiency, achievable in part through fertilisers, is necessary to sustain increasing output on existing farmland [1, 10, 11]. Fertilisers enable agricultural intensification by boosting yields and enabling high-yield crop varieties to respond effectively to nitrogen (N), phosphorus (P), and potassium (K) inputs, thereby pushing global production toward higher per-hectare outputs. The long-run trajectory of fertiliser use reflects this dynamic, with rising demand concentrated in regions with large agricultural sectors such as Asia, including China and India, where consumption patterns have historically driven global totals [12-14].

In this context, food security, a multidimensional concept encompassing availability, access, utilisation, and stability of food supply, benefits directly from higher, more reliable crop yields and improved nutritional quality that fertilisers can support when used efficiently [13-16]. The stabilisation of annual food availability hinges on predictable fertiliser supply and pricing, because fertiliser costs constitute a major portion of input costs for farmers; sharp price increases driven by energy costs, supply chain disruptions, or geopolitical events can prompt reductions in application rates and subsequent yield and output losses, which in turn affect food security and consumer prices [13, 16-19]. Energy is a central input in fertiliser production, particularly for nitrogen fertilisers produced via the Haber-Bosch process, which relies heavily on natural gas; consequently, fertiliser markets are closely linked to energy markets, with price fluctuations propagating through production costs and supply chains and potentially altering global distribution in response to geopolitical events and policy measures [14, 17, 20, 21].

At the same time, there is a growing awareness of the environmental footprint of fertiliser use. Excessive or inefficient application can drive nutrient runoff, water pollution, greenhouse gas emissions, and soil degradation, motivating a shift toward practices that improve nutrient use efficiency, adoption of precision agriculture technologies, and the integration of organic and inorganic nutrient sources to balance productivity with sustainability [1, 15, 21-23].

Looking forward, global fertiliser demand is projected to keep rising in response to population growth, dietary shifts (including increased meat and dairy consumption), and the productivity needs of developing regions; however, achieving this growth in an environmentally sustainable manner requires policies and market mechanisms that maintain economic viability for farmers while reducing environmental externalities [16, 22, 24, 25].

Fertilisers as drivers of agricultural intensification and productivity are underpinned by agronomic principles and global market dynamics. Modern high-yield crop varieties are bred to respond strongly to N, P, and K inputs, making fertiliser availability a key determinant of potential yields. Without adequate fertiliser use, these crops cannot realise their full genetic potential, which translates into lower yields and reduced overall agricultural productivity. The dependence of main food crops on fertiliser inputs is highlighted in both agronomic reviews and empirical analyses, which show substantial yield gains associated with appropriate N, P, and K fertilisation and the role of fertilisers in closing yield gaps in many regions [1, 14, 26]. The three primary fertilisers, N, P, and K, remain central to global fertiliser consumption patterns, with nitrogen typically representing the largest share of the applied nutrients due to its essential role in crop growth and the substantial energy intensity associated with its production; phosphorus and potassium are critical for various physiological and developmental processes in plants. These dynamics are reflected in sector analyses that emphasise the essential role of NPK fertilisers in sustaining high-yield production, while also acknowledging the environmental risks and the need for efficient management strategies to mitigate losses and pollution [1, 13, 14, 26].

Economic and policy dimensions reinforce the link between fertiliser demand and food security. Fertilisers account for a major portion of farmers' input costs, and their affordability and reliability influence planting decisions, application rates, and ultimately crop yields. Energy-price shocks and geopolitical tensions can disrupt fertiliser supply chains and elevate prices, provoking adjustments in fertiliser usage that reverberate through agricultural output and food prices. Empirical work across different countries demonstrates asymmetric transmission between global and domestic fertiliser prices, with exchange rates and energy costs shaping domestic affordability and access to fertilisers. This body of evidence highlights the importance of stable, affordable fertiliser supply for maintaining food security, and it underscores the potential policy leverage of energy and trade instruments in stabilising fertiliser markets [12, 17-21].

The interaction between energy markets and fertiliser production costs is particularly salient for nitrogen fertilisers, while geopolitical events (such as conflicts involving major fertiliser-producing nations) can disrupt supply and elevate prices, with knock-on effects for agricultural sectors and food affordability [12, 17-20]. Similarly, country-specific analyses show how exchange-rate movements and energy costs feed into domestic fertiliser pricing, with implications for fertiliser affordability and agricultural decision-making at the farm level in various economies [18, 20].

Sustainability concerns drive the evolving fertiliser landscape. The environmental consequences of fertiliser use, nitrate leaching, eutrophication, nitrous oxide emissions, soil degradation, and broader ecosystem impacts. have catalysed shifts toward more sustainable practices and technologies. Innovations in nutrient use efficiency, precision agriculture, and the integration of organic nutrient sources aim to maintain or increase productivity while reducing environmental harms. This includes enhanced efficiency fertilisers that tailor nutrient release to crop needs, coatings and inhibitors to reduce losses, and nutrient recovery from by-products or waste streams to support circular economy approaches. While fertilisers remain indispensable for achieving higher yields and food security, policy and industry discourse increasingly emphasise balancing productivity with environmental stewardship and resource sustainability [1, 14, 15, 22, 26].

Across a forward-looking lens, projections suggest continued growth in fertiliser demand driven by demographic and dietary trends and the need to uplift productivity in developing regions. Yet this growth must be stewarded through policies and market arrangements that ensure economic viability for farmers, promote efficient nutrient use, protect water quality and ecosystems, and align with broader climate and sustainability objectives. In this sense, fertiliser use remains a critical, but increasingly scrutinised, instrument in the global strategy to secure food security for a growing population while advancing environmental sustainability and energy resilience [11, 13, 14, 16, 24].

Case Example - War with Iran Threatens Global Food Security

Following joint U.S.–Israeli strikes on Iran on 28 February 2026, and Iran's retaliatory actions across the Middle East, the Strait of Hormuz has effectively been shut down [27]. This has placed severe pressure on one of the world's most critical chokepoints for oil, natural gas, and fertiliser trade. Recent attacks on three commercial vessels, along with the suspected deployment of sea mines, significantly increased the risks to shipping, creating uncertainty about when normal transit operations might resume. In response, the

United States targeted Iranian mine-laying vessels and proposed the use of military escorts to support commercial shipping [27].

Among the three major fertiliser nutrient groups, nitrogen, phosphorus, and potassium, nitrogen is the most widely used, accounting for around 59 percent of global fertiliser consumption in 2023, compared to 21 percent for phosphate and 20 percent for potash. Nitrogen is essential throughout the plant growth cycle, particularly for protein formation and photosynthesis, making it critical for crop yield [27]. Approximately 45 percent of global nitrogen fertiliser use is dedicated to staple crops such as wheat, rice, and maize, which together provide over 40 percent of the world's caloric intake [27].

This makes grain and cereal production highly vulnerable to disruptions in nitrogen fertiliser supply. The risk is compounded by the geographic concentration of fertiliser production. The Middle East is a major exporter of liquefied natural gas (LNG), which is a key feedstock for producing synthetic nitrogen fertilisers, as well as fertiliser products such as urea and ammonia. A large proportion of these exports pass through the Strait of Hormuz. Overall, the strait facilitates around 20 percent of global LNG exports and between 20 and 30 percent of global fertiliser exports, including approximately 35 percent of global urea trade [27].

Phosphate fertiliser supply is also exposed to disruption. Around 20 percent of global phosphate fertiliser trade originates from countries affected by the conflict or reliant on the strait, with Saudi Arabia and Israel together accounting for roughly 17 percent of global exports. In addition, sulphur, an important by-product of oil and gas processing, is a key input in phosphate fertiliser production. Approximately 45 percent of global sulphur trade is impacted by the current disruptions. If energy flows through the strait remain restricted, sulphur production is likely to decline alongside reduced fuel output, further limiting phosphate fertiliser manufacturing capacity worldwide [27].

At the same time, other major fertiliser-producing countries are not in a position to fully compensate for the reduced supply from the Middle East. Russia, which accounts for about 20 percent of global fertiliser trade and 14 percent of nitrogen fertiliser exports, is currently facing domestic export constraints and disruptions to production facilities [27]. Industry reports suggest that Russian producers are prioritising domestic supply. Meanwhile, China, a major phosphate fertiliser exporter, maintained export restrictions since 2021 to stabilise its domestic market. Although these restrictions are due to expire, recent global disruptions may lead China to extend them in order to safeguard domestic supply, further tightening international fertiliser markets [27].

Sustainability Challenges in Fertiliser Production

Fertiliser production is essential for modern agriculture, but it presents significant sustainability challenges across environmental, economic, and resource dimensions. As global demand for food continues to grow, the fertiliser industry must balance increased production with the need to reduce its environmental footprint and ensure long-term resource availability [28-31].

One of the most significant challenges is the high energy intensity of fertiliser production, particularly for nitrogen fertilisers. The production of ammonia, the building block of most nitrogen fertilisers, relies heavily on natural gas through processes such as steam methane reforming. This results in substantial carbon dioxide (CO_2) emissions, making fertiliser production a major contributor to global greenhouse gas emissions [28, 32, 33]. Reducing the carbon footprint of ammonia production is therefore a key sustainability priority, driving interest in alternatives such as green ammonia produced using renewable energy [28, 32, 33].

Another major issue is the environmental impact of fertiliser use, which is closely linked to how fertilisers are produced and applied. Excess fertiliser can be lost to the environment through leaching, runoff, and volatilisation. This can lead to water pollution, particularly eutrophication, where excess nutrients in water bodies cause algal blooms that deplete oxygen and harm aquatic life. Additionally, nitrogen fertilisers can release nitrous oxide (N_2O), a potent greenhouse gas that contributes to climate change [28-31].

The finite nature of key raw materials also presents a long-term sustainability concern. Phosphate rock, a critical input for phosphorus fertilisers, is a non-renewable resource concentrated in a few countries. Over-reliance on limited reserves raises concerns about future supply security and geopolitical dependence. Similarly, potash resources are geographically concentrated, which can create supply risks and market volatility [28, 30, 33].

Waste generation and by-products from fertiliser production pose further challenges. For example, the production of phosphoric acid generates large quantities of phosphogypsum, a by-product that can contain impurities such as heavy metals and radioactive elements. Managing and safely storing this waste is a significant environmental and regulatory issue. Improper disposal can lead to soil and water contamination [31, 32, 34-37].

Water usage is another important consideration. Fertiliser production processes require substantial amounts of water for cooling, chemical reactions, and processing. In regions where water resources are scarce, this can place additional pressure on local water supplies and ecosystems [28, 29, 32, 33].

There are also challenges related to nutrient use efficiency (NUE). A significant proportion of applied fertiliser is not absorbed by plants and is instead lost to the environment. This represents both an economic inefficiency and an environmental risk. Improving NUE through better fertiliser formulations, precision application, and soil management practices is critical for sustainable agriculture [28, 29, 32, 33].

From an economic and social perspective, price volatility and accessibility are key sustainability issues. Fertiliser production is closely tied to energy markets, meaning that fluctuations in fuel prices can significantly impact fertiliser costs. In many developing regions, high fertiliser prices can limit access for farmers, reducing crop yields and affecting food security [28, 32, 33].

In response to these challenges, the fertiliser industry is increasingly focusing on sustainable innovations and practices. These include the development of low-carbon production methods, such as green ammonia; the use of controlled-release and stabilised fertilisers to improve efficiency; recycling nutrients from organic waste streams; and adopting circular economy approaches. Precision agriculture technologies are also being used to optimise fertiliser application and reduce environmental impacts [28-30, 33].

Chapter 2

Soil Science and Plant Nutrition

This chapter builds directly on the foundational concepts established in the previous chapter, where the focus was placed on the origins, development, and global significance of fertilisers in supporting agricultural productivity and food security. Having examined how fertilisers emerged historically and why they are essential in modern farming systems, the discussion now shifts from a broad, system-level perspective to a more detailed exploration of the soil environment in which fertilisers operate.

In the previous chapter, fertilisers were introduced as a critical tool for replenishing nutrients removed through crop production and sustaining global food systems. However, the effectiveness of fertilisers cannot be fully understood without examining the medium through which they function—soil. Soil is not simply an inert growing medium; it is a dynamic, living system composed of physical, chemical, and biological components that interact continuously. These interactions determine how nutrients are stored, transformed, and made available to plants.

This chapter therefore focuses on the structure, composition, and behaviour of soils, providing the essential scientific context required to understand fertiliser performance. It explores soil texture, structure, and phase relationships, as well as the organisation of soil into distinct horizons. It also examines how soil properties such as pH, salinity, and biological activity influence nutrient availability and movement. Together, these factors explain why the same fertiliser can behave very differently depending on soil conditions.

A key theme throughout this chapter is the concept that nutrient availability is not solely determined by the amount of nutrient present, but by the complex interactions within the soil system. Concepts such as nutrient cycling, soil fertility, and deficiency symptoms are revisited in greater depth, linking plant nutrition directly to soil processes. This provides a more integrated understanding of how fertilisers interact with soil chemistry and biology to support plant growth.

By developing a detailed understanding of soil systems, this chapter establishes the scientific basis for effective fertiliser management. It highlights the importance of aligning fertiliser strategies with soil conditions to maximise nutrient use efficiency, minimise environmental losses, and support sustainable agricultural production. In doing so, it creates a clear progression from the "why" of fertiliser use explored in the previous chapter to the "how" and "where" that underpin practical application in real-world agricultural systems.

Soil Composition and Structure

Composition

Soil is a complex and dynamic natural system that forms the foundation of terrestrial plant life. It is composed of a mixture of mineral particles, organic matter, water, and air, all of which interact to support plant growth. The relative proportions of these components determine soil fertility, water-holding capacity, aeration, and overall productivity. In agricultural contexts, understanding soil composition is essential for effective fertiliser application, as it influences how nutrients are stored, transformed, and made available to plants.

The mineral component of soil originates from the weathering of rocks and is typically classified into three main particle sizes: sand, silt, and clay. Sand particles are the largest and provide good drainage and aeration but have low nutrient-holding capacity. Silt particles are intermediate in size and contribute to moderate water retention and fertility. Clay particles are the smallest and have a high surface area, allowing them to retain water and nutrients effectively. The proportion of these particles determines the soil's texture, which influences how easily roots can penetrate and how well the soil can retain moisture and nutrients.

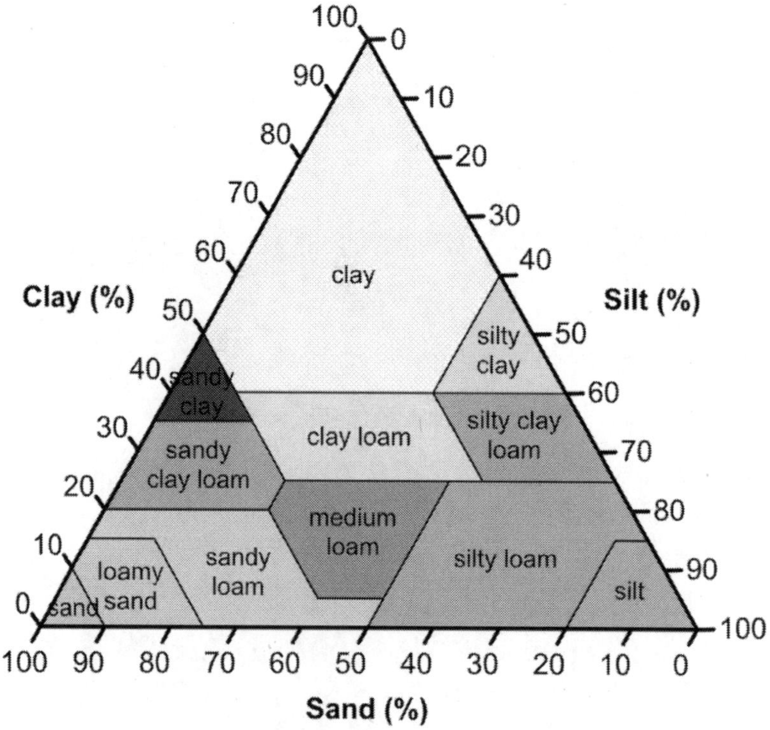

Figure 8: Soil types by clay, silt and sand composition. Zephyris at English Wikipedia, CC BY-SA 3.0, via Wikimedia Commons.

The diagram shown as Figure 8 is known as a soil texture triangle, a widely used tool in agriculture, soil science, and fertiliser management. It is designed to classify soils based on the relative proportions of three key components: sand, silt, and clay. By understanding the balance of these particles, farmers and agronomists can better predict how a soil will behave in terms of water retention, nutrient availability, and overall productivity.

The triangle is structured so that each of its three sides represents 100% of a particular soil component. The bottom axis corresponds to sand content, the right side represents silt, and the left side represents clay. Any point within the triangle reflects a specific combination of these three components, and together they always total 100%. This allows for precise classification of soil types based on their composition.

To read the diagram, the percentage of sand, silt, and clay in a soil sample must first be determined. From each axis, lines are followed inward that correspond to these percentages. The point at which the three lines intersect identifies the soil's textural class. For example, a soil containing 40% sand, 40% silt, and 20% clay is classified as loam, which is considered ideal for many crops. In contrast, a soil with a high clay content, such as 60%, will fall near the top of the triangle and be classified as clay.

The triangle divides soils into several major categories, each with distinct physical and agricultural characteristics. Sandy soils, located toward the bottom left of the triangle, consist of large particles that allow for excellent drainage and aeration. However, they have low nutrient retention, meaning fertilisers can leach away quickly. As a result, these soils often require smaller, more frequent fertiliser applications to maintain nutrient availability.

Silty soils, found toward the bottom right, have a smoother texture and moderate water-holding capacity. They are generally more fertile than sandy soils and can retain nutrients more effectively. However, they are more susceptible to erosion, which can lead to nutrient loss if not properly managed.

Clay soils, located at the top of the triangle, are composed of very fine particles. These soils have a high capacity to retain water and nutrients, but they often suffer from poor drainage and limited aeration. While nutrients are held effectively, they can sometimes become chemically bound and less available to plants. This can make nutrient management more complex in clay-rich soils.

At the centre of the triangle are the loam soils, which represent a balanced mixture of sand, silt, and clay. These soils are considered ideal for agriculture because they combine good drainage with strong nutrient and water retention. Loam soils support healthy root development and efficient nutrient uptake, making them highly productive and responsive to fertiliser application.

Understanding the soil texture triangle is essential in fertiliser management because soil type directly influences how nutrients behave in the soil. Clay soils tend to hold nutrients, while sandy soils are prone to losing them through leaching. Soil texture also affects water retention, which in turn influences how fertilisers dissolve and become available to plants. Additionally, the physical structure of the soil impacts root growth and the plant's ability to access nutrients.

By identifying where a soil falls within the triangle, agronomists and farmers can make more informed decisions about fertiliser use. This includes selecting appropriate fertiliser types, adjusting application rates, improving nutrient use efficiency, and reducing environmental impacts such as runoff and leaching. Ultimately, the soil texture triangle serves as a foundational tool for optimising both crop production and sustainable fertiliser practices.

In addition to minerals, soil contains organic matter, which consists of decomposed plant and animal residues. This material plays a crucial role in soil fertility by improving structure, increasing water retention, and supplying essential nutrients as it breaks down. Organic matter also supports soil microorganisms, which are vital for nutrient cycling processes such as nitrogen fixation and decomposition. Soils rich in organic matter are generally more productive and resilient.

The water and air components of soil occupy the spaces, or pores, between soil particles. These pore spaces are critical for plant growth. Water in the soil dissolves nutrients, making them available for uptake by plant roots, while air provides oxygen necessary for root respiration and microbial activity. A balanced ratio of water and air is essential; too much water can lead to poor aeration and root suffocation, while too little water limits nutrient availability and plant growth.

Structure

Soil structure refers to the arrangement of soil particles into aggregates or clusters, which affects how water, air, and roots move through the soil. Well-structured soils have stable aggregates that create a network of pores, allowing for good drainage, aeration, and root development. Poor soil structure, such as compacted or heavily crusted soils, restricts root growth, reduces water infiltration, and limits nutrient uptake. Soil structure is influenced by factors such as organic matter content, biological activity, and management practices.

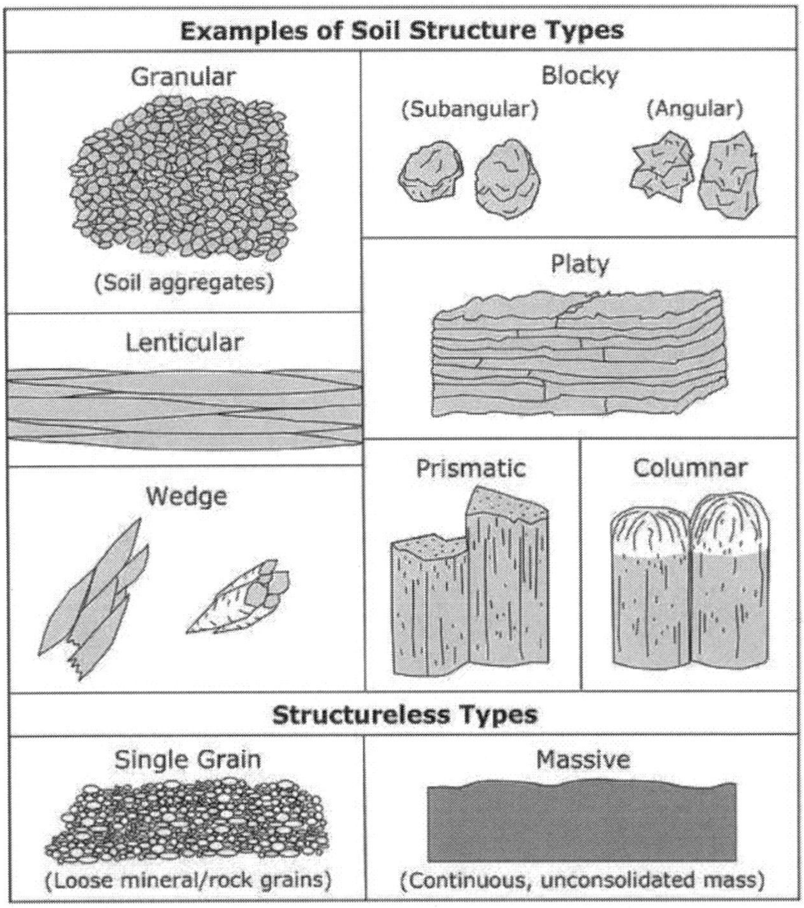

Figure 9: USDA-NRCS diagram of soil structure types including granular, subangular and angular blocky, lenticular, platy, wedge, prismatic, columnar, single grain, and massive. U.S. Department of Agriculture, Public Domain, via Picryl.

Figure 9 illustrates the different types of soil structure, which refers to how soil particles (sand, silt, and clay) are arranged into aggregates or clusters. Unlike soil texture, which describes particle size, soil structure determines how soil behaves in terms of water movement, aeration, root growth, and nutrient availability. This is particularly important in agriculture and fertiliser management, as structure directly affects how efficiently plants can access water and nutrients.

The first group shown is granular structure, where soil particles form small, rounded aggregates. This structure is common in topsoil and is considered highly desirable for agriculture. Granular soils allow good air circulation, water infiltration, and root penetration. They also support strong microbial activity, which enhances nutrient cycling. Because of these properties, fertilisers applied to granular soils are typically well utilised by plants.

The diagram also shows blocky structures, which are divided into subangular and angular types. These consist of irregular, block-like aggregates. Subangular blocky soils have rounded edges, while angular blocky soils have sharper corners. These structures are often found in subsoils and provide moderate drainage and aeration. They can support plant growth reasonably well, although they are not as ideal as granular soils. Fertiliser movement in these soils is generally acceptable but may be somewhat restricted compared to more open structures.

Platy structure is characterised by thin, flat layers of soil that lie horizontally. This structure often forms due to compaction or pressure, such as from heavy machinery or repeated cultivation. Platy soils restrict water movement and root penetration because the layers act as barriers. As a result, fertiliser uptake can be reduced, and water may accumulate above the layers, leading to poor plant growth.

The diagram includes prismatic and columnar structures, which consist of vertically elongated aggregates. Prismatic soils have flat tops, while columnar soils have rounded tops. These structures are typically found in deeper soil layers and can influence drainage patterns. Water tends to move vertically along the cracks between columns, but root growth may be limited within the dense aggregates. Nutrient movement can be uneven, affecting fertiliser efficiency.

Wedge and lenticular structures are less common and often associated with soils that experience shrinking and swelling, such as clay-rich soils. These structures can change shape depending on moisture conditions, which can disrupt root systems and affect nutrient uptake.

At the bottom of the diagram are structureless soils, which lack defined aggregates. These include:

- Single grain structure, where soil particles are loose and not bound together, typical of sandy soils. These soils drain quickly but have poor nutrient retention, meaning fertilisers can be easily lost through leaching.
- Massive structure, where soil forms a dense, solid mass with no visible structure. This type restricts root growth, water movement, and aeration, making it difficult for plants to access nutrients.

Overall, the diagram highlights that soil structure plays a critical role in determining soil quality and productivity. Well-structured soils, such as granular or moderately blocky types, support efficient water and nutrient movement, leading to better plant growth. Poorly structured or structureless soils can limit fertiliser effectiveness, reduce root development, and ultimately lower crop yields. Understanding soil structure is therefore essential for making informed decisions about soil management, cultivation practices, and fertiliser application strategies.

Another important aspect of soil is its chemical composition, including pH and nutrient content. Soil pH affects the availability of nutrients to plants; for example, highly acidic or alkaline soils can limit the uptake of essential elements like phosphorus or micronutrients. The presence of nutrients such as nitrogen, phosphorus, potassium, calcium, and magnesium determines the soil's fertility level. Fertilisers are often applied to correct deficiencies and optimise nutrient availability based on soil composition.

Overall, soil composition and structure are fundamental to plant growth and agricultural productivity. They determine how effectively water and nutrients are retained and supplied to plants, how roots develop, and how resilient the soil is to environmental stresses. A thorough understanding of these factors enables better soil management practices, more efficient fertiliser use, and improved crop yields.

Soil Phases

Soil is not a uniform solid material but a three-phase system consisting of solids, liquids, and gases. The solid phase is made up of mineral particles such as sand, silt, and clay, along with organic matter. The liquid phase consists of water that contains dissolved nutrients essential for plant growth, while the gas phase is made up of air occupying the spaces between soil particles. These three phases exist together within the soil and are constantly interacting. The balance between them plays a critical role in determining how effectively soil can support plant growth and agricultural productivity.

A soil phase diagram, such as shown as Figure 10, is a simplified representation used to show how these three phases are distributed within a given volume of soil. It helps visualise the proportion of soil that is made up of solids, the amount of space filled with water, and the portion occupied by air. In the attached diagram, this concept is illustrated in a basic form, with the blue section representing the liquid phase (water) and the brown section representing the solid phase (soil particles). Although simplified, the diagram effectively demonstrates that soil is composed of different physical components sharing the same space.

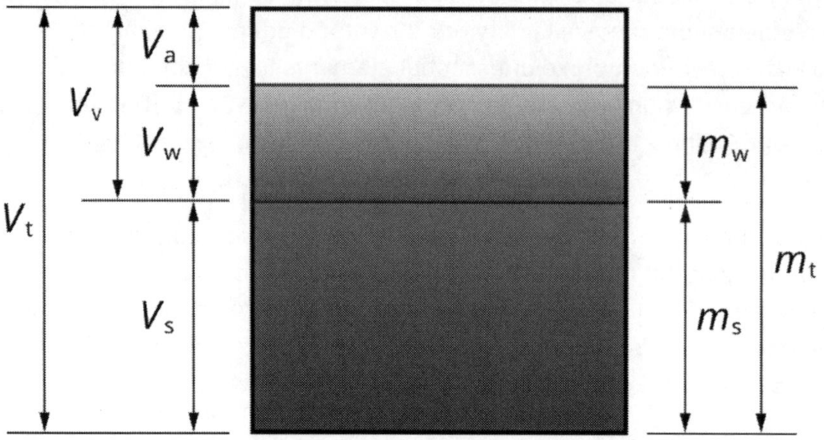

Figure 10: Soil phase diagram showing soil composition. V is for volume, M is for mass. Subscripts s, w, and a stand for soil particles, water and air respectively. Subscripts v and t stand for voids and total respectively. Derivative work: 5d7ygtr09hSjhan81, CC BY-SA 3.0, via Wikimedia Commons.

One of the key insights provided by a soil phase diagram is the concept of porosity, or pore space. Soil is not solid throughout; instead, it contains numerous voids between particles. These pores can be filled with either water or air, and their size and distribution influence how water moves through the soil, how well the soil is aerated, and how easily roots can grow. High porosity generally supports better plant growth by allowing adequate water retention and oxygen supply.

The diagram also helps explain water content, which refers to the proportion of pore space occupied by water. This is a critical factor in plant health. If too much of the pore space is filled with water, the soil can become waterlogged, reducing oxygen availability and harming root systems. Conversely, if too little water is present, plants may experience drought stress, and the movement of nutrients to roots is limited.

Although not always explicitly shown in simplified diagrams, the remaining pore space is occupied by air, which is essential for soil aeration. Oxygen in the soil is required for root respiration and for the activity of microorganisms that play a key role in nutrient cycling. Without sufficient air, biological processes slow down, and plant growth is negatively affected.

The relative proportions of solids and pore space also indicate the level of soil compaction or density. Soils with a high proportion of solids and limited pore space are compacted, which restricts root penetration, reduces water infiltration, and limits fertiliser effectiveness. In contrast, soils with well-balanced pore space are loose and well-structured, allowing for efficient movement of water, air, and nutrients.

Another important aspect illustrated by soil phases is nutrient movement and fertiliser efficiency. Nutrients are dissolved in the soil water and transported to plant roots through the liquid phase. If water balance is poor, nutrient uptake is reduced. Excess water can lead to nutrient losses through leaching, particularly in sandy soils, while insufficient water limits nutrient mobility.

Understanding soil phases is therefore essential in agriculture, particularly for fertiliser management. Nutrient availability depends on the presence of water, root health depends on adequate air supply, and nutrient storage and retention are influenced by the solid components, especially clay and organic matter. An ideal agricultural soil typically contains approximately 45% mineral solids, 5% organic matter, 25% water, and 25% air. This balance provides the optimal conditions for plant growth, ensuring efficient nutrient uptake, healthy root development, and sustainable crop production.

Nutrient Cycles (Nitrogen, Phosphorus, Potassium)

Nutrient cycles describe how essential elements move through the soil, plants, atmosphere, water, and living organisms. In agriculture, understanding these cycles is critical because it explains how nutrients become available to crops, how they are lost, and how fertilisers can be used to maintain soil fertility. The three primary nutrients, nitrogen, phosphorus, and potassium, each follow distinct pathways in the environment.

Nitrogen Cycle

The nitrogen cycle is the most complex of the three because nitrogen exists in multiple chemical forms and moves between the atmosphere, soil, and living organisms. Although nitrogen gas (N_2) makes up about 78% of the atmosphere, plants cannot use it directly.

The cycle begins with nitrogen fixation, where atmospheric nitrogen is converted into usable forms such as ammonia (NH_3). This occurs naturally through soil bacteria or artificially through industrial processes used in fertiliser production. Once in the soil, ammonia is converted into nitrite (NO_2^-) and then nitrate (NO_3^-) through nitrification, a process carried out by specialised bacteria.

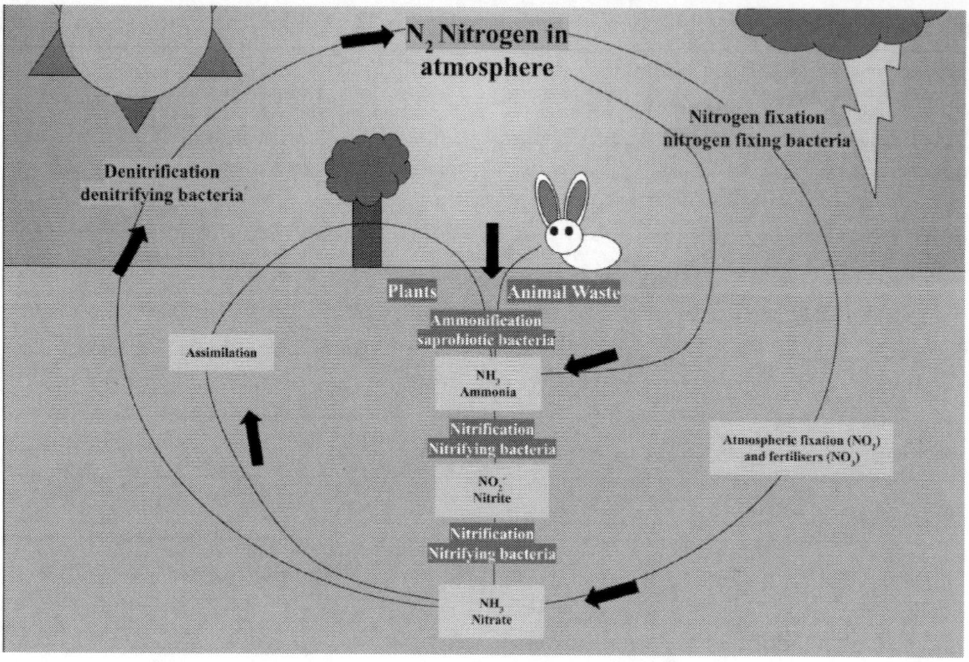

Figure 11: Simplified version of the nitrogen biogeochemical cycle with bacteria present alongside the processes that convert them via chemicals. EssauMejia, CC BY-SA 4.0, via Wikimedia Commons.

Plants absorb nitrogen mainly in the form of nitrate or ammonium (NH_4^+). This nitrogen is used to produce proteins, enzymes, and chlorophyll, making it essential for plant growth. When plants and animals die, organic nitrogen is returned to the soil through decomposition, where microorganisms convert it back into ammonium in a process called mineralisation.

Finally, some nitrogen is lost back to the atmosphere through denitrification, where bacteria convert nitrate into nitrogen gas. Nitrogen can also be lost through leaching (movement with water) or volatilisation (loss as gas), making it a highly dynamic and often limiting nutrient in agriculture.

Phosphorus Cycle

The phosphorus cycle is simpler than the nitrogen cycle because it does not involve a gaseous phase. Phosphorus primarily moves through soil, water, and living organisms.

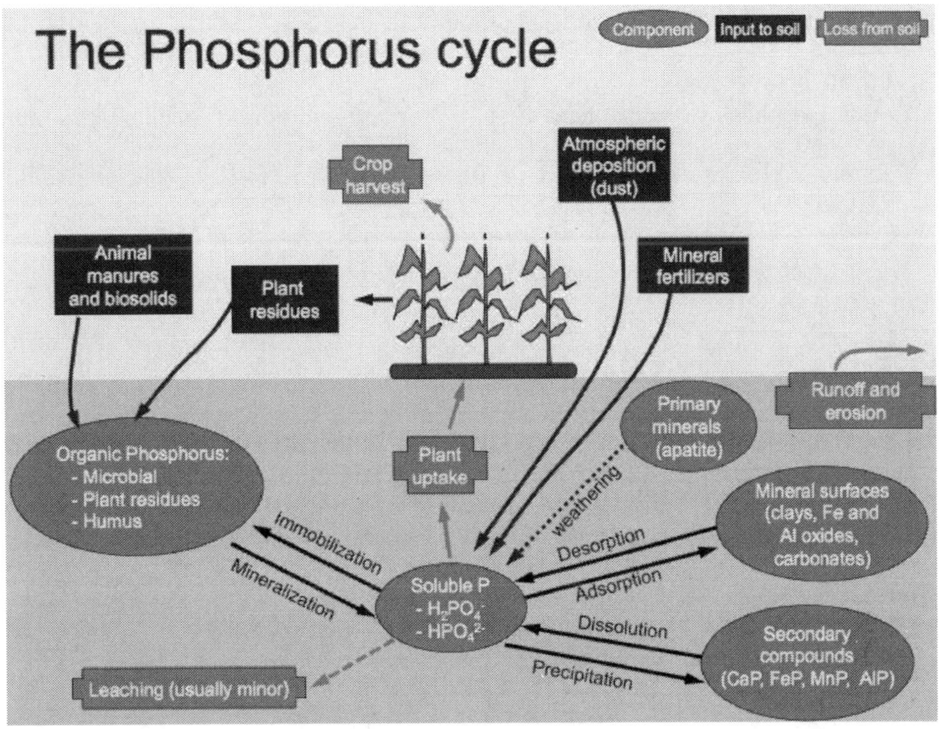

Figure 12: Phosphorus Cycle. Welcome1To1The1Jungle at English Wikipedia, CC BY 3.0, via Wikimedia Commons.

The cycle begins with the weathering of rocks, which releases phosphate ions (PO_4^{3-}) into the soil. Plants absorb phosphorus from the soil solution and use it for energy transfer (ATP), root development, and reproduction. Animals obtain phosphorus by consuming plants.

When plants and animals die, phosphorus is returned to the soil through decomposition. However, much of the phosphorus in soil becomes chemically bound to minerals such as calcium, iron, or aluminium, making it less available to plants. This process is known as fixation.

Phosphorus can also be lost from soils through erosion and runoff, especially into water bodies, where it can contribute to eutrophication. Because phosphorus is relatively immobile in soil, it tends to accumulate in the upper layers, and fertiliser placement becomes important for effective uptake.

Potassium Cycle

The potassium cycle differs from nitrogen and phosphorus because potassium does not form part of organic compounds in plants. Instead, it exists as a positively charged ion (K^+) and plays a key role in water regulation, enzyme activation, and stress resistance.

Potassium originates from the weathering of minerals in rocks. In soil, it exists in three main forms:

- Solution potassium → dissolved in soil water and immediately available to plants
- Exchangeable potassium → held on clay particles and organic matter, available for uptake
- Fixed or structural potassium → trapped within mineral structures and released slowly

Plants absorb potassium in its ionic form (K^+), and it moves easily within plant tissues. After plants die or shed material, potassium is quickly returned to the soil because it is not bound in complex organic structures.

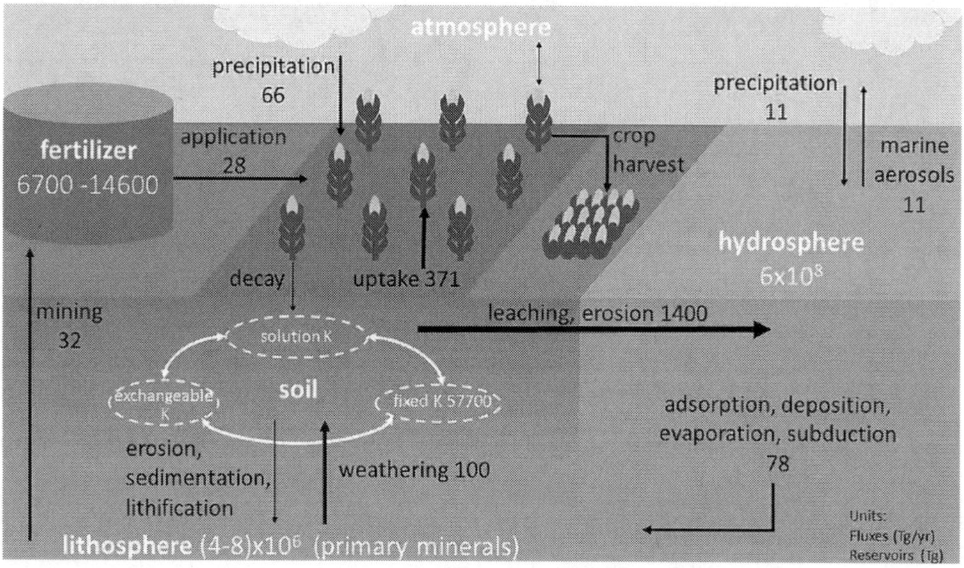

Figure 13: Figure depicting the biogeochemical cycle of potassium. Compost Camel, CC BY-SA 4.0, via Wikimedia Commons.

Unlike nitrogen, potassium is not lost to the atmosphere, but it can be lost through leaching, especially in sandy soils, or removed through crop harvesting. The balance between available and fixed potassium is influenced by soil type, particularly clay content.

Nutrient Cycles in Agriculture

Understanding nutrient cycles is fundamental to effective fertiliser management and sustainable agricultural practices. Nitrogen, phosphorus, and potassium each behave differently in the soil, and these differences directly influence how they should be applied and managed. Nitrogen is highly mobile and can be easily lost through processes such as leaching, volatilisation, and denitrification, which means it must be applied carefully and at the right time to match crop demand. In contrast, phosphorus is less mobile in the soil but can become chemically bound and unavailable to plants, making correct placement near the root zone critical for effective uptake. Potassium is generally retained within the soil, particularly in clay-rich soils, but it can become depleted over time through crop removal and leaching, requiring regular replenishment to maintain soil fertility.

Farmers and agronomists use their understanding of these nutrient cycles to make informed decisions about fertiliser use. This includes optimising application rates, timing, and methods to ensure that nutrients are available when plants need them most. By aligning fertiliser strategies with nutrient behaviour, they can improve nutrient use efficiency (NUE), ensuring that a greater proportion of applied nutrients is taken up by crops rather than lost to the environment. This approach also helps reduce negative environmental impacts, such as nutrient leaching into groundwater or runoff into waterways, which can lead to pollution and ecosystem damage. Ultimately, managing nutrient cycles effectively supports long-term soil fertility, sustainable crop production, and food security.

Nutrient Cycles in Fertiliser Manufacture

Understanding nutrient cycles is directly relevant to fertiliser manufacture because these cycles define how nutrients behave in the environment, how they become available to plants, and how they are lost. Fertiliser production is not simply about supplying nutrients—it is about designing products that align with the natural cycling processes of nitrogen, phosphorus, and potassium to maximise plant uptake and minimise losses.

In the case of the nitrogen cycle, its complexity and high mobility strongly influence fertiliser manufacturing processes. Nitrogen exists in multiple forms and is easily lost through leaching, volatilisation, and denitrification. As a result, fertiliser manufacturers produce nitrogen in forms that plants can readily absorb, such as ammonia, ammonium, and nitrate. Industrial processes, such as the Haber-Bosch process, are specifically designed to convert atmospheric nitrogen (N_2), which plants cannot use, into ammonia (NH_3), a usable form. Furthermore, because nitrogen is prone to loss, manufacturers develop enhanced-efficiency fertilisers, such as slow-release formulations and stabilised nitrogen products, to better match plant uptake and reduce environmental losses.

The phosphorus cycle also has a significant influence on fertiliser production. Unlike nitrogen, phosphorus does not have a gaseous phase and is relatively immobile in soil. However, it is often chemically fixed by soil minerals, making it unavailable to plants. This limitation drives the need for fertiliser manufacturing processes that convert raw phosphate rock into more soluble and plant-available forms, such as superphosphate or phosphoric acid-based fertilisers. Manufacturers also consider particle size, solubility, and chemical composition to ensure that phosphorus remains available in the soil long enough for plant uptake. In addition, because phosphorus losses are often associated with runoff and erosion, fertiliser design increasingly focuses on improving placement efficiency and reducing environmental impact.

The potassium cycle influences fertiliser production in a different way. Potassium is not part of organic compounds and exists primarily as a soluble ion (K^+). It is generally retained in the soil but can become locked in mineral structures or depleted through crop removal. Fertiliser manufacturing therefore focuses on producing potassium in readily soluble forms, such as potassium chloride (KCl) or potassium sulphate (K_2SO_4), which can replenish soil reserves and maintain adequate levels for plant growth. The relatively stable nature of potassium means that fertiliser strategies are often focused on maintaining long-term soil balance rather than managing rapid losses.

Potassium Humate Used in Fertilizers

Potassium humate is a widely used component in fertiliser formulations due to its ability to improve soil fertility, enhance nutrient availability, and support overall plant growth. It is a water-soluble organic compound derived from humic substances, typically extracted from natural sources such as leonardite, lignite, peat, or sapropel. Rich in humic acids, potassium humate plays an important role in sustainable agriculture by improving soil health while reducing reliance on synthetic inputs [38].

One of the primary benefits of potassium humate is its ability to improve soil structure and physical properties. It acts as a soil conditioner by binding soil particles into stable aggregates, which enhances porosity, aeration, and root penetration. This improved structure allows roots to access water and nutrients more effectively. In sandy soils, potassium humate increases water-holding capacity, helping to reduce moisture loss and improve drought resistance. In heavy clay soils, it reduces compaction and improves drainage, creating a more favourable environment for root development and microbial activity [38].

Potassium humate also plays a significant role in enhancing nutrient retention and availability. Due to its high cation exchange capacity (CEC), it can bind essential nutrients such as nitrogen, phosphorus, potassium, and micronutrients, reducing their loss through leaching. It also chelates micronutrients, making them more available for plant uptake. In addition, potassium humate contributes potassium directly to plant nutrition, supporting key processes such as photosynthesis, enzyme activation, and water regulation. Its buffering effect helps maintain nutrient availability in both acidic and alkaline soils [38].

Figure 14: Liquid Potassium Humate. UlmicA, CC BY-SA 4.0, via Wikimedia Commons.

As a plant growth stimulant, potassium humate promotes root development, improves seed germination, and enhances overall plant vigour. It supports increased chlorophyll production, which improves photosynthesis and leads to greater biomass and yield. Crops treated with potassium humate often show improvements in quality, including better fruit size, colour, sugar content, and shelf life. These benefits make it particularly valuable in both field crops and high-value horticultural systems [38].

Another important function of potassium humate is its ability to stimulate soil microbial activity. It provides a carbon source for beneficial microorganisms, which play a key role in nutrient cycling and organic matter decomposition. This leads to improved soil fertility over time and helps suppress harmful pathogens. Enhanced microbial activity also contributes to the formation of humus, further improving soil structure and long-term productivity [38].

Potassium humate contributes to soil pH regulation and remediation by buffering extreme pH conditions and improving nutrient availability. It can bind heavy metals and reduce their bioavailability, making it useful in rehabilitating degraded or contaminated soils. This property supports safer crop production and reduces environmental risks associated with soil pollution [38].

In terms of plant resilience, potassium humate helps improve resistance to environmental stress, including drought, salinity, and temperature extremes. It enhances water-use efficiency and supports the plant's natural defence mechanisms, enabling crops to better withstand adverse conditions and recover from stress events [38].

A key advantage of potassium humate in fertiliser systems is its ability to reduce dependence on synthetic fertilisers. By improving nutrient use efficiency, it allows for lower application rates of conventional fertilisers without compromising crop yield. It also enhances the performance of common fertilisers such as urea and NPK blends by reducing nutrient losses and improving uptake [38].

Potassium humate is also well suited to sustainable and organic farming systems. As a biodegradable and environmentally friendly product, it supports soil organic matter buildup and contributes to carbon sequestration. Its use aligns with the growing demand for sustainable agricultural practices and reduced environmental impact [38].

In practical terms, potassium humate is highly versatile and can be applied across a wide range of agricultural systems, including field crops, horticulture, orchards, greenhouses, and hydroponics. It can be used in soil applications, foliar sprays, drip irrigation systems, and seed treatments, making it adaptable to different farming methods and crop requirements [38].

Overall, potassium humate enhances fertiliser efficiency, improves soil health, and supports sustainable crop production. Its multifunctional role makes it a valuable component in modern fertiliser manufacturing and agricultural systems, particularly as the industry moves toward more environmentally responsible and resource-efficient practices [38].

More broadly, knowledge of nutrient cycles allows fertiliser manufacturers to design products that improve nutrient use efficiency (NUE). By understanding how nutrients move, transform, and are lost in the soil–plant system, manufacturers can develop fertilisers that release nutrients at controlled rates, remain available under different soil conditions, and match crop demand throughout the growing season. This includes innovations such as coated fertilisers, inhibitors, and customised nutrient blends.

Nutrient cycles also inform the selection of raw materials and production technologies. For example, the reliance of nitrogen fertilisers on natural gas highlights the link between energy systems and fertiliser supply. Similarly, the finite nature of phosphate rock and the geological origin of potash influence global supply chains and manufacturing strategies. Understanding these cycles helps manufacturers anticipate resource constraints and develop more sustainable production methods, including recycling nutrients from waste streams.

Soil Fertility and Deficiency Symptoms

Soil fertility refers to the ability of soil to supply essential nutrients to plants in adequate amounts and in forms that are readily available for uptake. A fertile soil supports healthy plant growth by providing the right balance of macronutrients (such as nitrogen, phosphorus, and potassium), secondary nutrients (such as calcium, magnesium, and sulphur), and micronutrients (such as iron, zinc, and manganese). In addition to nutrient content, soil fertility is influenced by physical properties such as structure, texture, aeration, and water-holding capacity, as well as biological activity, including microorganisms that help cycle nutrients into plant-available forms.

When soil fertility is well balanced, plants are able to grow vigorously, develop strong root systems, and produce high yields. However, when one or more nutrients are deficient, plant growth becomes limited, even if other nutrients are present in sufficient amounts. This concept is often described by the "law of the minimum," where the nutrient in shortest supply determines overall plant performance. As a result, identifying and correcting nutrient deficiencies is essential for maintaining productive agricultural systems.

Deficiency symptoms occur when plants do not receive enough of a particular nutrient to meet their physiological needs. These symptoms are often visible in leaves, stems, or overall plant growth patterns, and they can vary depending on the specific nutrient involved. For example, nitrogen deficiency typically causes yellowing (chlorosis) of older leaves, as nitrogen is mobile within the plant and is moved to younger tissues. This results in reduced leaf area and stunted growth, directly impacting photosynthesis and yield.

Phosphorus deficiency often appears as dark green or purplish discoloration of leaves, particularly in younger plants. It can also lead to poor root development and delayed maturity, affecting flowering and seed formation. Because phosphorus is involved in energy transfer within the plant, its deficiency limits the plant's ability to grow and reproduce efficiently.

Potassium deficiency is usually seen as yellowing or browning along the edges of leaves, known as leaf scorch. It can weaken plant resistance to disease, reduce water-use efficiency, and negatively affect crop quality. Since potassium plays a key role in regulating water balance and enzyme activity, its deficiency can make plants more vulnerable to environmental stress.

Secondary nutrient deficiencies also present distinct symptoms. Calcium deficiency often affects growing points, leading to poor root and shoot development, while magnesium deficiency typically causes interveinal chlorosis (yellowing between leaf veins) in older leaves, as magnesium is a central component of chlorophyll. Sulphur deficiency, on the other hand, resembles nitrogen deficiency but usually affects younger leaves first, as sulphur is less mobile within the plant.

Micronutrient deficiencies, although required in smaller amounts, can have significant impacts on plant health. For instance, iron deficiency causes interveinal chlorosis in young leaves, zinc deficiency can lead to stunted growth and reduced leaf size, and boron deficiency may result in poor flower and fruit development. These deficiencies are often influenced by soil pH, as certain nutrients become less available under highly acidic or alkaline conditions.

The following are a range of that illustrate how nutrient deficiencies appear in real cropping systems and how they affect productivity.

Nitrogen (N) Deficiency – Cereal Crops (Wheat, Rice, Maize): In cereal crops such as wheat, maize, and rice, nitrogen deficiency is commonly observed as yellowing of older (lower) leaves, progressing upward as the deficiency worsens. Because nitrogen is mobile, it is relocated to younger tissues, leaving older leaves depleted. Crops become pale, thin, and stunted, with reduced tillering in wheat and smaller cob formation in maize. This directly reduces photosynthesis and ultimately leads to significant yield losses, particularly in grain production systems.

Phosphorus (P) Deficiency – Early Crop Establishment (Corn, Canola): In crops such as corn and canola, phosphorus deficiency is often seen in early growth stages. Plants may develop a dark green to purplish coloration, especially on leaves and stems, due to impaired energy transfer. Root systems remain poorly developed, limiting nutrient and water uptake. This results in delayed maturity, poor flowering, and reduced seed formation, making phosphorus critical during crop establishment.

Figure 15: Maize seedling showing phosphorus deficiency symptoms. Although sufficient P fertilizer had been applied, the soil had a high acid saturation, and aluminium toxicity restricted root growth and P uptake. Alandmanson, CC0, via Wikimedia Commons.

Potassium (K) Deficiency – Horticulture (Tomatoes, Potatoes, Fruit Trees): In horticultural crops like tomatoes, potatoes, and fruit trees, potassium deficiency appears as yellowing and browning along leaf edges (leaf scorch). Plants may have weak stems, poor fruit quality, and increased susceptibility to disease. In fruit crops, this can result in smaller fruits, reduced sugar content, and poor shelf life, significantly affecting market value.

Calcium (Ca) Deficiency – Vegetables (Tomatoes, Lettuce):

Calcium deficiency commonly affects rapidly growing tissues. In tomatoes, it causes blossom end rot, where the fruit develops dark, sunken lesions. In leafy vegetables like lettuce, it leads to tip burn, damaging leaf edges. Since calcium is immobile, symptoms appear in new growth, affecting crop quality and reducing marketability.

Figure 16: Heart rot of cabbage caused by calcium deficiency. Scot Nelson, Public Domain, via Flickr.

Figure 17: Calcium deficiency of banana - Symptom: Drooping and slightly chlorotic flag leaf (in severe cases the leaf is white in colour with a necrotic tip). Scot Nelson, Public Domain, via Flickr.

Magnesium (Mg) Deficiency – Vine Crops (Grapes, Citrus): Magnesium deficiency is common in crops like grapes and citrus. It appears as interveinal chlorosis in older leaves, where the tissue between veins turns yellow while veins remain green. This reduces chlorophyll production and photosynthesis, leading to lower sugar accumulation in fruits and reduced crop quality.

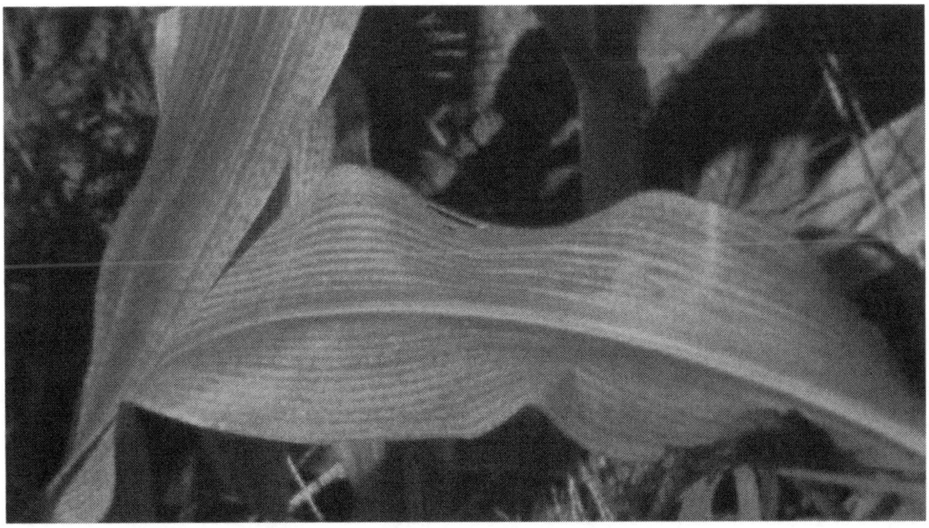

Figure 18: Maize leaves showing magnesium-deficiency symptoms; in the acid soils of a small-scale farmer in Vulindlela, KwaZulu-Natal, South Africa. Acid soils often have low Mg reserves, and Mg uptake is suppressed at high soil acid saturation. Magnesium deficiency symptoms are therefore often a symptom of severe soil acidity. Alandmanson, CC BY 4.0, via Wikimedia Commons.

Iron (Fe) Deficiency – High pH Soils (Soybeans, Fruit Crops): Iron deficiency is common in alkaline soils and affects crops like soybeans and fruit trees. It causes interveinal chlorosis in young leaves, as iron is not easily transported within the plant. Severe cases result in almost white leaves and reduced growth, limiting photosynthesis and yield potential.

Zinc (Zn) Deficiency – Broadacre Crops (Wheat, Maize): Zinc deficiency in crops like wheat and maize leads to stunted growth, shortened internodes, and small leaves. In maize, it may appear as striping or banding on leaves. This reduces plant vigour and limits yield, especially in soils with low organic matter or high pH.

Figure 19: Macadamia shoots showing zinc deficiency symptoms. The youngest leaves show chlorosis (yellowing), dwarfing and malformation. Near Margate, KwaZulu-Natal. Alandmanson, CC BY-SA 4.0, via Wikimedia Commons.

Boron (B) Deficiency – Reproductive Crops (Canola, Sunflower, Fruit Trees): Boron deficiency affects reproductive development. In crops like canola and sunflower, it can lead to poor flowering, empty seed pods, and reduced pollination success. In fruit trees, it may cause deformed fruits and poor fruit set, directly impacting yield and quality.

Understanding soil fertility and recognising deficiency symptoms are critical for effective fertiliser management. By diagnosing nutrient deficiencies early, farmers and agronomists can apply the appropriate fertilisers to correct imbalances, improve nutrient availability, and restore plant health. This not only enhances crop productivity but also supports sustainable soil management by ensuring that nutrients are used efficiently and responsibly.

Soil Fertility and Fertiliser Manufacturing

A foundational view of fertiliser manufacture is that soil fertility and nutrient deficiency are the principal drivers of product purpose, formulation, and delivery strategies. Across soils and cropping systems, a clear understanding of which nutrients are limiting and in what forms crops can assimilate them under given conditions provides the essential direction for designing fertiliser products that deliver nutrients in accessible, plant-available forms. Fertiliser production aims to correct specific imbalances in soil nutrient supply, ensuring crops obtain the essential elements in forms and at rates aligned with plant uptake capabilities and soil chemistry, thereby maximizing crop response while mitigating losses [1, 39, 40]. The broader literature consistently emphasizes that NUE and environmental considerations are integral to product development, reinforcing that fertilisers are not generic nutrient surrogates but targeted interventions tailored to soil–crop–environment interactions [1, 41, 42].

A central conceptual bridge between soil fertility and fertiliser manufacture is the law of the minimum, which asserts that plant growth is constrained by the nutrient in shortest supply rather than by the abundance of other nutrients. This principle has direct implications for product formulation: fertiliser blends must be designed to address the specific limiting nutrients of a soil or cropping system, rather than supplying nutrients indiscriminately. Consequently, the development of tailored blends, ranging from conventional NPK formulations to micronutrient-enriched products, emerges as a practical response to field deficiencies, ensuring that nutrient additions target the most limiting elements and thereby optimize economic and agronomic outcomes [1, 39, 43]. Further to this is the recognition that nutrient deficiency symptoms, observed as visible plant responses, provide empirical cues that guide product design toward nutrients and release patterns that align with the onset and duration of demand.

Nitrogen deficiency in cereals, for example, motivates the production of highly available N sources (such as urea and ammonium nitrate) to meet rapid uptake during vegetative growth, while phosphorus deficiency seen in early development supports soluble phosphate fertilisers designed to place P near the root zone for immediate uptake; potassium deficiency symptoms underpin the manufacture of potassic fertilisers to restore adequate K nutrition. The linkage between deficiency symptoms and product strategy is well established in the literature, underscoring that symptomology informs which nutrients are prioritized, in what quantities, and in which chemical forms [39, 44-46].

Soil fertility considerations further shape the chemical form and solubility characteristics of fertilisers. Nutrients must be delivered in forms that are compatible with soil conditions and plant uptake mechanisms, accounting for soil processes such as fixation, pH-dependent availability, and microbial interactions. For phosphorus, for instance, fertilisers are produced in water-soluble forms to overcome soil fixation and ensure the availability of phosphate near the root zone, whereas nitrogen fertilisers are manufactured in multiple forms (ammonium, nitrate, and urea) to accommodate different soil types and climatic conditions.

The micronutrient sector has responded to deficiency challenges (e.g., iron and zinc) with chelated fertilisers to maintain plant-available micronutrients even in challenging soil environments such as high pH soils. This reflects a systems perspective: product formulation is not only about nutrient content but also about the chemistry and stability of nutrient forms under diverse soil conditions and the kinetics of release and uptake [44, 47-49]. Moreover, soil properties such as pH, texture, and biological activity influence how fertilisers perform, guiding strategies such as adding calcium or magnesium to ameliorate acidic soils or using micronutrient forms that retain availability in alkaline soils; this aligns fertiliser development with broader soil health and microbial interactions that support nutrient cycling and influence the design of biofertiliser products [41, 47, 50, 51].

A further dimension is the drive for efficiency technologies that respond to soil fertility constraints, particularly nutrient losses through leaching, fixation, volatilisation, and denitrification. Fertiliser manufacturers increasingly pursue enhanced efficiency fertilisers (EEFs) that improve nutrient use efficiency and reduce environmental losses by tailoring release patterns and incorporating inhibitors or coatings. Slow-release and coated fertilisers, stabilised formulations using urease and nitrification inhibitors, and polymer-coated products are all advanced strategies that synchronise nutrient release with plant demand and soil processes. The 4R nutrient stewardship framework (right source, right rate, right time, right place) remains a guiding principle for engineering EEFs that minimize losses while maintaining yields, and this paradigm is reinforced across multiple sources that discuss regulatory and environmental drivers alongside agronomic performance data for these products [1, 42, 45, 52-54]. The evidence base includes field trials showing that coatings and inhibitors can reduce N losses through volatilisation and leaching while maintaining or enhancing yields in certain contexts, though results are sometimes context-dependent and require careful consideration of soil type, climate, and management practices [41, 42, 45, 53, 54].

4R Nutrient Stewardship Framework

The 4R Nutrient Stewardship Framework is a globally recognised approach to fertiliser management that aims to optimise crop productivity while minimising environmental impact. Developed and promoted by organisations such as the International Fertilizer Association and The Fertilizer Institute, the framework is built around four core principles: applying the Right Source, at the Right Rate, at the Right Time, and in the Right Place. Together, these principles provide a science-based system for improving nutrient use efficiency (NUE), supporting sustainable agriculture, and maintaining long-term soil fertility.

The first principle, Right Source, refers to selecting the appropriate type of fertiliser to match the specific nutrient requirements of the crop and the characteristics of the soil. Different fertilisers supply nutrients in different chemical forms, which affects how readily plants can absorb them. For example, nitrogen may be supplied as urea, ammonium, or nitrate, each behaving differently in the soil. Similarly, phosphorus fertilisers must be chosen based on soil pH and fixation potential. The right source ensures that nutrients are available in forms that plants can efficiently utilise, while also considering compatibility with other inputs and environmental conditions.

The second principle, Right Rate, focuses on applying the correct amount of nutrients to meet crop demand without over- or under-application. This requires an understanding of soil fertility, crop nutrient requirements, yield targets, and existing nutrient levels. Applying too little fertiliser can limit plant growth and reduce yields, while excessive application can lead to nutrient losses through leaching, runoff, or volatilisation. Determining the right rate often involves soil testing, plant tissue analysis, and the use of decision-support tools to balance productivity with environmental protection.

The third principle, Right Time, emphasises synchronising nutrient application with the crop's growth stages and nutrient uptake patterns. Crops require different nutrients at different stages of development—for example, nitrogen is critical during vegetative growth, while phosphorus is important during early root development. Applying fertilisers at the wrong time can result in nutrients being lost before plants can use them. By timing applications to coincide with peak demand, farmers can improve nutrient uptake efficiency and reduce losses to the environment.

The fourth principle, Right Place, involves applying nutrients where they are most accessible to plant roots and least likely to be lost. This includes decisions about fertiliser placement methods, such as broadcasting, banding, or fertigation. For example, placing phosphorus fertiliser close to the root zone improves uptake because phosphorus is relatively immobile in soil. Similarly, incorporating nitrogen fertilisers into the soil can reduce volatilisation losses. Proper placement ensures that nutrients are effectively delivered to the crop while minimising environmental risks.

The 4R framework is highly relevant to both fertiliser manufacture and agricultural practice. For manufacturers, it drives the development of advanced fertiliser products, such as slow-release formulations, stabilised nitrogen fertilisers, and customised nutrient blends that align with 4R principles. For farmers and agronomists, it provides a practical decision-making framework that integrates soil science, crop physiology, and environmental management.

Importantly, the 4R approach supports broader goals such as sustainable intensification, environmental protection, and food security. By improving nutrient use efficiency, it helps reduce greenhouse gas emissions, minimise water pollution from nutrient runoff, and conserve finite resources such as phosphate rock. At the same time, it enables farmers to maintain or increase yields, ensuring that agricultural systems can meet the demands of a growing global population.

In parallel, innovations in fertiliser formulation increasingly hinge on matching release kinetics to crop phenology and soil nutrient dynamics. Biopolymer coatings, slow-release matrices, and double-coating strategies have been developed to slow nutrient release, with the aim of aligning N, P, and K availability with critical growth stages. Reviews of enhanced efficiency fertilisers highlight the opportunities and challenges of coating technologies, biologically derived polymers, and complex encapsulation techniques that can tailor release rates while mitigating environmental externalities. A recurring theme is that improved NUE depends on achieving better synchrony between nutrient availability and plant demand, which is achieved through smarter release profiles, improved formulation science, and site-specific management integrated with 4R principles [1, 39, 42, 46, 54]. Several empirical investigations underscore that the economics and agronomic benefits of slow-release and stabilised products are highly context-specific, with some studies reporting substantial NUE gains in certain systems while others observe more modest or no advantages under particular conditions, emphasizing the need for site- and crop-specific evaluation when deploying these products [41-43, 51].

Environmental implications are central to the rationale for advancing fertiliser technology. Reducing nitrous oxide (N_2O) emissions and nitrate leaching through smarter N management and stabilised fertilisers is a major objective in contemporary fertiliser design. Studies comparing different N sources and formulations, including stabilized urea, ammonium-based compound fertilisers, and NPK blends with varying nitrate to ammonium ratios, demonstrate that fertiliser composition can influence N_2O emissions and NUE, with findings showing reductions in emissions under certain formulations and management regimes, though outcomes depend on soil type, moisture, and management practices. This literature supports the claim that environmental performance is a core consideration in product development and selection, linking soil fertility, nutrient dynamics, and emission outcomes to fertiliser design decisions [41, 53-56]. In addition, organic amendments and green manures emerge as complementary approaches that can improve NUE and reduce losses when integrated with inorganic inputs, highlighting that sustainable fertiliser strategies often involve combining organic and inorganic sources to optimize nutrient release patterns and soil health, further informing product development strategies that accommodate diverse agronomic systems [39, 43, 57, 58].

Finally, the practical realization of these concepts requires crop- and system-specific product lines and management recommendations. The wide range of deficiency examples across agricultural systems, such as nitrogen deficiency in cereals, phosphorus deficiency in seedlings, or calcium deficiency in horticultural crops, illustrates that fertiliser manufacture must be contextual and tailored, producing specialist products for different crops, growth stages, and environments to ensure that nutrient supply matches plant requirements as closely as possible. This crop-specificity is reinforced by analytic approaches to NUE that partition uptake and utilisation efficiency (NUpE, NUtE) and consider the role of soil organic matter, microbial activity, and soil nutrient pools in determining how fertilisers behave in situ. Thus, soil fertility knowledge underpins not only product formulation but also the assessment of NUE and environmental performance, guiding ongoing improvements in fertiliser technology, agronomic practice, and policy frameworks [39, 41, 42, 44, 55, 59].

Soil Profile Layers (O, A, B, C, R Horizons)

A soil profile is a vertical section of soil that shows how it is organised into distinct layers, known as horizons. These layers develop over time through processes such as weathering, organic matter accumulation, and the movement of water and nutrients. The main horizons—O, A, B, and C—each have unique characteristics and play different roles in supporting plant growth and influencing fertiliser behaviour.

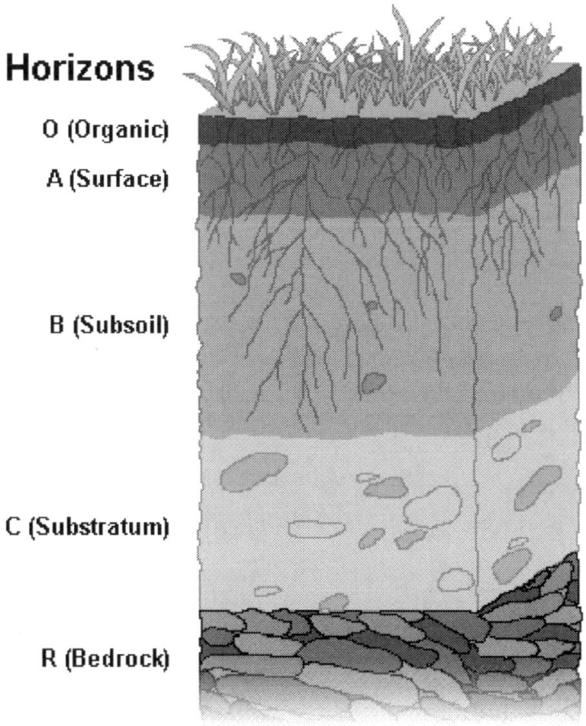

Figure 20: The soil profile with 5 horizons. Jokullmusic, CC BY-SA 4.0, via Wikimedia Commons.

O Horizon (Organic Layer)

The O horizon is the topmost layer and is primarily composed of organic material such as decomposing leaves, plant residues, and animal matter. This layer is rich in humus, which improves soil fertility by supplying nutrients and enhancing water retention. It is especially prominent in forested areas but may be thin or absent in cultivated agricultural soils.

The O horizon plays a key role in nutrient cycling, as microorganisms break down organic matter and release nutrients into forms that plants can absorb. It also helps protect the soil surface from erosion and contributes to soil structure. In fertiliser management, this layer is important because it can influence how nutrients are retained or released into the soil below.

A Horizon (Topsoil)

The A horizon, commonly known as topsoil, is the most important layer for plant growth. It consists of a mixture of mineral particles (sand, silt, and clay) and organic matter. This layer is typically darker in colour due to its higher organic content and is where most root activity, microbial activity, and nutrient uptake occur.

Topsoil is highly fertile because it contains essential nutrients and supports strong biological activity. It is also the primary zone where fertilisers are applied and where nutrient transformations take place. The structure and composition of the A horizon influence water retention, aeration, and the efficiency of fertiliser use. Maintaining healthy topsoil is critical for sustainable agriculture.

B Horizon (Subsoil)

The B horizon, or subsoil, lies beneath the topsoil and is characterised by the accumulation of materials leached from the A horizon above. These materials may include clay, iron, aluminium, and other minerals. As a result, the B horizon is often denser and lighter in organic matter compared to the topsoil.

Although it contains fewer nutrients than the A horizon, the B horizon still plays an important role in water storage and root penetration. Deep-rooted plants may access nutrients and moisture from this layer. However, its denser structure can sometimes restrict root growth and water movement. Fertiliser management must consider the B horizon, particularly in relation to nutrient leaching and long-term soil fertility.

C Horizon (Parent Material)

The C horizon consists of partially weathered parent material, such as rock fragments and mineral deposits from which the soil originally formed. This layer has very little organic matter and limited biological activity. It serves as the foundation for the upper soil layers and influences the soil's overall texture and mineral composition.

While the C horizon does not directly support plant growth, it is important because it determines the long-term nutrient supply and soil characteristics. Over time, weathering of this layer releases minerals that contribute to soil fertility. In some cases, the properties of the C horizon can affect drainage and root depth, indirectly influencing crop performance.

Importance of Soil Horizons in Agriculture

Understanding soil profile layers is essential for effective agricultural management and fertiliser use, as each horizon plays a distinct role in supporting plant growth. The O and A horizons are the most biologically active zones, where organic matter decomposition, nutrient cycling, and fertiliser interactions occur. These upper layers are critical for supplying nutrients to plants and are the primary focus for fertiliser application.

The B horizon, located beneath the topsoil, plays an important role in water storage and the movement of nutrients through the soil profile. It can act as a reservoir for moisture and certain minerals, but its denser structure may also influence root penetration and the downward movement of fertilisers.

The C horizon forms the underlying parent material of the soil and determines its long-term mineral composition and development. Although it has limited direct involvement in plant growth, it influences soil texture, drainage, and the natural supply of minerals over time.

By understanding how these soil layers function, farmers and agronomists can make more informed decisions about fertiliser placement, soil improvement strategies, and overall crop management. This knowledge supports improved nutrient use efficiency, helps maintain soil health, and contributes to more sustainable agricultural production systems.

R Horizon (Bedrock)

In addition to the O, A, B, and C horizons, many soil profiles also include a deeper layer known as the R horizon, which represents bedrock. The R horizon is the lowest layer in the soil profile and consists of solid, unweathered or minimally weathered rock, such as granite, basalt, or limestone. Unlike the layers above it, the R horizon does not contain soil particles, organic matter, or biological activity, and it is generally impermeable to roots and water.

The R horizon plays an indirect but important role in soil formation and agricultural potential. Over long geological timeframes, the gradual weathering of bedrock contributes to the development of the overlying C horizon and ultimately the entire soil profile. The type of bedrock present influences the mineral composition, texture, and fertility of the soil above. For example, soils formed over limestone tend to be more alkaline and rich in calcium, while those formed over granite are often more acidic and lower in certain nutrients.

From an agricultural perspective, the depth to the R horizon is critical. Shallow soils with bedrock close to the surface can restrict root growth, limit water storage capacity, and reduce overall crop productivity. In contrast, deeper soils with a well-developed profile above the R horizon provide more space for root development, better moisture retention, and greater nutrient availability.

Although fertilisers are not applied directly to the R horizon, understanding its presence and characteristics is important for land management. It influences drainage, soil depth, and long-term fertility, all of which affect how fertilisers behave in the soil system. In summary, the R horizon forms the geological foundation of the soil profile and plays a key role in shaping the properties and productivity of agricultural soils.

pH, Salinity, and Nutrient Availability

Soil pH, salinity, and nutrient availability are closely interconnected factors that determine how effectively plants can access and use nutrients. Together, they influence soil chemistry, biological activity, and overall crop productivity. Understanding these relationships is essential for both agronomy and fertiliser management.

Soil pH and Its Role in Nutrient Availability

Soil pH is a measure of how acidic or alkaline the soil is, typically on a scale from 0 to 14. Most crops grow best in a slightly acidic to neutral range (approximately pH 6–7), where the majority of nutrients are most available. Soil pH directly affects the chemical form of nutrients and their solubility, which determines whether plants can absorb them.

At low pH (acidic soils), nutrients such as phosphorus, calcium, and magnesium become less available, while elements like aluminium and manganese may become more soluble and potentially toxic to plants. This can restrict root growth and reduce nutrient uptake. In contrast, at high pH (alkaline soils), micronutrients such as iron, zinc, and manganese become less available, often leading to deficiency symptoms like interveinal chlorosis in young leaves.

Because of these effects, pH is often considered a master variable in soil chemistry. Fertiliser strategies must take pH into account, for example, using lime to raise pH in acidic soils or applying sulphur-based amendments to lower pH in alkaline conditions. Additionally, fertilisers themselves can influence soil pH over time, particularly nitrogen fertilisers, which may acidify the soil.

Soil Salinity and Its Impact on Plant Growth

Soil salinity refers to the concentration of soluble salts in the soil solution. These salts may include sodium (Na^+), chloride (Cl^-), calcium (Ca^{2+}), and magnesium (Mg^{2+}), among others. While some salts are essential nutrients, excessive salinity can create significant challenges for plant growth.

High salinity affects plants primarily through osmotic stress, where the presence of salts reduces the plant's ability to take up water from the soil. Even when water is present, plants may experience drought-like conditions because water uptake is inhibited. In addition, high concentrations of specific ions, such as sodium or chloride, can be toxic and interfere with normal physiological processes.

Figure 21: Tree roots breathing from the surface to avoid soil salinity. Sharifshabbir, CC BY-SA 4.0, via Wikimedia Commons.

Salinity can also disrupt nutrient balance by competing with essential nutrients for uptake. For example, excess sodium can interfere with potassium absorption, while high salt levels can reduce the availability of calcium and magnesium. This imbalance can lead to poor plant growth, reduced yields, and degraded soil structure, particularly in irrigated or poorly drained soils.

Managing salinity often involves improving drainage, using high-quality irrigation water, and applying soil amendments such as gypsum to displace sodium ions. Fertiliser selection must also consider salt content, as some fertilisers can contribute to salinity if applied excessively.

Nutrient Availability in Soil Systems

Nutrient availability refers to the extent to which essential nutrients are present in forms that plants can absorb. Even if nutrients are present in the soil, they may not be available due to chemical fixation, poor solubility, or unfavourable soil conditions.

The availability of nutrients is influenced by several factors, including pH, salinity, soil texture, organic matter content, and microbial activity. For example, phosphorus can become fixed in soils by binding with calcium in alkaline conditions or with iron and aluminium in acidic conditions. Similarly, micronutrients may become unavailable if soil pH is outside the optimal range.

Water also plays a critical role, as nutrients are transported to plant roots in the soil solution. If the soil is too dry, nutrient movement is limited; if it is too wet, nutrients may be lost through leaching or runoff. Soil microorganisms further influence nutrient availability by breaking down organic matter and converting nutrients into plant-available forms.

Interaction Between pH, Salinity, and Nutrients

Soil pH, salinity, and nutrient availability are closely interconnected factors that collectively determine how effectively plants can access and utilise nutrients. Soil pH influences the chemical form and solubility of nutrients, affecting whether they are available for plant uptake. At the same time, salinity impacts the plant's ability to absorb water and maintain a balanced uptake of nutrients. Together, these factors control the overall efficiency of nutrient use within the soil–plant system.

In acidic soils, high concentrations of soluble aluminium can become toxic to plant roots. This toxicity damages root systems, limiting their ability to absorb water and nutrients, even when those nutrients are present in the soil. As a result, plant growth is restricted despite apparently adequate nutrient levels.

In alkaline soils, the issue is often the opposite. Nutrients may be present in sufficient quantities, but certain micronutrients, such as iron, zinc, and manganese, become chemically unavailable due to high pH conditions. This can lead to deficiency symptoms in plants, such as chlorosis, even though the soil contains adequate total nutrient reserves.

In saline soils, high concentrations of dissolved salts create osmotic stress, making it difficult for plants to absorb water. In addition, excess ions such as sodium can compete with essential nutrients like potassium and calcium, further disrupting nutrient uptake. This dual effect reduces both water availability and nutrient absorption, leading to poor plant performance.

These interactions demonstrate that effective fertiliser management must go beyond simply supplying nutrients. It requires a comprehensive understanding of the soil environment, including pH and salinity conditions, to ensure that applied nutrients remain available and can be efficiently taken up by plants.

Relevance to Fertiliser Management

Understanding soil pH, salinity, and nutrient availability is essential for designing effective fertiliser strategies. Fertilisers cannot be applied in isolation; they must be selected and managed in a way that reflects the specific chemical and physical conditions of the soil. This ensures that nutrients remain in forms that are accessible to plants and are not lost or rendered unavailable due to adverse soil conditions.

In acidic soils, for example, nutrient availability can be limited and toxic elements such as aluminium may become more soluble. To address this, lime is often applied prior to fertilisation to raise the soil pH, improve nutrient availability, and create a more favourable environment for root development and microbial activity.

In alkaline soils, certain micronutrients such as iron, zinc, and manganese may become chemically unavailable, even when present in adequate amounts. In these situations, chelated micronutrient fertilisers are commonly used, as they help keep these nutrients in plant-available forms and prevent deficiency symptoms.

In saline environments, high concentrations of soluble salts can interfere with both water and nutrient uptake. To manage this, farmers may use low-salt fertilisers and implement careful irrigation practices to minimise salt accumulation and maintain a balanced soil environment.

By effectively managing pH, salinity, and nutrient availability, farmers can significantly improve nutrient use efficiency, reduce losses to the environment, and optimise crop growth and yields.

Relevance to Fertiliser Manufacturing

The relationships between soil pH, salinity, and nutrient availability are highly significant to fertiliser manufacture because they determine not only what nutrients must be produced, but also the chemical form, formulation, and delivery method of fertiliser products. Fertiliser manufacturing is fundamentally driven by the need to ensure that nutrients remain available under a wide range of soil conditions, rather than simply supplying raw nutrient elements.

One of the most important influences is soil pH, often referred to as a master variable in soil chemistry. Because pH controls nutrient solubility and chemical form, fertiliser manufacturers must design products that remain effective in both acidic and alkaline environments. For example, in acidic soils where phosphorus becomes fixed by iron and aluminium, manufacturers produce highly soluble phosphate fertilisers (such as superphosphates) to improve availability. In alkaline soils, where micronutrients like iron and zinc become unavailable, manufacturers develop chelated micronutrient fertilisers that protect these elements from precipitation and ensure plant uptake. In addition, fertiliser products such as lime (calcium carbonate) and sulphur-based amendments are manufactured specifically to modify soil pH and improve overall nutrient availability.

Soil pH considerations also influence the chemical composition of nitrogen fertilisers. Different nitrogen forms (ammonium, nitrate, and urea) behave differently in soil and can either acidify or stabilise pH over time. As a result, manufacturers carefully select nitrogen forms and may include stabilisers or inhibitors to control how nitrogen transforms in the soil, ensuring efficient uptake and reduced losses.

Soil salinity is another critical factor shaping fertiliser manufacture. Since fertilisers themselves are salts, their contribution to soil salinity must be carefully managed. Manufacturers therefore design fertilisers with controlled salt indices, particularly for sensitive crops or saline environments. For example, potassium sulphate (K_2SO_4) may be preferred over potassium chloride (KCl) in salt-sensitive crops. Similarly, low-salt or specialty fertilisers are developed for use in greenhouse systems, hydroponics, and arid regions where salinity is a major constraint. This ensures that fertiliser application does not exacerbate osmotic stress or ion toxicity in plants.

Salinity considerations also drive the production of soil amendment products, such as gypsum (calcium sulphate), which is used to displace sodium ions and improve soil structure in saline and sodic soils. These products are often manufactured and marketed alongside traditional fertilisers as part of integrated soil management solutions.

The concept of nutrient availability further shapes fertiliser manufacturing by determining how nutrients must be formulated to remain accessible in the soil environment. Many nutrients, particularly phosphorus and micronutrients, can become chemically fixed or unavailable depending on soil conditions. To address this, manufacturers produce fertilisers with enhanced solubility, controlled-release coatings, or chemical stabilisation. For example, coated fertilisers and slow-release formulations are designed to regulate nutrient release over time, ensuring that nutrients remain available when plants need them most.

In addition, fertiliser manufacturers increasingly incorporate additives and technologies that improve nutrient use efficiency (NUE) under varying soil conditions. These include urease and nitrification inhibitors to reduce nitrogen losses, as well as humic substances and biostimulants that improve nutrient uptake and soil health. Such innovations are directly linked to the challenges posed by pH, salinity, and nutrient dynamics in soils.

The interaction between pH, salinity, and nutrient availability also drives the development of customised and blended fertilisers. Rather than producing one-size-fits-all products, manufacturers create region-specific or crop-specific fertiliser blends that account for local soil conditions. This ensures that nutrients are supplied in the right balance and form to overcome soil constraints and maximise plant uptake.

Finally, these factors are central to improving the sustainability of fertiliser production and use. By understanding how soil conditions affect nutrient behaviour, manufacturers can design products that reduce nutrient losses, minimise environmental impacts such as leaching and runoff, and support efficient resource use. This is particularly important in the context of global food security, where maximising the effectiveness of fertilisers is essential for increasing agricultural productivity without degrading soil and water resources.

Interaction between Soil Biology and Fertilisers

The interaction between soil biology and fertilisers is a fundamental aspect of modern agriculture, as it directly influences nutrient cycling, availability, and overall soil health [60-62]. Soil is not just a physical medium but a living system containing microorganisms such as bacteria, fungi, protozoa, and earthworms, and these organisms play a critical role in transforming nutrients into forms that plants can absorb. Their activity is closely linked to how fertilisers behave in the soil, shaping nutrient fate and plant uptake dynamics [60-62]. A growing body of work emphasises that fertiliser strategies should be designed with microbial processes in mind, so that nutrient supply aligns with microbial mineralisation, immobilisation, solubilisation, and turnover rates to maximise plant availability while preserving soil life [62-64].

One of the most important roles of soil biology is in nutrient cycling. Microorganisms decompose organic matter, releasing nutrients such as nitrogen, phosphorus, and sulphur in plant-available forms. For example, bacteria mineralise organic nitrogen to ammonium and then other specialised bacteria carry out nitrification, converting ammonium to nitrate, thereby linking organic matter turnover to plant-accessible N forms [61, 65]. Fertilisers interact with these processes by supplying additional nutrients that can either stimulate or disrupt microbial activity, depending on application method and form. When managed correctly, fertilisers can enhance microbial processes and improve nutrient availability by supporting a dynamic soil food web that mediates nutrient mineralisation and stabilization within soil organic matter pools [61-63]. The literature also notes that the form and timing of nutrient inputs influence microbial-mediated transformations, with organic amendments and certain slower-release fertilisers more likely to sustain microbial-driven nutrient cycling than immediate surges from highly soluble inorganic nutrients [63, 66].

Fertilisers also influence microbial population dynamics and diversity. Moderate and balanced fertiliser use can promote beneficial microorganisms, such as nitrogen-fixing bacteria (e.g., Rhizobium) and mycorrhizal fungi, which enhance nutrient uptake by plants [63, 64, 66]. Mycorrhizal fungi extend the root system and improve the uptake of phosphorus and micronutrients, creating a symbiotic network that can be particularly beneficial under nutrient-limited conditions or with certain phosphorus management regimes [62-64]. However, excessive or imbalanced fertiliser use, particularly high levels of nitrogen, can suppress these beneficial organisms and reduce biological diversity, potentially compromising soil structure, disease suppression, and long-term soil health [62, 63, 66]. The dichotomy in responses highlights the need for fertiliser strategies that balance inorganic inputs with organic amendments and microbial inoculants to sustain a diverse microbial community essential for resilience [63, 66].

The interaction between fertilisers and soil biology is particularly important for phosphorus availability. Many soil microorganisms, including phosphate-solubilising bacteria and fungi, help convert insoluble forms of phosphorus into forms that plants can absorb. Fertilisers that support microbial activity, such as those containing organic matter or humic substances, can enhance this process by providing carbon substrates and energy sources that fuel microbial solubilisation and mineralisation pathways [60-62]. Conversely, over-application of phosphorus fertilisers may reduce the need for microbial mediation, potentially weakening these biological relationships, though context-dependent outcomes exist and require field-scale validation [61-63]. Phosphorus-solubilising microorganisms (PSM) are frequently recognized as key contributors to P availability, with mechanisms including organic acid production and mineral dissolution that release P for plant use; however, field responses are variable and influenced by soil pH, organic matter, and existing microbial communities [61, 62]. Reviews systematically discuss the conditions under which PSM yield agronomically meaningful gains versus scenarios where plant P nutrition is not substantially enhanced by inoculation alone [61].

Soil biology also plays a key role in improving nutrient use efficiency (NUE). Microorganisms help retain nutrients in the soil by incorporating them into organic forms, reducing losses through leaching or volatilisation. Microbial biomass acts as a temporary reservoir, releasing nutrients gradually as organisms die and decompose, which can align with crop uptake patterns when fertiliser inputs are distributed in a crop-appropriate schedule. Fertiliser strategies that support microbial activity, such as split applications, use of slow-release products, or integration with organic amendments, can improve NUE and reduce environmental losses [61-63]. This perspective is supported by syntheses arguing that the turnover of microbial biomass contributes to P and N cycling in ways that may be as important as directly supplying nutrients through chemical fertilisers, prompting a shift in fertiliser design toward microbiome-friendly formulations [61].

Another important interaction is related to soil structure and fertility. Soil organisms contribute to the formation of stable soil aggregates through the production of organic compounds and physical activity (e.g., earthworm burrowing). Improved soil structure enhances aeration, water infiltration, and root growth, all of which influence how effectively fertilisers are utilised. Healthy biological activity therefore increases the efficiency of fertiliser inputs by creating conditions that favour root access to nutrients and microbial transformations. In this way, soil biota contribute to nutrient retention and distribution within the soil profile, linking biological health with fertiliser efficiency and plant performance [61, 63, 66].

Fertilisers can also affect soil pH and chemical conditions, which in turn influence microbial activity. Repeated use of certain nitrogen fertilisers, particularly ammonium-based forms, can acidify the soil, potentially reducing microbial diversity and activity. Changes in pH can alter microbial community composition, favouring some organisms while inhibiting others, which underscores the need to manage fertiliser regimens to maintain a stable environment conducive to soil biology and nutrient cycling processes [60-63]. The biogeochemical consequences of pH shifts include altered mineral solubility and microbial enzyme activity, with implications for P and N availability and for soil organic matter turnover [61, 63].

In addition, there is growing interest in the integration of biofertilisers and biological inputs with conventional fertilisers. Biofertilisers contain living microorganisms that enhance nutrient availability, such as nitrogen-fixing bacteria or phosphate-solubilising microbes.

When combined with traditional fertilisers, these products can improve nutrient efficiency, reduce chemical inputs, and support sustainable agricultural systems by promoting synergistic effects between chemical nutrients and microbial processes [60, 67, 68]. The concept of biofertilisers has been advanced in policy-oriented and scientific discourse as a means to reorganise fertiliser regimes toward more ecologically harmonious systems, with attention to legal definitions, quality control, and standardised efficacy testing to realise their agronomic potential [68].

The interaction between soil biology and fertilisers also has important environmental implications. Healthy microbial activity can reduce nutrient losses by stabilising nutrients in the soil and improving plant uptake, whereas poor management of fertilisers can lead to nutrient imbalances, reduced biological activity, and increased risks of pollution such as nitrate leaching or nitrous oxide emissions. Field and meta-analytic work emphasises that the anticipated environmental benefits of microbiome-informed fertiliser practices depend on context, including soil type, climate, crop system, and management history; thus, predictive success requires careful site-specific testing and adaptive management [61, 63]. In environments where phosphorus is abundant but poorly available, for example, microbial solubilisation can be a pivotal mechanism, but the magnitude of response is often contingent on soil biology and physical-chemical constraints; hence researchers advocate for integrated approaches that couple microbial inoculants with balanced fertiliser inputs and organic matter management to realize both agronomic and environmental gains [61, 62].

Possible integration of biofertilisers with conventional fertilisers is particularly relevant for phosphorus management. Biofertilisers that contain phosphate-solubilising bacteria or phosphatase-producing organisms can improve P availability in concert with P fertilisers, potentially reducing total P inputs while maintaining crop yields in certain systems. Reviews and empirical studies discuss how inoculants can enhance P uptake when paired with organic amendments and mycorrhizal inoculation, though outcomes under field conditions are variable and highly soil-specific [61, 62, 67]. In oil palm and other crops, combined strategies involving rock phosphate-enriched compost, microbial inoculants, and mycorrhizal associations have been explored as a route to sustainable P management and reduced chemical dependence, illustrating the practical application of these principles in productive perennial systems [69-71].

Environmental and sustainability considerations further motivate research into alternative, low-input fertiliser strategies. Phosphate-solubilising microorganisms are highlighted as a promising component of sustainable agriculture because they can increase P availability while potentially reducing reliance on finite rock phosphate resources. Foundational reviews emphasise the mechanisms by which PSOMs solubilise inorganic phosphate and mineralize organic P, and discuss the ecological and agronomic conditions required for reliable field performance. However, they also caution that PSM effects are not universally robust in field settings and call for better mechanistic understanding and standardized evaluation frameworks to translate laboratory potential into consistent crop outcomes [61, 62]. Etiologies of variability are often traced to soil pH, organic matter, microbial community structure, and plant root exudation patterns, which together govern the net benefit of PSM in any given system [61].

Chapter 3

Classification of Fertilisers

Chapter 2 established the foundational concepts of soil science, including soil composition, structure, nutrient cycling, soil fertility, and the environmental factors that influence nutrient availability. It emphasised that plant growth is not determined solely by the presence of nutrients, but by how those nutrients interact within the soil–plant system. Building on this understanding, the focus now shifts from the soil environment to the tools used to manage and optimise it—fertilisers.

This chapter introduces the classification of fertilisers, providing a structured framework for understanding the wide range of products used in modern agriculture. As outlined in this chapter, fertilisers can be grouped based on their origin, nutrient composition, physical form, release characteristics, and method of application. These classifications are not merely theoretical; they underpin practical decision-making in fertiliser manufacture, selection, and application.

The chapter begins by distinguishing between organic and inorganic fertilisers, highlighting their different roles in supplying nutrients and influencing soil health. It then explores how fertilisers are formulated, whether as straight, compound, or blended products, and explains how these formulations affect nutrient distribution and efficiency in the field. Further sections examine nutrient groupings, including macronutrients, secondary nutrients, and micronutrients, and consider how fertilisers are delivered in different physical forms such as solids, liquids, and gases.

In addition, the chapter addresses more advanced fertiliser technologies, including controlled-release and slow-release products, precision application methods such as fertigation and foliar spraying, and emerging innovations such as biofertilisers, enhanced efficiency fertilisers, and nanofertilisers. These developments reflect a broader shift toward improving nutrient use efficiency and reducing environmental impacts.

Overall, this chapter builds on the soil-focused principles of the previous chapter by examining how fertilisers are designed, classified, and applied to meet crop nutrient requirements. It provides the essential link between soil science and fertiliser technology, equipping readers with the knowledge needed to make informed decisions in both agricultural practice and fertiliser manufacture.

Classification Overview

Fertilisers can be classified in a number of ways depending on their origin, composition, nutrient content, physical form, release characteristics, and method of application. This classification framework is essential for agronomists, manufacturers, and farmers, as it enables informed decision-making when selecting fertiliser products suited to specific soil conditions, crop requirements, and management systems. By understanding how fertilisers are grouped, it becomes easier to match nutrient supply with plant demand and optimise agricultural productivity.

One of the most common ways to classify fertilisers is by their origin, or source of nutrients. Fertilisers are broadly divided into organic and inorganic (synthetic or mineral) types. Organic fertilisers are derived from natural materials such as plant residues, animal manures, compost, and other biological sources. They release nutrients gradually through microbial decomposition, contributing not only to nutrient supply but also to improvements in soil structure, organic matter content, and biological activity. However, their nutrient concentrations are generally lower and more variable compared to synthetic alternatives. In contrast, inorganic fertilisers are manufactured through industrial processes or obtained from mined mineral deposits. These fertilisers provide nutrients in concentrated, readily available forms, allowing for precise and predictable nutrient management. Common examples include urea, ammonium nitrate, superphosphate, and potassium chloride. While highly efficient, their use requires careful management to minimise nutrient losses and environmental impacts.

Fertilisers are also classified according to their nutrient content, particularly the number and combination of nutrients they supply. Straight fertilisers contain a single primary nutrient, such as nitrogen in urea or phosphorus in single superphosphate, and are typically used to correct specific deficiencies. Compound or complex fertilisers contain two or more nutrients that are chemically combined within each granule, providing a balanced nutrient supply. NPK fertilisers, such as 15-15-15 formulations, are common examples. Blended fertilisers, on the other hand, are physical mixtures of different fertiliser materials. Although each granule may contain different nutrients, the overall mixture is designed to deliver a targeted nutrient ratio.

NPK Fertilisers

NPK fertilisers are compound fertilisers that supply the three primary macronutrients required for plant growth: nitrogen (N), phosphorus (P), and potassium (K). These nutrients are essential for different physiological functions in plants, and together they form the foundation of most fertiliser programs in agriculture.

The numbers shown on an NPK fertiliser label, such as 15-15-15, represent the percentage by weight of each nutrient in the product. In this case, a 15-15-15 fertiliser contains:

- 15% nitrogen (N)
- 15% phosphorus (expressed as P_2O_5)

- 15% potassium (expressed as K_2O)

The remaining percentage consists of other materials, such as fillers or conditioning agents, which help with handling and application. A 15-15-15 formulation is therefore considered a balanced fertiliser, as it provides equal proportions of all three primary nutrients.

Each component of NPK fertilisers plays a distinct role in plant development. Nitrogen is primarily responsible for vegetative growth, promoting the development of leaves, stems, and overall plant biomass. It is a key component of chlorophyll and proteins, making it essential for photosynthesis and growth. Phosphorus supports energy transfer within the plant, particularly through compounds such as ATP, and is critical for root development, flowering, and seed formation. Potassium regulates water balance, enzyme activation, and stress resistance, improving plant health, disease tolerance, and crop quality.

NPK fertilisers are typically produced as compound (complex) fertilisers, where each granule contains all three nutrients in the specified ratio. This ensures uniform nutrient distribution during application. Alternatively, they may be blended fertilisers, where different granules supply different nutrients but are mixed to achieve the desired overall ratio.

A formulation such as 15-15-15 is commonly used as a general-purpose fertiliser, particularly in situations where soil nutrient levels are relatively balanced or where a broad-spectrum nutrient supply is required. It is widely applied in field crops, horticulture, and home gardening. However, in more advanced agricultural systems, fertiliser formulations are often tailored to specific crop needs and soil test results, meaning that other ratios (such as 20-10-10 or 12-24-12) may be more appropriate depending on nutrient demand.

Another important classification is based on the type of nutrients supplied. Fertilisers may provide macronutrients, secondary nutrients, or micronutrients, depending on plant requirements. Macronutrient fertilisers supply nitrogen (N), phosphorus (P), and potassium (K), which are required in the largest quantities and are most often limiting in agricultural systems. Secondary nutrient fertilisers provide calcium (Ca), magnesium (Mg), and sulphur (S), which are needed in moderate amounts. Micronutrient fertilisers supply trace elements such as iron (Fe), zinc (Zn), copper (Cu), manganese (Mn), boron (B), and molybdenum (Mo). Although required in small quantities, these micronutrients are essential for proper plant growth and development.

Fertilisers can also be grouped according to their physical form, which influences how they are handled, stored, and applied. Solid fertilisers, including granules, prills, powders, and pellets, are the most widely used, particularly in broadacre agriculture. Liquid fertilisers are available as solutions or suspensions and are commonly applied through irrigation systems (fertigation) or as foliar sprays. Gaseous fertilisers, such as anhydrous ammonia, are injected directly into the soil and are typically used in large-scale farming operations.

The rate at which nutrients are released provides another basis for classification. Quick-release fertilisers dissolve rapidly in soil water, making nutrients immediately available to plants. While this supports rapid plant growth, it also increases the risk of nutrient losses through leaching or volatilisation. In contrast, slow-release or controlled-release fertilisers are designed to release nutrients gradually over time. This can be achieved through specialised coatings, chemical formulations, or incorporation into organic materials. These products improve nutrient use efficiency and help reduce environmental impacts.

Fertilisers are further classified by their method of application. Broadcast fertilisers are spread evenly across the soil surface, while banded fertilisers are placed in concentrated zones near plant roots to enhance nutrient uptake. Foliar fertilisers are applied directly to plant leaves, allowing for rapid absorption of nutrients, particularly micronutrients. Fertigation involves delivering fertilisers through irrigation systems, enabling precise and efficient nutrient management.

In practice, these classification systems are interconnected, and a single fertiliser may fall into multiple categories simultaneously. For example, a fertiliser may be synthetic, compound, granular, slow-release, and applied through broadcasting. Understanding these overlapping classifications provides a comprehensive framework for selecting and managing fertilisers effectively.

Overall, fertiliser classification offers a structured way to understand the wide variety of products available and their roles in supporting soil fertility, crop productivity, and sustainable agricultural systems.

Organic vs Inorganic Fertilisers

Organic and inorganic fertilisers represent two broad approaches to supplying nutrients for plant growth, differing primarily in their source, nutrient release patterns, and effects on soil systems. Understanding the distinction between these fertiliser types is essential for designing effective and sustainable nutrient management strategies.

Organic fertilisers are derived from natural biological materials, such as animal manures, compost, crop residues, bone meal, and green manures. Their nutrients are contained within organic compounds and must be broken down by soil microorganisms before becoming available to plants. This process, known as mineralisation, results in a slow and gradual release of nutrients, which helps reduce the risk of leaching and nutrient loss. In addition to supplying nutrients, organic fertilisers play a critical role in improving soil health. They increase soil organic matter, enhance soil structure, improve water-holding capacity, and stimulate beneficial microbial activity. However, their nutrient content is generally lower and more variable, making it more difficult to precisely match nutrient supply with crop demand. As a result, larger quantities are often required to achieve the same nutrient input as synthetic fertilisers.

In contrast, inorganic fertilisers, also referred to as synthetic or mineral fertilisers, are produced through industrial processes or extracted from naturally occurring mineral deposits. These fertilisers supply nutrients in readily available, inorganic forms that plants can absorb immediately. For example, nitrogen may be supplied as nitrate (NO_3^-) or ammonium (NH_4^+), phosphorus as phosphate (PO_4^{3-}), and potassium as potassium ions (K^+).

The way fertilisers supply nutrients such as nitrate (NO_3^-), ammonium (NH_4^+), phosphate (PO_4^{3-}), and potassium (K^+) is fundamentally a chemical process involving dissolution, ion formation, and plant uptake through roots. These nutrients are not taken up as solid fertiliser granules, but rather as dissolved ions in the soil solution.

When a fertiliser is applied to soil, the first step is dissolution in soil water. Most inorganic fertilisers are highly soluble, meaning that when moisture is present (from rainfall or irrigation), they dissolve and release their nutrients into the soil solution. For example, a fertiliser such as urea or ammonium nitrate dissolves and releases nitrogen-containing compounds, while superphosphate releases phosphate ions and potassium chloride releases potassium ions.

Once dissolved, these nutrients exist as charged particles (ions) in the soil water. Nitrogen can be present as nitrate (NO_3^-), which carries a negative charge, or ammonium (NH_4^+), which carries a positive charge. Phosphorus is present as phosphate ions (PO_4^{3-} or related forms depending on pH), and potassium exists as the positively charged potassium ion (K^+). These ionic forms are critical because plant roots can only absorb nutrients in these soluble, ionic states.

The next step involves movement of nutrients through the soil to the plant roots. This occurs through three main mechanisms:

- Mass flow, where nutrients move with water toward the roots as plants take up water
- Diffusion, where nutrients move from areas of high concentration to low concentration near the root surface
- Root interception, where growing roots physically contact nutrients in the soil

Once the ions reach the root surface, they are taken up through specialised transport proteins in root cell membranes. For example, nitrate and ammonium are absorbed through different transport systems, while potassium uptake is regulated by ion channels. This process is selective, meaning plants actively control which nutrients they absorb and in what quantities.

It is important to note that soil chemistry strongly influences how these ions behave. For instance, nitrate (NO_3^-) is highly mobile in soil water and can be easily leached, whereas ammonium (NH_4^+) can be held on soil particles due to its positive charge. Phosphate (PO_4^{3-}) tends to react with soil minerals and become less available, while potassium (K^+) can be held on clay surfaces and released gradually.

Because of their high nutrient concentration and predictable composition, inorganic fertilisers allow for precise nutrient management, making them highly effective for maximising crop yields. However, their rapid availability also means they are more susceptible to losses through leaching, runoff, or volatilisation if not managed carefully. Over-reliance on inorganic fertilisers without proper soil management can also lead to soil degradation, reduced biological activity, and environmental issues such as water pollution.

A key difference between the two lies in their impact on soil systems. Organic fertilisers contribute to long-term soil fertility by building organic matter and supporting biological processes, while inorganic fertilisers primarily focus on supplying nutrients directly to plants. For this reason, many modern agricultural systems adopt an integrated approach, combining both organic and inorganic fertilisers to balance immediate nutrient needs with long-term soil health.

In practical terms, the choice between organic and inorganic fertilisers depends on factors such as crop type, soil condition, production scale, and sustainability goals. Organic fertilisers are often preferred in regenerative and organic farming systems, where soil health and environmental outcomes are prioritised. Inorganic fertilisers are widely used in intensive agriculture, where high productivity and precise nutrient control are required.

Straight, Compound, and Blended Fertilisers

Straight, compound, and blended fertilisers are three key categories used to describe how nutrients are formulated and delivered in fertiliser products. The distinction between them is important for both fertiliser manufacture and practical agricultural use, as it affects nutrient distribution, application accuracy, and overall efficiency.

Straight fertilisers are the simplest type and contain only one primary nutrient. They are used when a specific nutrient deficiency has been identified in the soil or crop. For example, urea supplies nitrogen, single superphosphate supplies phosphorus, and potassium chloride supplies potassium. Because they deliver a single nutrient, straight fertilisers allow for precise correction of individual deficiencies and are often used in targeted fertiliser programs. However, they require careful management, as multiple products may need to be applied separately to meet the full nutrient requirements of a crop.

Straight fertilisers are manufactured through targeted industrial or mining processes designed to produce a single, concentrated nutrient in a plant-available form. Each type of straight fertiliser follows a different production pathway depending on the nutrient being supplied—nitrogen (N), phosphorus (P), or potassium (K). The goal in all cases is to convert raw materials into highly soluble, stable, and transportable products that can be easily applied in agriculture.

For nitrogen fertilisers, the process typically begins with the industrial production of ammonia (NH_3) using the Haber–Bosch process. In this process, nitrogen gas (N_2) from the air is combined with hydrogen (usually derived from natural gas) under high temperature and pressure in the presence of a catalyst to form ammonia. This ammonia becomes the foundation for many nitrogen fertilisers. For example, urea is produced by reacting ammonia with carbon dioxide (CO_2), forming a solid product that contains about 46% nitrogen. The molten urea is then cooled and formed into granules or prills for ease of handling and application. Other nitrogen fertilisers, such as ammonium nitrate or ammonium sulphate, are produced through similar chemical reactions involving ammonia.

For phosphorus fertilisers, production begins with the mining of phosphate rock, which contains phosphorus in a form that is not readily available to plants. To make it usable, the rock is chemically treated with acids. In the case of single superphosphate (SSP), phosphate rock is reacted with sulphuric acid. This process converts insoluble phosphate minerals into water-soluble monocalcium phosphate, which plants can absorb. The reaction also produces gypsum as a by-product. The resulting material is then cured, granulated, and dried to produce a stable fertiliser product suitable for agricultural use.

For potassium fertilisers, the process is generally based on mining and refining natural mineral deposits, rather than chemical synthesis. Potassium is commonly sourced from ores such as sylvite (potassium chloride, KCl) or carnallite. These ores are extracted from underground deposits or evaporite basins and then processed through crushing, flotation, and crystallisation to separate potassium salts from other minerals. The purified potassium chloride is then granulated or compacted into a uniform product. Because potassium in these minerals is already in a soluble ionic form (K^+), minimal chemical transformation is required compared to nitrogen and phosphorus fertilisers.

After production, all straight fertilisers undergo finishing processes to improve their usability. These include granulation or prilling to create uniform particle sizes, drying to reduce moisture content, and coating to prevent caking or dust formation during storage and transport. Quality control ensures that the nutrient concentration meets specified standards and that the product performs reliably in the field.

Compound fertilisers (also called complex fertilisers) contain two or more nutrients that are chemically combined within each granule. This means that every granule has the same nutrient composition, ensuring uniform nutrient distribution when applied to the field. Common examples include NPK fertilisers such as 15-15-15 or 20-10-10. These fertilisers are manufactured through chemical processes that integrate nutrients into a single product, making them convenient and efficient for application. Because each granule is balanced, compound fertilisers are particularly useful for achieving consistent crop nutrition across large areas.

Compound (or complex) fertilisers are manufactured through integrated chemical processes that combine multiple nutrients—typically nitrogen (N), phosphorus (P), and potassium (K)—into a single, uniform granule. The defining feature of these fertilisers is that each granule contains the same nutrient ratio, which is achieved by chemically reacting raw materials rather than simply mixing them.

The manufacturing process begins with the selection of raw materials, which typically include ammonia (NH_3) for nitrogen, phosphoric acid (H_3PO_4) or phosphate rock for phosphorus, and potassium salts such as potassium chloride (KCl) or potassium sulphate (K_2SO_4) for potassium. These materials are carefully proportioned according to the desired formulation (e.g., 15-15-15 or 20-10-10), ensuring that the final product meets specific nutrient specifications.

One of the most common production methods is the granulation process, where liquid and solid raw materials are combined in a reactor or granulator. In processes such as ammoniation–granulation, ammonia is reacted with phosphoric acid to form ammonium phosphate compounds (such as monoammonium phosphate or diammonium phosphate). At the same time, potassium salts are introduced into the mixture. This creates a slurry or melt in which all nutrients are chemically integrated. As the material is agitated in a rotating drum or pan granulator, small particles form and grow into uniform granules, each containing the full NPK composition.

Another method involves nitrophosphate processes, where phosphate rock is treated with nitric acid instead of sulphuric acid. This produces a mixture containing nitrogen and phosphorus, to which potassium is added later. The resulting material is then neutralised, concentrated, and granulated. This process allows for efficient incorporation of multiple nutrients in a single production stream.

During granulation, the material is carefully controlled in terms of temperature, moisture, and particle size to ensure consistent granule formation. Once the granules reach the desired size, they are dried to remove excess moisture, which improves storage stability and prevents caking. The product is then screened to separate correctly sized granules from fines or oversized particles, which are recycled back into the process.

After drying and screening, the fertiliser may be coated with anti-caking agents or conditioners to improve flowability and handling. The final product is a uniform, free-flowing granular fertiliser in which each granule contains the same ratio of nitrogen, phosphorus, and potassium.

The key advantage of this manufacturing approach is nutrient uniformity. Because the nutrients are chemically combined during production, every granule delivers a balanced nutrient supply when applied in the field. This ensures even nutrient distribution across the soil, which is particularly important in large-scale agriculture where consistent crop performance is required.

Blended fertilisers are physical mixtures of different fertiliser materials that are combined to achieve a desired nutrient ratio. Unlike compound fertilisers, the nutrients in blended fertilisers are not chemically combined, so individual granules may contain different nutrients. For example, one granule may contain nitrogen while another contains phosphorus or potassium. While the overall blend delivers the correct nutrient proportions, there can be some variability in nutrient distribution, especially if the blend is not well mixed or if particle sizes differ. Blended fertilisers are widely used because they are flexible, cost-effective, and can be customised to suit specific soil and crop requirements.

Blended fertilisers are manufactured through a mechanical mixing process, rather than chemical reaction. The objective is to combine different fertiliser materials—each supplying specific nutrients—into a uniform physical mixture that delivers a targeted nutrient ratio (e.g., NPK grades such as 20-10-10 or custom blends based on soil tests).

The process begins with the selection of raw fertiliser components, which are typically straight fertilisers such as urea (nitrogen), diammonium phosphate or superphosphate (phosphorus), and potassium chloride or potassium sulphate (potassium). Additional materials, including secondary or micronutrient fertilisers, may also be included depending on crop and soil requirements. Each component is chosen based on its nutrient content, physical characteristics, and compatibility with other materials.

Before blending, the individual fertilisers are often screened and sized to ensure that the particles are of similar size and density. This step is critical because differences in particle size or weight can lead to segregation, where components separate during handling, transport, or application. Matching particle characteristics helps maintain a consistent nutrient distribution throughout the blend.

The actual blending is carried out in specialised equipment such as rotary drum mixers, ribbon blenders, or paddle mixers. These machines gently but thoroughly mix the different fertiliser components to create a homogeneous mixture. The goal is not to chemically combine the nutrients, but to ensure that the final product contains the correct proportion of each component throughout the batch.

Once blending is complete, the product may undergo quality control checks to verify nutrient composition, uniformity, and physical properties. The fertiliser is then typically bagged or stored in bulk for distribution. In some cases, the blend may be treated with anti-caking agents or coatings to improve flowability and reduce dust during handling.

An important consideration in blended fertiliser manufacture is compatibility between components. Some fertilisers can react with each other or absorb moisture when mixed, leading to caking, nutrient loss, or reduced effectiveness. For example, certain ammonium-based fertilisers may react with materials containing lime or become hygroscopic (absorbing moisture from the air). Manufacturers must therefore carefully select compatible materials and control storage conditions.

The key advantage of blended fertilisers lies in their flexibility and customisation. Blends can be tailored to specific soil test results, crop nutrient requirements, and regional conditions, making them highly adaptable. They are also generally more cost-effective

than compound fertilisers, as they do not require complex chemical manufacturing processes.

However, because nutrients are not chemically combined, each granule in a blended fertiliser contains only one or a few nutrients, rather than a complete nutrient package. This means that uniform application is essential to ensure even nutrient distribution in the field. Proper blending, handling, and spreading techniques are therefore critical to achieving optimal results.

In practice, the choice between straight, compound, and blended fertilisers depends on the level of precision required, the specific nutrient needs of the crop, and economic considerations. Straight fertilisers are ideal for targeted nutrient correction, compound fertilisers provide uniform and convenient nutrient delivery, and blended fertilisers offer flexibility and customisation. Together, these categories form the basis of most fertiliser products used in modern agriculture.

Controlled-Release and Slow-Release Fertilisers

Controlled-release and slow-release fertilisers are designed to supply nutrients to plants gradually over time, rather than delivering them all at once like conventional (quick-release) fertilisers. Their primary purpose is to improve nutrient use efficiency (NUE), reduce losses to the environment, and better match nutrient supply with plant demand.

Although the terms are sometimes used interchangeably, there is a technical distinction between the two.

Slow-release fertilisers are materials that naturally release nutrients at a slower rate due to their chemical composition or low solubility. The release of nutrients depends largely on biological or chemical processes in the soil, such as microbial activity, temperature, and moisture.

A common example is urea-formaldehyde, where nitrogen is chemically bound in complex molecules. Soil microorganisms gradually break these compounds down, releasing nitrogen over time. Other examples include natural organic fertilisers (such as compost or manure), where nutrients become available through decomposition.

Because their release depends on environmental conditions, slow-release fertilisers can be less predictable. For instance, in warm, moist soils with high microbial activity, nutrients may be released more quickly, while in cooler or drier conditions, release may be delayed.

Controlled-release fertilisers are more advanced products that are engineered to release nutrients at a predetermined and predictable rate. This is typically achieved through physical coatings or encapsulation technologies.

In these fertilisers, nutrient granules (such as urea or NPK compounds) are coated with materials like polymers, sulphur, or resins. Water penetrates the coating, dissolves the nutrient inside, and then the dissolved nutrients diffuse slowly out through the coating. The release rate is controlled by factors such as:

- Coating thickness
- Temperature
- Moisture levels

Unlike slow-release fertilisers, controlled-release products are designed to provide a more consistent and predictable nutrient supply, often aligned with the growth stages of specific crops. For example, a fertiliser may be formulated to release nutrients over 2–6 months.

The main difference between the two lies in how the release rate is controlled:

- Slow-release fertilisers depend on natural processes (microbial breakdown, chemical reactions)
- Controlled-release fertilisers use engineered coatings to regulate nutrient release

As a result, controlled-release fertilisers generally offer greater precision and reliability, while slow-release fertilisers are often simpler and more cost-effective.

Both types of fertilisers provide several important advantages in agricultural systems. By releasing nutrients gradually, they help ensure that nutrients are available when plants need them most, rather than being lost shortly after application. This reduces losses through leaching (especially nitrogen), volatilisation, and runoff, which are common issues with conventional fertilisers.

They also reduce the risk of nutrient burn, which can occur when high concentrations of soluble fertilisers damage plant roots or foliage. In addition, fewer applications may be required, reducing labour and operational costs.

Controlled- and slow-release fertilisers play an important role in sustainable agriculture. By improving nutrient use efficiency, they reduce the amount of fertiliser required and minimise environmental impacts such as:

- Groundwater contamination from nitrate leaching
- Eutrophication of waterways due to nutrient runoff
- Greenhouse gas emissions associated with nitrogen losses

These benefits make them particularly valuable in environmentally sensitive areas and high-value cropping systems.

Liquid vs Solid Fertilisers

Liquid and solid fertilisers represent two major physical forms in which nutrients are supplied to crops. While both serve the same fundamental purpose—delivering essential nutrients such as nitrogen, phosphorus, and potassium—their form, handling, application methods, and behaviour in soil–plant systems differ significantly. Understanding these differences is important for selecting the most appropriate fertiliser strategy for specific crops, soils, and production systems.

Solid fertilisers are the most widely used form in agriculture and are typically produced as granules, prills, pellets, or powders. Common examples include urea, ammonium nitrate, diammonium phosphate (DAP), and potassium chloride (MOP). These products are manufactured through processes such as granulation or prilling to create uniform particles that are easy to store, transport, and apply.

One of the key advantages of solid fertilisers is their high nutrient concentration, which makes them cost-effective for bulk handling and large-scale application. They are well suited to broadacre farming systems, where fertilisers are applied using spreaders or seed drills. Solid fertilisers can be applied by broadcasting across the soil surface or banded near the root zone to improve nutrient uptake efficiency.

Once applied, solid fertilisers must dissolve in soil moisture before nutrients become available to plants. This means their effectiveness depends on factors such as rainfall, irrigation, and soil moisture levels. In dry conditions, nutrient availability may be delayed, while in wet conditions, nutrients may become available quickly but also be more prone to losses through leaching or runoff.

Another important consideration is nutrient distribution. While modern application equipment is highly efficient, uneven spreading can lead to variability in nutrient availability across a field. However, solid fertilisers are generally stable, have a long shelf life, and are less prone to handling issues compared to liquids.

Liquid fertilisers are supplied as solutions or suspensions, where nutrients are already dissolved or finely dispersed in water. Examples include urea-ammonium nitrate (UAN) solutions, liquid ammonium polyphosphate, and various micronutrient solutions. Because nutrients are already in solution, they are immediately available for plant uptake after application.

Liquid fertilisers offer a high degree of precision and flexibility. They can be applied through a variety of methods, including:

- Fertigation (through irrigation systems)
- Foliar spraying (directly onto leaves)
- Soil injection or dribbling

These three methods, fertigation, foliar spraying, and soil injection or dribbling, are commonly used to apply fertilisers in a way that improves nutrient delivery, efficiency, and crop response. Each method differs in how nutrients reach the plant and is suited to different crops, soil conditions, and management systems.

Fertigation (through irrigation systems)

Fertigation is the process of applying fertilisers through an irrigation system, allowing nutrients to be delivered directly to the root zone along with water. This method is widely used in drip irrigation, sprinkler systems, and centre pivots, particularly in horticulture and high-value cropping systems.

In fertigation, liquid or fully soluble fertilisers are injected into the irrigation water using specialised equipment such as dosing pumps or venturi injectors. As the water moves through the system, it carries dissolved nutrients evenly across the field and into the soil where plant roots can absorb them.

Figure 22: Fertigation tank. Athira thadathil gopinathan, CC BY-SA 4.0, via Wikimedia Commons.

One of the major advantages of fertigation is precision. Nutrients can be applied in small, frequent doses that match the crop's growth stages, improving nutrient use efficiency and reducing losses through leaching or runoff. It also ensures uniform distribution, especially in well-designed irrigation systems. However, fertigation requires good water quality, proper system maintenance, and careful management to avoid issues such as emitter clogging or uneven application.

Foliar Spraying (directly onto leaves)

Foliar spraying involves applying fertilisers directly to plant leaves in the form of a fine spray. In this method, nutrients are absorbed through the leaf surface and enter the plant tissues relatively quickly.

This approach is particularly useful for correcting nutrient deficiencies rapidly, especially for micronutrients such as iron, zinc, and manganese, which may be unavailable in the soil due to pH or other factors. Foliar application bypasses soil-related limitations and delivers nutrients directly to where they are needed.

Foliar fertilisation is most effective when applied under suitable conditions, such as cooler temperatures and high humidity, which enhance nutrient absorption and reduce evaporation. However, it is generally used as a supplementary method, rather than a primary nutrient source, because plants can only absorb limited amounts of nutrients through their leaves. Incorrect concentrations or application during hot conditions can lead to leaf burn (phytotoxicity).

Soil Injection or Dribbling

Soil injection and dribbling are methods used to apply liquid fertilisers directly to the soil, often near or below the soil surface, close to the plant root zone.

Soil injection involves placing fertiliser beneath the soil surface using specialised equipment, such as knives or injectors. This method is commonly used for applying nitrogen fertilisers like anhydrous ammonia or liquid nitrogen solutions. By placing nutrients below the surface, soil injection reduces losses from volatilisation (gas loss) and positions nutrients where roots can access them more efficiently.

Dribbling, on the other hand, involves applying liquid fertiliser in narrow bands on the soil surface, typically alongside crop rows. While not injected, the fertiliser is still placed close to the root zone, improving uptake compared to broad surface application.

Both methods are valued for their ability to target nutrient placement, reduce environmental losses, and improve efficiency. They are particularly useful in row crops and large-scale farming systems. However, they require specialised equipment and careful operation to ensure accurate placement and avoid crop damage.

This makes them particularly useful in intensive cropping systems, horticulture, and controlled environments such as greenhouses. Liquid fertilisers allow for accurate dosing and timing, enabling nutrients to be supplied in smaller, more frequent applications that match crop growth stages.

Because they are applied in solution, liquid fertilisers provide uniform nutrient distribution, especially when applied through irrigation systems. They are also ideal for delivering micronutrients, which are often required in small, precise amounts.

However, liquid fertilisers have some limitations. They typically have lower nutrient concentration per unit volume compared to solids, which increases transport and storage costs. They may also require specialised equipment for storage, mixing, and application. Additionally, improper application (e.g., foliar spraying at high concentrations) can lead to leaf burn or phytotoxicity.

The primary difference between liquid and solid fertilisers lies in how nutrients are delivered and made available to plants. Solid fertilisers require dissolution in soil water, making them dependent on environmental conditions, whereas liquid fertilisers are already in plant-available form, allowing for rapid uptake.

Solid fertilisers are generally preferred for large-scale, cost-effective nutrient application, while liquid fertilisers are favoured for precision agriculture, high-value crops, and situations requiring rapid nutrient correction. Liquid systems also enable integration with modern technologies such as automated irrigation and precision dosing.

From a management perspective, the choice between liquid and solid fertilisers depends on several key factors. These include the type and value of the crop being grown, as high-value crops often justify more precise and flexible nutrient delivery methods. Soil conditions and the availability of irrigation also play a critical role, particularly where fertigation or liquid applications can be integrated into existing water management systems. In addition, the selection is influenced by the equipment and infrastructure available on the farm, as different fertiliser forms require specific handling and application systems. Finally, the desired level of precision in nutrient application is an important consideration, with liquid fertilisers generally offering greater control over timing and dosage compared to solid fertilisers.

In many modern farming systems, liquid and solid fertilisers are not used in isolation but rather in combination. For example, a farmer may apply a solid base fertiliser at planting to supply bulk nutrients, followed by liquid fertiliser applications during the growing season to fine-tune nutrient supply. This integrated approach maximises efficiency and supports optimal crop performance.

Biofertilisers and Emerging Products

Biofertilisers and emerging fertiliser products represent a growing area of innovation in agriculture, aimed at improving nutrient efficiency, soil health, and environmental sustainability. Unlike conventional fertilisers that primarily supply nutrients directly, these products often work by enhancing natural biological and chemical processes in the soil–plant system.

Biofertilisers are products that contain living microorganisms which promote plant growth by increasing the availability or uptake of nutrients. These microorganisms may include bacteria, fungi, or algae that interact with plant roots and soil systems. For example, nitrogen-fixing bacteria such as *Rhizobium* form symbiotic relationships with legume roots, converting atmospheric nitrogen (N_2) into plant-available forms.

> *Rhizobium* is a group of soil bacteria best known for its ability to form symbiotic relationships with legume plants (such as peas, beans, lentils, clover, and soybeans). This relationship is highly significant in agriculture because *Rhizobium* can convert atmospheric nitrogen (N_2), which plants cannot use, into plant-available forms of nitrogen, such as ammonia (NH_3). This biological process is known as biological nitrogen fixation.
>
> When *Rhizobium* bacteria are present in the soil and compatible legume plants are grown, the bacteria infect the plant roots and stimulate the formation of root nodules. Inside these nodules, the bacteria live in a protected environment where they receive carbohydrates (energy) from the plant. In return, they fix nitrogen from the atmosphere into ammonia, which the plant can use to produce proteins, enzymes, and chlorophyll. This mutual relationship significantly reduces or even eliminates the need for synthetic nitrogen fertilisers in legume cropping systems.

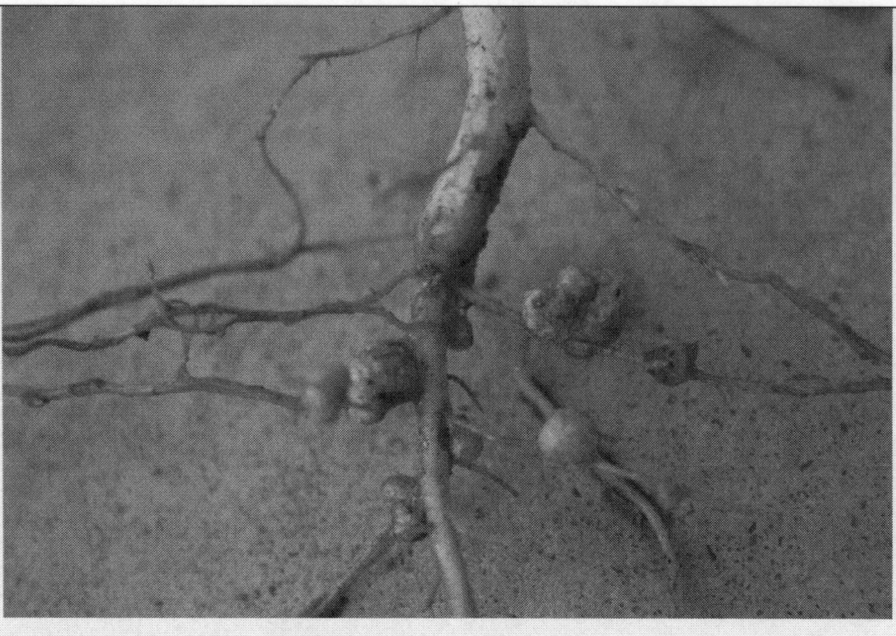

Figure 23: Rhizobia nodules attached to roots of Vigna unguiculata (Cowpea/ Black-eyed pea). Stdout, CC BY-SA 3.0, via Wikimedia Commons.

The effectiveness of *Rhizobium* depends on host specificity, meaning that particular strains of *Rhizobium* are compatible with specific legume species. For example, the strain that works with soybeans is different from the one that works with clover. Therefore, selecting the correct strain is essential for successful nitrogen fixation.

Rhizobium can be included in fertiliser systems through biofertiliser formulations, which contain live bacterial cultures. These are typically applied in several ways. One of the most common methods is seed inoculation, where seeds are coated with a *Rhizobium* culture before planting. This ensures that the bacteria are present in close contact with the emerging roots, allowing rapid colonisation and nodule formation.

Another method is soil application, where *Rhizobium* is applied directly to the soil in liquid or granular form. In some cases, *Rhizobium* may be incorporated into carrier materials such as peat, compost, or lignite, which help protect the bacteria and maintain their viability during storage and application.

More advanced formulations include coated fertilisers or granules that contain both nutrients and microbial inoculants. In these products, *Rhizobium* may be encapsulated or combined with organic materials to protect it from environmental stress and ensure survival until it reaches the soil. However, care must be taken when combining live microbes with chemical fertilisers, as high salt concentrations or certain chemicals can reduce microbial viability.

The inclusion of *Rhizobium* in fertiliser systems offers several important benefits. It reduces dependence on synthetic nitrogen fertilisers, lowers production costs, and decreases environmental impacts such as greenhouse gas emissions and nitrate leaching. It also contributes to improved soil fertility by increasing nitrogen levels for subsequent crops in rotation systems.

Other microbes, such as *Azotobacter* and *Azospirillum*, fix nitrogen independently in the soil. Phosphate-solubilising bacteria and fungi help convert insoluble phosphorus into forms that plants can absorb, while mycorrhizal fungi extend the root system and improve the uptake of nutrients such as phosphorus and micronutrients.

Azotobacter and *Azospirillum* are two important groups of nitrogen-fixing bacteria used in agriculture as biofertilisers. Unlike *Rhizobium*, which forms symbiotic relationships with legumes, these bacteria are free-living or loosely associated with plant roots. This makes them suitable for use across a wide range of crops, including cereals, vegetables, and grasses, where they contribute to improved nutrient availability and plant growth.

Azotobacter is a free-living soil bacterium capable of fixing atmospheric nitrogen independently, without requiring a host plant. It converts nitrogen gas (N_2) into ammonia (NH_3), which can then be utilised by plants. In addition to nitrogen fixation, *Azotobacter* provides several indirect benefits that enhance plant growth and soil health. It produces growth-promoting substances such as auxins, gibberellins, and cytokinins, which stimulate root development and improve plant vigour. It also enhances overall soil microbial activity, contributing to improved soil fertility. Furthermore, *Azotobacter* can produce protective compounds that help suppress certain soil-borne pathogens. Because it does not form nodules like *Rhizobium*, it operates primarily in the rhizosphere, the zone of soil surrounding plant roots, where it contributes to nutrient cycling in a more generalised manner.

Azospirillum, in contrast, is a plant-associated bacterium that forms a close, though not fully symbiotic, relationship with plant roots. It is particularly associated with grasses and cereal crops such as maize, wheat, and rice. *Azospirillum* contributes to plant growth through several mechanisms. It fixes atmospheric nitrogen, supplying modest but meaningful amounts of nitrogen to the plant. It also produces plant growth hormones that promote root elongation and branching, leading to a more extensive root system. In addition, it enhances water and nutrient uptake, especially under stress conditions such as drought or poor soil fertility. Its primary benefit lies not only in nitrogen supply but also in its ability to stimulate root development, which improves overall plant performance.

Both *Azotobacter* and *Azospirillum* are incorporated into fertiliser strategies as biofertilisers and can be applied in a variety of practical ways. The most common method is seed inoculation, where seeds are coated with a formulation containing the bacteria prior to planting. This ensures that the microorganisms are in close contact with emerging roots, allowing rapid colonisation and early activity. They can also be applied directly to the soil, either as liquid suspensions or mixed with carrier materials such as peat, compost, or lignite. These carriers help protect the bacteria and maintain their viability during storage and application.

In modern agricultural systems, these microorganisms may also be included in integrated fertiliser products, where they are combined with organic amendments or low concentrations of mineral fertilisers. However, compatibility must be carefully managed, as high salt levels or certain chemical fertilisers can reduce microbial survival. Another application method involves fertigation, where liquid biofertilisers are delivered through irrigation systems, ensuring even distribution within the root zone. Through these approaches, *Azotobacter* and *Azospirillum* contribute to more sustainable and efficient nutrient management in agricultural systems.

The primary benefit of biofertilisers is that they enhance natural nutrient cycling processes, reducing the need for synthetic fertilisers. They also improve soil structure, increase microbial diversity, and support long-term soil fertility. However, their effectiveness can vary depending on environmental conditions such as soil type, temperature, moisture, and pH, as well as the compatibility between the microorganism and the crop.

In addition to traditional biofertilisers, a range of emerging fertiliser products is being developed to address the challenges of modern agriculture, including nutrient losses, environmental impacts, and the need for sustainable intensification. One important category is enhanced efficiency fertilisers (EEFs), which include products such as urease inhibitors and nitrification inhibitors. These compounds slow down the conversion of nitrogen in the soil, reducing losses through volatilisation, leaching, and denitrification, and improving nitrogen use efficiency.

Enhanced efficiency fertilisers (EEFs) are designed to improve how nitrogen behaves in the soil, ensuring that more of the applied fertiliser is actually taken up by plants rather than being lost to the environment. To understand how they work, it is important to recognise that nitrogen in soil is highly dynamic and can rapidly change form through biological and chemical processes.

When a conventional nitrogen fertiliser such as urea is applied, it is quickly converted in the soil. First, urea is broken down by an enzyme called urease into ammonia (NH_3). Some of this ammonia can be lost to the atmosphere as a gas, a process known as volatilisation. The remaining nitrogen is converted into ammonium (NH_4^+), which is then further transformed by soil bacteria into nitrate (NO_3^-) through a process called nitrification. While nitrate is easily taken up by plants, it is also highly mobile in soil and can be lost through leaching (washed below the root zone) or denitrification (converted back into nitrogen gases under wet conditions). Enhanced efficiency fertilisers work by slowing down these conversion processes, keeping nitrogen in more stable forms for longer periods. This is achieved using specific additives.

Urease inhibitors are compounds that temporarily block the activity of the urease enzyme. By slowing the conversion of urea into ammonia, they reduce the amount of nitrogen lost through volatilisation. This allows more nitrogen to remain in the soil and become available for plant uptake over time.

Nitrification inhibitors, on the other hand, slow down the activity of soil bacteria responsible for converting ammonium (NH_4^+) into nitrate (NO_3^-). By keeping nitrogen in the ammonium form for longer, these inhibitors reduce losses through leaching and denitrification, as ammonium is less mobile in the soil and can bind to soil particles.

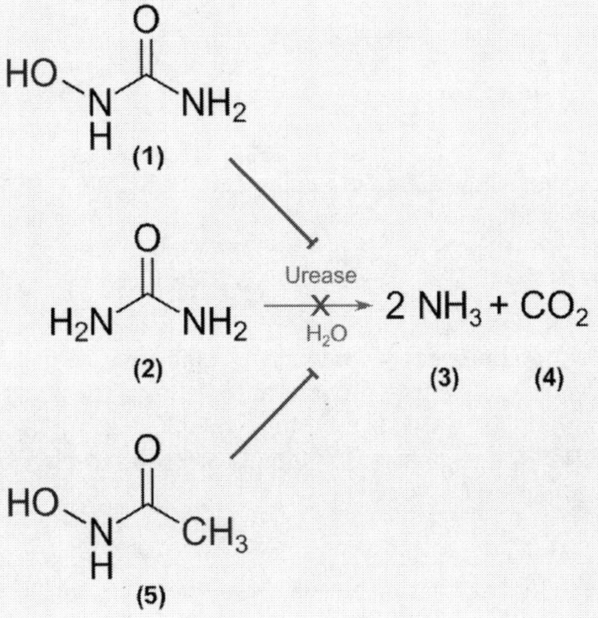

Figure 24: AHA & Hydroxyurea as urease inhibitors. Benff, CC BY-SA 4.0, via Wikimedia Commons.

Figure 24 illustrates how a urease inhibitor works to slow down the breakdown of urea in the soil. At the centre of the image is the enzyme urease, which normally catalyses the reaction between urea and water (H_2O). Under normal conditions, urease rapidly converts urea into ammonia (NH_3) and carbon dioxide (CO_2). This reaction happens very quickly in soil, and if it occurs near the surface, a significant portion of the nitrogen can be lost to the atmosphere as ammonia gas (volatilisation).

The "X" across the urease reaction indicates that this process is being blocked or inhibited. The chemical structures shown (with functional groups such as –OH and CH_3) represent a urease inhibitor molecule (commonly compounds like NBPT). These inhibitors interact with the urease enzyme and temporarily deactivate it.

The arrows pointing toward urease show that the inhibitor is binding to or interfering with the enzyme, preventing it from catalysing the reaction. As a result, the conversion of urea into ammonia is slowed down rather than occurring immediately.

By delaying this reaction, the fertiliser (urea) has more time to:

- Move into the soil with rainfall or irrigation
- Convert into more stable forms such as ammonium (NH_4^+)
- Be taken up by plant roots

This reduces nitrogen losses to the atmosphere and improves nitrogen use efficiency (NUE).

The overall effect of these technologies is to better synchronise nitrogen availability with plant demand. Instead of a rapid surge of available nitrogen followed by significant losses, EEFs provide a more gradual and sustained nutrient supply. This improves nitrogen use efficiency (NUE), meaning that a higher proportion of applied fertiliser is taken up by the crop.

From an environmental perspective, enhanced efficiency fertilisers (EEFs) play an important role in reducing negative impacts associated with nitrogen use. By slowing nitrogen transformations in the soil, they help minimise ammonia emissions to the atmosphere, reduce nitrate leaching into groundwater, and lower the release of nitrous oxide, a potent greenhouse gas.

Another area of innovation is nanofertilisers, where nutrients are engineered at the nanoscale to improve solubility, uptake, and controlled release. These products aim to deliver nutrients more precisely and in smaller quantities, although their widespread adoption is still under development and subject to regulatory and safety considerations.

Nanofertilisers represent a modern innovation in fertiliser technology where nutrients are engineered at the nanoscale (typically 1–100 nanometres) to enhance how they behave in the soil–plant system. At this extremely small size, materials have unique physical and chemical properties, such as a very high surface area and increased reactivity, which can significantly improve nutrient efficiency.

In conventional fertilisers, a portion of applied nutrients is often lost through leaching, volatilisation, or fixation in the soil. Nanofertilisers aim to address these inefficiencies by improving solubility and dispersion, allowing nutrients to remain more evenly distributed in the soil solution. Because of their small size, they can interact more effectively with plant roots and, in some cases, penetrate plant tissues more easily, leading to improved nutrient uptake.

Another key advantage is controlled or targeted nutrient release. Nanofertilisers can be designed with coatings or carrier materials that release nutrients slowly over time or in response to environmental triggers such as moisture, temperature, or root activity. This helps synchronise nutrient availability with plant demand, reducing waste and improving nutrient use efficiency.

In addition, nanotechnology allows for the development of smart delivery systems, where nutrients are encapsulated or bound to nanomaterials that protect them from premature loss in the soil. For example, phosphorus can be protected from fixation, and nitrogen can be released gradually, minimising environmental losses.

Nanofertilisers are manufactured using a combination of chemical, physical, and biological (green synthesis) processes to produce nutrient materials at the nanoscale and to incorporate them into delivery systems that enhance efficiency and control release. The approach used depends on the nutrient involved (e.g., nitrogen, phosphorus, potassium, or micronutrients) and the intended application.

One common approach is top-down physical processing, where bulk fertiliser materials are mechanically reduced to nanoscale particles. This can be achieved through techniques such as high-energy ball milling, grinding, or ultrasonication. For example, conventional phosphate or micronutrient compounds can be broken down into nanoparticles, increasing their surface area and improving solubility. While relatively straightforward, this method requires precise control to ensure uniform particle size and stability.

Another major method is bottom-up chemical synthesis, where nanoparticles are built atom by atom or molecule by molecule through controlled chemical reactions. This includes processes such as precipitation, sol–gel synthesis, hydrothermal synthesis, and chemical reduction. In these systems, nutrients are converted into nano-sized particles under controlled conditions of temperature, pressure, and pH. For instance, nano-sized zinc oxide or iron oxide particles can be synthesised and used as micronutrient fertilisers with enhanced bioavailability.

A more sustainable approach is biological or "green" synthesis, which uses microorganisms, plant extracts, or organic compounds to produce nanoparticles. In this method, bacteria, fungi, or plant-derived compounds act as reducing and stabilising agents, converting nutrient salts into nanoparticles. This approach is gaining attention because it reduces the use of hazardous chemicals and aligns with environmentally sustainable production practices.

In addition to producing nano-sized nutrients, a key part of nanofertiliser manufacture involves encapsulation and carrier systems. Nutrients are often embedded within or coated onto nanomaterials such as polymers, clay nanoparticles, silica, or carbon-based materials. These carriers act as protective shells, preventing nutrient loss and enabling controlled release. For example, nitrogen fertilisers may be encapsulated in polymer nanocoatings that slowly degrade, releasing nutrients over time in response to soil moisture or microbial activity.

Another technique involves nano-emulsions and nano-dispersions, particularly for liquid fertilisers. In these systems, nutrients are dispersed in very fine droplets or particles within a liquid medium, improving uniformity and uptake when applied through fertigation or foliar spraying.

Finally, advanced systems include stimuli-responsive or "smart" nanofertilisers, where release mechanisms are engineered to respond to environmental triggers such as pH, temperature, or root exudates. These systems require precise material engineering and often combine multiple technologies, including polymers and nanocomposites.

Biostimulants are also an important emerging category. These products may contain substances such as seaweed extracts, humic acids, amino acids, or beneficial microbes that stimulate plant growth and stress tolerance without directly supplying large amounts of nutrients. They enhance processes such as root development, nutrient uptake, and resistance to environmental stresses like drought or salinity.

Additionally, there is growing interest in recycled and circular fertilisers, which are derived from waste streams such as food waste, sewage sludge, or livestock manure processed into stable, nutrient-rich products. These fertilisers contribute to resource efficiency and reduce reliance on finite raw materials such as phosphate rock.

Overall, biofertilisers and emerging products reflect a shift towards more sustainable and integrated nutrient management systems. Rather than simply adding nutrients to the soil, these approaches focus on improving how nutrients are cycled, retained, and utilised within the soil–plant system. When used alongside conventional fertilisers, they can enhance productivity, reduce environmental impacts, and support the long-term resilience of agricultural systems.

Chapter 4

Raw Materials and Feedstocks

This chapter marks a critical transition from foundational concepts of soils, nutrients, and fertiliser types into the industrial and resource-based realities that underpin fertiliser production. While the previous chapter examined how fertilisers are classified, formulated, and applied within agricultural systems, this chapter shifts the focus upstream—toward the raw materials, feedstocks, and industrial processes that make those fertilisers possible.

Understanding fertilisers at the application level is only part of the picture. To fully appreciate how modern agriculture is sustained, it is necessary to examine where these nutrients originate, how they are extracted, and how they are transformed into usable products. This chapter therefore builds directly on earlier discussions of nutrient cycles, soil fertility, and fertiliser types by exploring the material and industrial foundations of nitrogen, phosphorus, and potassium supply chains.

The chapter begins with natural gas, the dominant feedstock for ammonia production and the cornerstone of nitrogen fertiliser manufacture. It then moves into phosphate rock, examining its geological origins, beneficiation, and chemical transformation into plant-available phosphorus. Potash mining and processing are also explored, highlighting how potassium is extracted from ancient evaporite deposits and prepared for agricultural use. These sections collectively emphasise that, unlike nitrogen, which can be synthesised from the atmosphere, phosphorus and potassium are finite mineral resources that must be mined and processed.

In addition to these primary nutrient sources, the chapter also examines the role of sulphur and other industrial inputs, which enable critical processes such as acidulation and phosphoric acid production. These inputs demonstrate the interconnected nature of fertiliser manufacturing, where outputs from one industry—such as petroleum refining—become essential inputs for another.

The chapter concludes by expanding the discussion to global supply chains and geopolitical considerations, recognising that fertiliser production does not occur in isolation. Instead, it is embedded within complex international systems shaped by resource distribution, trade networks, energy markets, and political dynamics. These factors influence not only the cost and availability of fertilisers but also broader issues of food security and agricultural sustainability.

Natural Gas (Ammonia Production)

Process

From the perspective of raw materials and feedstocks, natural gas is the primary input used in the industrial production of nitrogen fertilisers, particularly in the manufacture of ammonia (NH_3). It serves both as a source of hydrogen and as a major energy input, making it one of the most critical and cost-determining components in fertiliser production.

The process begins with the extraction and processing of natural gas, which is composed mainly of methane (CH_4). Before it can be used, impurities such as sulphur compounds are removed because they can poison catalysts used in later stages. The purified natural gas is then subjected to steam methane reforming (SMR), where methane reacts with high-temperature steam to produce hydrogen (H_2), carbon monoxide (CO), and carbon dioxide (CO_2). This step is fundamental, as hydrogen is the key building block for ammonia synthesis.

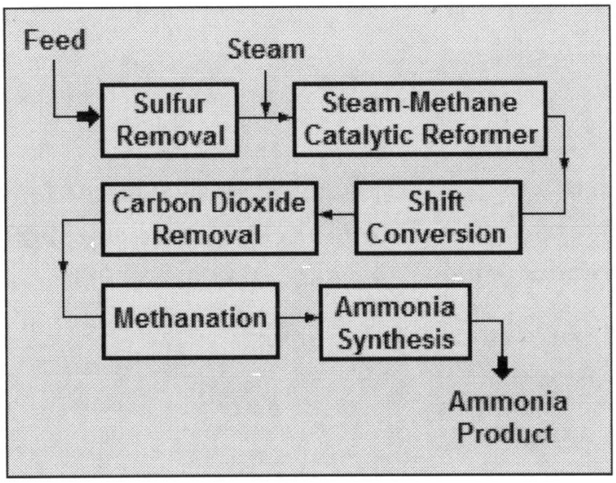

Figure 25: Block flow diagram of an ammonia synthesis industrial plant. Mbeychok at English Wikipedia, CC BY-SA 3.0, via Wikimedia Commons.

Following reforming, the gas mixture undergoes further processing to maximise hydrogen yield. The carbon monoxide produced is converted into additional hydrogen and carbon dioxide through the water–gas shift reaction. The carbon dioxide is then removed, leaving a hydrogen-rich gas stream. At the same time, nitrogen (N_2) is introduced, typically sourced from the air, to provide the second essential component for ammonia production.

The hydrogen and nitrogen gases are then combined in the Haber–Bosch process, where they react under high temperature and pressure in the presence of an iron-based catalyst to form ammonia. The overall reaction is:

$$N_2 + 3H_2 \rightarrow 2NH_3$$

Ammonia produced through this process is the foundation for a wide range of nitrogen fertilisers, including urea, ammonium nitrate, and ammonium sulphate. As a result, the availability and cost of natural gas have a direct impact on fertiliser production costs and global fertiliser markets.

From a feedstock perspective, natural gas is particularly valued because it is relatively abundant, energy-dense, and efficient for hydrogen production compared to alternative sources such as coal or oil. However, its use also has environmental implications. The reforming process generates significant carbon dioxide emissions, making ammonia production a major contributor to industrial greenhouse gas emissions.

In response, there is growing interest in alternative feedstocks and technologies, such as green ammonia, where hydrogen is produced using renewable energy through water electrolysis instead of natural gas. Despite these developments, natural gas remains the dominant feedstock for ammonia production worldwide due to its established infrastructure and cost efficiency.

Origin of Natural Gas

Natural gas is a fossil fuel formed over millions of years from the remains of ancient plants and microorganisms. These organic materials were buried under layers of sediment, where heat and pressure transformed them into hydrocarbons, primarily methane (CH_4).

It is typically found in three main geological settings:

- Conventional gas reservoirs, where gas accumulates in porous rock formations beneath impermeable layers
- Associated gas, found alongside crude oil in oil fields
- Unconventional sources, such as shale gas, tight gas, and coal seam gas, which require advanced extraction techniques like hydraulic fracturing

Major global producers include countries such as the United States, Russia, Qatar, Iran, and Australia, all of which play significant roles in supplying gas for industrial uses, including fertiliser production.

Availability for Fertiliser Production

Natural gas is the dominant feedstock for ammonia production and underpins the majority of global nitrogen fertiliser manufacture. It provides both the hydrogen required for ammonia synthesis and the energy needed to drive the process, making it central to modern fertiliser production systems [72-74].

In terms of availability, global natural gas reserves are considered sufficient to support current consumption levels for several decades. The fertiliser industry accounts for only a relatively small share of total gas use, consuming approximately 3–5% of global supply, primarily for ammonia production. Despite this modest share, natural gas plays a critical role in fertiliser manufacturing, with around 70–80% of global ammonia production relying on it as the primary hydrogen source [72, 73].

This means that, although fertiliser production is highly dependent on natural gas, it represents only a small proportion of overall demand. Larger volumes of natural gas are used in electricity generation, residential and industrial heating, and broader industrial energy applications.

Figure 26: Ammonia production installation. Tseno Tanev (цено та..., CC BY-SA 3.0, via Wikimedia Commons.

However, it is important to recognise that regional availability has a greater impact than global supply. Fertiliser production is typically concentrated in areas where natural gas is abundant and relatively inexpensive, such as the Middle East and North America [72, 75-77]. In contrast, in regions where gas supply is limited or prices are high, such as parts of Europe, fertiliser production can become economically unviable, leading to reduced output or increased reliance on imports.

This situation can lead to reduced output or higher vulnerability to price shocks, reflecting the sensitivity of ammonia-based fertiliser manufacturing to regional gas supply and pricing conditions rather than merely global gas stock levels [72, 75-77]. In the context of global supply chains, researchers emphasize that while global reserves may be deemed sufficient, the distribution of gas resources, regional market structures, and regulatory frameworks collectively shape fertiliser production viability and pricing across regions [72, 75-77].

Risks to Supply

Despite the overall global availability of natural gas, a range of factors can affect its reliability and cost for fertiliser production. One of the most significant risks is price volatility. Natural gas prices can fluctuate widely due to changes in market demand, seasonal variations, and global economic or political events. Because natural gas can account for up to 70–90% of the cost of ammonia production, sudden price increases can make fertiliser production uneconomical, leading to reduced output or temporary plant shutdowns.

Geopolitical risks also play a major role in supply stability. Natural gas resources are often concentrated in specific regions, and supply chains can be disrupted by political instability, trade restrictions, or conflicts. Events such as pipeline disruptions or international sanctions can significantly impact the availability of gas, which in turn affects fertiliser production and global supply.

Infrastructure constraints further influence access to natural gas. The transportation of gas depends on pipelines or liquefied natural gas (LNG) facilities, both of which require substantial investment and maintenance. In regions where infrastructure is limited or underdeveloped, access to natural gas can be restricted, reducing the capacity for local fertiliser production and increasing reliance on imports.

Environmental and regulatory pressures are also increasingly important. Ammonia production from natural gas is carbon-intensive, and growing concerns about climate change have led to stricter emissions regulations and the introduction of carbon pricing in some regions. These measures can increase production costs and encourage a shift toward lower-emission alternatives.

In addition, natural gas is in high demand across multiple sectors, including electricity generation, residential heating, and industrial use. During periods of peak demand, such as winter heating seasons or energy shortages, fertiliser production may be deprioritised, limiting supply to the agricultural sector.

Finally, the transition toward alternative technologies introduces longer-term uncertainty. The development of renewable energy and green ammonia—where hydrogen is produced through electrolysis rather than from natural gas—has the potential to reduce reliance on fossil fuels. While this transition is gradual, it creates uncertainty for future investment in traditional gas-based fertiliser production and may reshape the industry over time.

Phosphate Rock Sources and Beneficiation

From the perspective of raw materials and feedstocks, phosphate rock is the primary natural source of phosphorus used in fertiliser manufacture. Unlike nitrogen, which is synthesised from atmospheric gases, phosphorus must be mined from finite geological deposits, making it a critical and non-renewable input to global agriculture.

Figure 27: Phosphate rock mining continues at this quarry near South Point on Christmas Island, Australia. David Stanley from Nanaimo, Canada, CC BY 2.0, via Wikimedia Commons.

Phosphate rock originates from ancient marine environments where the remains of organisms accumulated and were transformed into phosphate-rich sedimentary deposits over millions of years. These deposits are typically found as sedimentary phosphorites, although some igneous sources also exist. Major global reserves are concentrated in regions such as Morocco (including Western Sahara), China, the United States, and Russia, with Morocco holding the largest share of known reserves. The uneven geographic distribution of phosphate rock makes supply security an important consideration in fertiliser production.

Figure 28: Phosphate-rich sedimentary rock (Phosphoria Formation, Permian; Deer Creek Valley area; Preuss Range, Idaho, USA). James St. John, CC BY 2.0, via Wikimedia Commons.

In its natural form, phosphate rock contains phosphorus primarily as apatite minerals (commonly calcium phosphate), but it is not immediately suitable for fertiliser use. The raw ore often contains impurities such as clay, sand, carbonates, iron, and aluminium compounds, which reduce its nutrient concentration and processing efficiency. As a result, phosphate rock must undergo beneficiation before it can be used as a feedstock for fertiliser manufacture.

Beneficiation refers to the series of physical and chemical processes used to upgrade the quality of phosphate ore by increasing its phosphorus content and removing unwanted materials. The process typically begins with crushing and grinding, which reduce the ore to a fine particle size to liberate phosphate minerals from surrounding impurities. This is followed by washing and screening, where water is used to remove fine clays and separate particles based on size.

One of the most important steps in beneficiation is froth flotation, a process that separates phosphate minerals from other materials based on differences in surface properties. In this process, chemicals known as collectors and frothers are added to a slurry of ground ore. Air bubbles are introduced, and the phosphate particles attach to the bubbles and rise to the surface, forming a froth that can be removed, while unwanted materials sink and are discarded. This step significantly increases the concentration of phosphorus in the final product.

In some cases, additional processes such as magnetic separation or calcination (heating) may be used to remove specific impurities, depending on the composition of the ore. The result of beneficiation is a phosphate concentrate, typically containing a much higher proportion of phosphorus (often expressed as P_2O_5), which is suitable for further chemical processing.

Beneficiation

Beneficiation of phosphate rock is the process of upgrading raw mined ore into a concentrated, usable feedstock for fertiliser manufacture. Because natural phosphate deposits contain a mixture of valuable phosphate minerals (mainly apatite) and unwanted materials such as clay, sand, carbonates, iron, and aluminium compounds, beneficiation is essential to increase the phosphorus (P_2O_5) content, improve processing efficiency, and reduce transport costs.

The process typically begins with mining and initial handling. Phosphate rock is extracted either by open-pit mining (common for sedimentary deposits) or underground mining. The raw ore is then transported to a beneficiation plant, where it is stockpiled and prepared for processing. At this stage, the material is highly heterogeneous, with varying particle sizes and impurity levels.

The first major step in beneficiation is crushing and grinding. Large rocks are reduced in size using crushers (e.g., jaw crushers and cone crushers), followed by grinding mills that further reduce the material to fine particles. The purpose of this step is to liberate phosphate minerals from the surrounding gangue (waste material), allowing them to be separated more effectively in later stages. The particle size is carefully controlled to optimise separation efficiency.

Following size reduction, the material undergoes washing and desliming. Water is used to remove fine clay particles and slimes, which can interfere with separation processes such as flotation. This step often involves hydrocyclones or classifiers, which separate particles based on size and density. Removing clays is particularly important because they can consume reagents and reduce the efficiency of downstream processing.

One of the most critical stages of beneficiation is froth flotation, which is widely used to separate phosphate minerals from impurities. In this process, the finely ground ore is mixed with water to form a slurry, and chemical reagents are added. These include:

- Collectors, which selectively attach to phosphate particles and make them hydrophobic (water-repellent)
- Frothers, which help create stable air bubbles

- Modifiers, which adjust pH and suppress unwanted minerals

Air is then introduced into the slurry, creating bubbles. The hydrophobic phosphate particles attach to these bubbles and rise to the surface, forming a froth that can be skimmed off. The remaining waste materials (tailings) sink and are removed. This process significantly increases the concentration of phosphate in the final product.

In some deposits, additional separation techniques may be used depending on the nature of the impurities. For example, magnetic separation can remove iron-bearing minerals, while calcination (thermal treatment) may be used to remove organic matter or carbonates. The choice of method depends on the mineralogy of the ore and the desired product quality.

After separation, the concentrated phosphate material is thickened, filtered, and dried. Thickening removes excess water, while filtration further reduces moisture content. Drying ensures that the product is stable for storage and transport. The final product, known as phosphate concentrate, typically has a much higher P_2O_5 content (often 28–35% or higher) compared to the original ore.

Quality control is an important part of beneficiation. The concentrate is analysed to ensure it meets specifications for phosphorus content, particle size, and impurity levels. This is critical because the efficiency of subsequent fertiliser production processes (such as acidulation) depends on the quality of the feedstock.

From an operational perspective, beneficiation also generates tailings and waste streams, which must be managed carefully. Tailings often contain clays, sand, and residual chemicals from flotation, and are typically stored in tailings dams. Water used in the process is often recycled to reduce environmental impact and improve efficiency.

Once beneficiated, the phosphate rock is used as a feedstock in fertiliser manufacturing processes, most commonly through acidulation.

Acidulation

Acidulation is the key chemical process used to convert insoluble phosphate rock into plant-available phosphorus fertilisers. In its natural form, phosphate rock contains phosphorus mainly as calcium phosphate (apatite), which is only sparingly soluble in soil and therefore not readily available to plants. Acidulation transforms this material into water-soluble or plant-accessible forms, making it suitable for agricultural use.

The process involves reacting finely ground phosphate rock with a strong acid—most commonly sulphuric acid (H_2SO_4) or phosphoric acid (H_3PO_4). The choice of acid determines the type of fertiliser produced.

When phosphate rock is treated with sulphuric acid, the result is single superphosphate (SSP). In this reaction, insoluble calcium phosphate is converted into monocalcium phosphate, which is water-soluble and can be taken up by plants. At the same time, gypsum (calcium sulphate) is formed as a by-product. This gives SSP its additional value as a source of both sulphur and calcium. The simplified reaction is:

$$Ca_3(PO_4)_2 + 2H_2SO_4 \rightarrow Ca(H_2PO_4)_2 + 2CaSO_4$$

When phosphate rock is reacted with phosphoric acid, the product is triple superphosphate (TSP). This process produces a much more concentrated phosphorus fertiliser because it does not generate large amounts of gypsum. The resulting product contains a higher proportion of soluble phosphorus, making it more efficient for transport and application.

In modern fertiliser production, acidulation is also used to produce phosphoric acid itself, which serves as an intermediate for manufacturing other fertilisers such as monoammonium phosphate (MAP) and diammonium phosphate (DAP). In this case, phosphate rock is first reacted with sulphuric acid to produce phosphoric acid and gypsum. The phosphoric acid is then further processed to create a range of high-analysis fertilisers.

From a processing perspective, acidulation typically occurs in a controlled reactor where finely ground phosphate rock is mixed with acid under specific conditions of temperature, concentration, and residence time. The reaction produces a slurry, which is then allowed to cure (complete the chemical reaction), followed by granulation, drying, and cooling to produce a stable fertiliser product.

Acidulation is critical because it determines the availability of phosphorus to plants. Without this process, most of the phosphorus in phosphate rock would remain locked in insoluble forms. However, it also has important environmental and operational considerations. The process generates by-products such as phosphogypsum, which must be managed safely, and it requires significant quantities of acid, linking fertiliser production to other industrial processes such as sulphur processing.

In this step, the concentrated phosphate rock is reacted with acids such as sulphuric acid or phosphoric acid to produce soluble fertilisers like single superphosphate (SSP), triple superphosphate (TSP), or phosphoric acid, which is then used to manufacture compound fertilisers such as MAP and DAP.

Single superphosphate (SSP), triple superphosphate (TSP), and phosphoric acid are all derived from phosphate rock, but they differ in how they are processed, their nutrient concentration, and how they are used in fertiliser systems.

Single Superphosphate (SSP) is produced by reacting phosphate rock with sulphuric acid. This process converts insoluble phosphate minerals into a form that plants can absorb, primarily monocalcium phosphate, while also producing gypsum (calcium sulphate) as a by-product. As a result, SSP typically contains around 16–20% P_2O_5, along with useful amounts of calcium and sulphur. Because of its lower phosphorus concentration and additional nutrients, SSP is often used in situations where both phosphorus and sulphur are required, particularly in pasture systems and soils deficient in sulphur.

Triple Superphosphate (TSP) is a more concentrated phosphorus fertiliser produced by reacting phosphate rock with phosphoric acid instead of sulphuric acid. This process yields a product that is much higher in phosphorus content, typically around 44–48% P_2O_5, and contains little or no gypsum. As a result, TSP provides a more concentrated source of phosphorus compared to SSP, making it more efficient to transport and apply when high phosphorus inputs are needed. However, it does not supply sulphur, which may need to be added separately if required.

Figure 29: Triple phosphate. Rasbak, CC BY-SA 3.0, via Wikimedia Commons.

Phosphoric Acid (H_3PO_4) is an intermediate product rather than a direct fertiliser in most cases. It is produced by treating phosphate rock with sulphuric acid in a process that generates both phosphoric acid and gypsum. The phosphoric acid can then be further processed to produce a range of fertilisers, including TSP, monoammonium phosphate (MAP), and diammonium phosphate (DAP). In some cases, it is also used directly in liquid fertiliser formulations. Phosphoric acid is highly concentrated and plays a central role in modern fertiliser manufacturing because it allows precise formulation of phosphorus-containing products.

SSP is a lower-concentration fertiliser that also supplies sulphur and calcium, TSP is a high-concentration phosphorus fertiliser with greater efficiency in nutrient delivery, and phosphoric acid is a key intermediate used to manufacture a wide range of phosphate fertilisers. Together, they represent different stages and options in the processing of phosphate rock for agricultural use.

From a feedstock perspective, phosphate rock differs from natural gas in several important ways. It is a finite, geographically concentrated resource, subject to depletion and geopolitical considerations. Its quality varies widely depending on the deposit, which affects processing costs and fertiliser efficiency. Additionally, beneficiation and processing generate waste materials, such as tailings and phosphogypsum, which must be managed carefully to minimise environmental impacts.

Origin of Phosphate Rock

Phosphate rock is a naturally occurring mineral resource formed over millions of years from the accumulation of marine sediments rich in phosphorus, which originate primarily from the remains of marine organisms such as plankton, shells, and bones that settled on ancient seabeds. Over geological time these materials were compacted and chemically transformed into phosphorite deposits that contain high concentrations of phosphate minerals, mainly apatite. This sedimentary signal is consistent with modern understanding that marine sedimentary phosphorites constitute the bulk of commercial phosphate rock today [78-80]. The genesis often involves biogenic and diagenetic processes in shallow, oxygenated to suboxic marine settings, with upwelling-driven productivity being a recurring mechanism in many phosphorite belts linked to the Tethyan and similar paleogeographic configurations [78, 80, 81].

The formation (genesis) of phosphate rock is closely linked to both biological (biogenic) and post-depositional (diagenetic) processes that occur in specific marine environments. In simple terms, this means that phosphate deposits originate from the accumulation of biological material—such as microscopic marine organisms—and are later transformed into solid mineral deposits through chemical and physical changes within the sediment.

In many cases, this process takes place in shallow marine environments where water conditions range from oxygen-rich (oxygenated) to low-oxygen (suboxic). These conditions are important because they influence how organic matter is preserved and how phosphorus is released and concentrated in the sediment. In oxygenated waters, biological activity is high, supporting the growth of plankton and other organisms. When these organisms die, their remains settle on the seafloor, forming organic-rich sediments.

A key driver of phosphate formation in these environments is upwelling, a process where deep, nutrient-rich ocean water rises to the surface. This upwelled water contains high concentrations of nutrients, including phosphorus, which stimulate intense biological productivity. As a result, large amounts of organic material are produced, die, and accumulate on the seabed. Over time, the decomposition of this organic matter releases phosphorus into the surrounding sediments.

Following deposition, diagenetic processes transform these phosphorus-rich sediments into phosphate minerals. During diagenesis, chemical reactions within the sediment cause phosphorus to combine with calcium and other elements, forming stable minerals such as apatite. This transformation gradually converts loose sediment into solid phosphate rock (phosphorite).

These processes are commonly associated with ancient ocean systems, particularly those linked to the Tethyan realm, a large prehistoric ocean that existed between the continents of Gondwana and Laurasia. Many of the world's major phosphate deposits are found in regions that were once part of this and similar paleogeographic environments. The consistent presence of upwelling zones and high biological productivity in these regions made them ideal for the formation of extensive phosphorite belts.

In summary, phosphate rock forms through a combination of biological production, sediment accumulation, and chemical transformation in shallow marine environments. Upwelling-driven productivity plays a central role by supplying nutrients that fuel biological growth, while diagenetic processes convert accumulated organic material into concentrated phosphate deposits over geological time.

Most commercially viable phosphate rock is found in sedimentary deposits, though igneous phosphate rocks occur in more limited quantities and often require beneficiation to achieve economic P2O5 contents. Sedimentary phosphorites of marine origin account for the overwhelming majority of world production (approximately 80–90% in various syntheses), whereas igneous deposits generally contribute smaller shares but can host different mineralogies and impurity profiles, sometimes requiring distinct processing approaches [79, 80, 82]. A robust regional pattern emerges from global reviews: Morocco and Western Sahara sit atop the world's largest sedimentary phosphorite reserves; other major producers include China, the United States, Russia, Jordan, Saudi Arabia, and several MENA countries, with notable regional clustering in the Middle East and North Africa and in central China [79, 80, 82].

The Moroccan phosphorite deposits illustrate the dominance of sedimentary marine phosphorites in North Africa, with Morocco hosting extensive sedimentary phosphates and associated igneous complexes that remain underexplored compared with the world-at-large; these deposits are primarily Late Cretaceous to Eocene in age and are linked to upwelling-driven phosphogenesis and regional paleogeography within the Tethys-related phosphogenic province [78, 80, 83]. In other major producing regions, sedimentary phosphorites are similarly linked to marine transgressions, upwelling centres, and diagenetic processes that concentrate apatite within pelagic ramps or nearshore shelves, often accompanied by skeletal fragments, peloids, and bioclasts as primary phosphatic components [83-86]. Igneous phosphates, while significant in global resources, appear in Russia, Brazil, South Africa, Canada, and Finland, among others, with ore textures including apatite-rich carbonatites and related alkaline complexes; their processing often emphasizes upgrading to high-grade concentrates to reach typical P2O5 contents observed in sedimentary phosphorites [80, 82].

Global distribution patterns show that phosphate rock resources are geographically uneven, with Morocco/Western Sahara, China, the United States, Russia, Jordan, Saudi Arabia, and other MENA countries forming a core set of reserves and production centres. Works reviewing global reserves, regional production, and future projections emphasize Morocco's central role, the rising importance of Jordan, Saudi Arabia, and Egypt in the regional phosphogenic belts, and the continued dominance of sedimentary phosphorites in global output; these sources also stress that data availability and reserve estimation remain uncertain in several regions, requiring cautious interpretation and continued exploration [79, 80, 82].

Availability for Fertiliser Production

Phosphate rock is the primary feedstock for all phosphorus fertilisers and is therefore essential for global food production. Unlike nitrogen, which can be synthesised from the atmosphere, phosphorus must be extracted from naturally occurring mineral deposits. This makes it a finite and non-renewable resource, placing greater importance on its efficient use and long-term management.

In terms of availability, global phosphate rock reserves are considered sufficient to support fertiliser production for many decades, and potentially over a century, depending on future consumption rates and agricultural demand. However, these reserves are not evenly distributed. A significant proportion—often estimated at more than 70%—is concentrated in Morocco and Western Sahara, making these regions critically important to global supply.

The vast majority of phosphate rock production is directed toward agriculture, with nearly 90% of mined material used in fertiliser manufacture. The remaining portion is utilised in other applications, including animal feed supplements and various industrial processes.

Despite the large overall reserve base, not all phosphate deposits are economically or technically viable to extract. The suitability of phosphate rock for fertiliser production depends heavily on ore quality, including its phosphorus concentration and the presence of impurities such as clay, carbonates, or heavy metals. Lower-grade ores require more intensive processing and beneficiation, increasing costs and influencing the overall efficiency of fertiliser production.

Risks to Supply

Despite the presence of relatively large global reserves, the supply of phosphate rock is subject to several important risks that can affect its availability, cost, and long-term sustainability. One of the most significant challenges is its geographic concentration. A large proportion of global reserves is located in a small number of regions, particularly Morocco and Western Sahara. This concentration creates supply security concerns, as any disruption in these areas could have widespread impacts on global fertiliser production and food systems.

Another key issue is resource depletion and declining ore quality. Phosphate rock is a non-renewable resource, and high-grade deposits are gradually being exhausted. As mining progresses, producers are increasingly required to process lower-grade ores, which contain less phosphorus and more impurities. This leads to higher energy use, more intensive beneficiation processes, and increased production costs.

Geopolitical and trade risks also play a significant role in supply stability. Political instability, export controls, and trade disputes can restrict access to phosphate rock, particularly for countries that rely heavily on imports. These risks can disrupt supply chains and create price volatility in global fertiliser markets.

Environmental and regulatory pressures further influence phosphate rock production. Mining and processing activities can result in land degradation, high water consumption, and the generation of waste materials such as tailings and phosphogypsum. As environmental regulations become more stringent, compliance costs may increase, and expansion of mining operations may be limited.

In addition, some phosphate deposits contain contaminants and impurities, including elements such as cadmium, uranium, and other heavy metals. These substances can pose environmental and health risks, and regulatory limits on their presence may restrict the use of certain deposits or require additional processing to meet safety standards.

Growing global demand for food also places increasing pressure on phosphate resources. As populations rise and agricultural production intensifies, the demand for phosphorus fertilisers is expected to increase, placing further strain on a finite resource and raising concerns about long-term supply and price stability.

Finally, phosphate rock is particularly critical because there is no direct substitute for phosphorus in plant nutrition. Unlike nitrogen, which can be synthesised from the atmosphere, phosphorus must be sourced from mineral deposits. This makes phosphate rock a strategically important and irreplaceable resource in global agriculture, underscoring the need for efficient use, recycling, and sustainable management.

Potash Mining and Processing

Potash refers to naturally occurring potassium-bearing minerals that are mined and processed to produce potassium fertilisers. Potassium is one of the three primary macronutrients (N–P–K) required for plant growth, and unlike nitrogen, it cannot be synthesised industrially. Instead, it must be extracted from geological deposits, making potash a critical and finite mineral feedstock for fertiliser production.

> Potash refers to a group of potassium-bearing salts that originate from the evaporation of ancient inland seas millions of years ago. As seawater evaporated in restricted basins, dissolved salts crystallised in sequence, forming layered deposits known as evaporites. Potassium-rich minerals, primarily sylvite (KCl) and carnallite, formed in the later stages of evaporation when more soluble salts remained.

Potash deposits are typically formed through the evaporation of ancient inland seas millions of years ago. As seawater evaporated, dissolved salts crystallised and settled in layers, forming extensive evaporite deposits. These deposits contain a range of potassium minerals, most commonly sylvite (potassium chloride, KCl), as well as carnallite ($KCl{\cdot}MgCl_2{\cdot}6H_2O$) and other mixed salts. Major global potash reserves are located in countries such as Canada (particularly Saskatchewan), Russia, Belarus, China, and Germany, with Canada being one of the largest producers [87-91].

> Like phosphorus, potassium cannot be manufactured synthetically and must be extracted from natural deposits, making it a finite mineral resource.
>
> In terms of availability:
>
> - Global potash reserves are considered abundant and sufficient for many decades to over a century at current rates of use
> - A significant proportion of economically recoverable reserves is concentrated in a few countries, particularly Canada, Russia, and Belarus
> - The majority of potash production, well over 90%, is used in fertiliser manufacture, with only small amounts used in industrial applications
>
> Compared to phosphate rock, potash reserves are generally more widely distributed and, in many cases, of relatively consistent quality, which supports long-term supply stability.

Mining of potash is carried out using two main methods, depending on the depth and characteristics of the deposit. Conventional underground mining is used where deposits are deep. In this method, shafts are sunk into the ground, and ore is extracted using room-and-pillar mining techniques. The ore is then transported to the surface for processing. Alternatively, solution mining is used for deeper or less accessible deposits. In this method, hot water is injected underground to dissolve potassium salts, and the resulting brine is pumped back to the surface, where the salts are recovered through evaporation and crystallisation.

Figure 30: Sigmundshall potash mine of K+S AG in the vicinity of Wunstorf. Vasyatka1, CC BY-SA 4.0, via Wikimedia Commons.

Once extracted, the raw potash ore undergoes processing and beneficiation to separate potassium minerals from impurities such as sodium chloride (common salt), clay, and other non-valuable materials. The process typically begins with crushing and grinding to liberate potassium-bearing minerals. This is followed by separation techniques, most commonly flotation, where chemical reagents are used to selectively separate sylvite from halite (NaCl). In some cases, crystallisation processes are used, particularly in solution mining operations, where differences in solubility are exploited to recover potassium salts.

After separation, the concentrated potassium product is dried, screened, and granulated to produce a uniform fertiliser product, typically potassium chloride (also known as muriate of potash, or MOP). Other forms, such as potassium sulphate (SOP), may be produced through additional chemical processing where chloride-sensitive crops are involved.

From a feedstock perspective, potash differs from nitrogen and phosphorus in that it is already present in a plant-available ionic form (K^+) within the mineral structure, meaning that less chemical transformation is required compared to phosphate rock or ammonia production. The main challenge lies in extraction and purification rather than chemical conversion.

However, potash supply is influenced by several factors. Deposits are geographically concentrated, and production is dominated by a few countries, creating potential supply risks. Mining and processing are energy-intensive and require significant infrastructure, particularly for underground or solution mining operations. Environmental considerations, such as waste salt disposal and water use, must also be managed carefully.

Risks to Supply

The global potash supply is concentrated in a few major producers, with Canada, Russia, and Belarus accounting for a large share of exports and, in some periods, a pivotal role in setting world prices. This concentration creates exposure to regional disruptions and policy actions in those suppliers, reinforcing market sensitivity to shocks in these jurisdictions [92-96]. Belarussian exports have been particularly impacted by EU and other sanctions regimes, illustrating how political actions can constrain supply channels even when global reserves remain ample [93, 97, 98]. Russia's role as a leading fertilizer exporter, along with its prominence as a natural gas supplier (a key input for nitrogen fertilizer), further links potash supply dynamics to broader energy and policy developments [95, 98]. Taken together, these sources corroborate that a small number of dominant suppliers can influence global supply stability and price trajectories, despite substantial global resource endowments [92-96].

Sanctions and export controls directed at major potash exporters (notably Russia and Belarus) have demonstrably tightened supply conditions and influenced global fertilizer markets. Multiple sources document how export restrictions and sanctions regimes altered trade flows and price dynamics in recent years, underscoring the vulnerability of fertilizer markets to geopolitical developments [89, 92, 93, 97, 98]. The Russia–Ukraine conflict intensified these effects through disruption of grain and fertilizer trade networks, elevated energy costs, and altered transport routes, all of which feed into fertilizer availability and prices globally [93, 96, 97, 99-101]. Some debates emphasize the resilience of certain supply chains and the capacity of countries to reallocate flows, but the prevailing evidence indicates meaningful price and availability responses to geopolitical actions and policy measures [89, 93, 97]. Overall, the literature converges on the view that fertilizer markets are highly exposed to geopolitical risk via sanctions, trade restrictions, and shifts in trade policy [93].

Sulphur and Other Industrial Inputs

Sulphur and other industrial inputs play a critical supporting role in fertiliser manufacture, particularly in the production of phosphate and some nitrogen fertilisers. While nutrients such as nitrogen, phosphorus, and potassium form the core of fertiliser products, these materials cannot be effectively processed or converted into plant-available forms without key industrial inputs such as sulphur, acids, water, air, and energy.

Sulphur is one of the most important industrial feedstocks in fertiliser production. It is primarily used to produce sulphuric acid (H_2SO_4), which is essential for processing phosphate rock. In the acidulation process, sulphuric acid reacts with phosphate rock to produce phosphoric acid and fertilisers such as single superphosphate (SSP). As a result, the availability of sulphur directly influences the production of phosphorus fertilisers.

Modern sulphur supply is largely derived as a by-product of oil and natural gas refining. During fuel processing, sulphur compounds are removed to meet environmental regulations, and the recovered sulphur is then converted into elemental sulphur. This means that sulphur supply is closely linked to the petroleum and gas industries rather than being mined directly in large quantities. In some regions, sulphur is also obtained from natural deposits or from metallurgical processes such as the smelting of sulphide ores.

In fertiliser manufacturing, elemental sulphur is typically burned in air to produce sulphur dioxide (SO_2), which is then further oxidised to sulphur trioxide (SO_3) and reacted with water to form sulphuric acid. This process is highly efficient and forms the backbone of the wet-process phosphoric acid industry.

The wet-process phosphoric acid industry is the dominant industrial method used to produce phosphoric acid for fertiliser manufacture. It forms the backbone of global phosphorus fertiliser production, accounting for the vast majority of phosphoric acid used worldwide. This process converts naturally occurring phosphate rock into a form that plants can readily use, making it essential for modern agriculture and food production.

At its core, the wet process involves the chemical reaction between phosphate rock and a strong mineral acid, typically sulphuric acid (H_2SO_4). Phosphate rock is primarily composed of calcium phosphate minerals, most commonly fluorapatite. When this rock is finely ground and reacted with sulphuric acid, it undergoes acidulation, producing phosphoric acid (H_3PO_4) and a by-product known as phosphogypsum (calcium sulphate, $CaSO_4{\cdot}2H_2O$). The simplified reaction can be represented as:

$$Ca_5(PO_4)_3F + 5H_2SO_4 + 10H_2O \rightarrow 3H_3PO_4 + 5CaSO_4 \cdot 2H_2O + HF$$

This reaction takes place in large, controlled reactors where temperature, mixing, and acid concentration are carefully managed to maximise conversion efficiency.

The production process typically begins with beneficiated phosphate rock, which has been upgraded to increase its phosphorus content and remove impurities such as clay and sand. The rock is then finely ground to increase its surface area, improving the efficiency of the chemical reaction. The ground rock is fed into a reactor, where it is mixed with sulphuric acid and water to form a slurry.

During the reaction, phosphoric acid is released into the liquid phase, while phosphogypsum forms as a solid by-product. The mixture is then transferred to a filtration system, where the phosphogypsum is separated from the phosphoric acid. The solid phosphogypsum is typically washed to recover any remaining acid and then disposed of in large stacks, although there is growing interest in finding alternative uses for this by-product.

The resulting phosphoric acid is usually produced in a relatively dilute form (around 25–35% P_2O_5 equivalent) and may be further concentrated by evaporation to increase its strength for downstream applications. This concentrated phosphoric acid is then used to manufacture a wide range of fertilisers, including diammonium phosphate (DAP), monoammonium phosphate (MAP), and triple superphosphate (TSP).

One of the key advantages of the wet-process method is its cost-effectiveness and scalability, making it suitable for large-scale fertiliser production. It can handle a wide range of phosphate rock qualities and is less energy-intensive than alternative methods such as the thermal process, which is mainly used for producing high-purity phosphoric acid for industrial applications.

However, the wet-process phosphoric acid industry also presents several challenges. The generation of phosphogypsum waste is a major environmental concern, as it is produced in large volumes and may contain impurities such as residual acids, heavy metals, and naturally occurring radioactive materials. Managing and storing this waste safely is a significant issue for the industry. In addition, the process can release gases such as hydrogen fluoride (HF) and requires strict environmental controls to minimise emissions.

The quality of the final phosphoric acid is also influenced by the composition of the phosphate rock. Impurities such as cadmium, uranium, and organic matter can carry through into the acid and must be managed, especially where regulatory limits apply.

From a feedstock perspective, the wet-process phosphoric acid industry highlights the interdependence between sulphur and phosphate rock. Large quantities of sulphur are required to produce sulphuric acid, meaning that phosphorus fertiliser production is closely linked to the availability of sulphur from petroleum refining and other industrial sources.

In addition to sulphur, several other industrial inputs are essential. Air is a critical feedstock in nitrogen fertiliser production, providing nitrogen (N_2) for ammonia synthesis through the Haber–Bosch process. Water is equally important, serving as a reactant (e.g., in steam reforming for hydrogen production), a solvent for fertiliser dissolution, and a medium for processing and transport within plants.

Energy is another key input, particularly in the form of natural gas, electricity, or other fuels. Energy is required for high-temperature and high-pressure processes, such as ammonia synthesis, as well as for crushing, grinding, drying, and granulating fertiliser products. The cost and availability of energy therefore have a significant impact on fertiliser production economics.

Other important inputs include acids and chemical intermediates, such as phosphoric acid and nitric acid, which are used to manufacture compound fertilisers like ammonium phosphate and ammonium nitrate. These materials are often produced within integrated industrial complexes, where outputs from one process serve as inputs to another, improving efficiency and reducing costs.

From a feedstock perspective, these industrial inputs are interconnected. For example, sulphur from petroleum refining is used to produce sulphuric acid, which is then used to process phosphate rock into fertiliser. Similarly, natural gas provides hydrogen for ammonia production, while air supplies nitrogen. This integration highlights the dependence of fertiliser production on broader industrial systems.

However, the supply of these inputs is subject to various risks. Because sulphur is largely a by-product of fossil fuel processing, its availability can fluctuate with changes in energy markets and refining activity. Environmental regulations affecting emissions and industrial processes can also influence supply and cost. Additionally, disruptions in energy supply or increases in energy prices can significantly impact fertiliser production.

Supply Chains and Geopolitical Considerations

Supply chains and geopolitical considerations are central to understanding how fertilisers are produced, distributed, and made available to farmers worldwide. Fertiliser production depends on a small number of critical raw materials—natural gas for nitrogen fertilisers, phosphate rock for phosphorus, potash for potassium, and sulphur for acid production. These resources are unevenly distributed across the globe, which means that fertiliser supply is inherently international and highly dependent on trade, infrastructure, and political stability.

A key feature of fertiliser supply chains is the geographic concentration of resources. For example, large reserves of phosphate rock are located in North Africa, particularly Morocco and Western Sahara, while potash production is dominated by countries such as Canada, Russia, and Belarus. Natural gas, essential for ammonia production, is abundant in regions like the Middle East, the United States, and Russia. Because few countries possess all these resources, most rely on imports at some stage of the fertiliser production process, creating global interdependence.

Fertiliser supply chains are also multi-stage and highly integrated. Raw materials are extracted, processed into intermediate products (such as ammonia or phosphoric acid), and then converted into finished fertilisers like urea, DAP, or NPK blends. These products are transported across continents via pipelines, rail networks, and maritime shipping routes before reaching distributors and farmers. Each stage of this chain must function efficiently; disruptions at any point can affect availability and price.

Geopolitical factors can significantly influence these supply chains. Conflicts, sanctions, and trade restrictions can limit access to key raw materials or finished fertilisers. For instance, restrictions on exports from major producers can reduce global supply and cause price spikes. Similarly, political instability in resource-rich regions can disrupt mining or production operations, affecting downstream fertiliser manufacturing.

Another important issue is resource security and national policy. Some countries may impose export controls or prioritise domestic supply to protect their agricultural sectors, especially during periods of global shortage. This form of resource nationalism can reduce availability on international markets and increase competition for remaining supplies. Countries that depend heavily on imports are particularly vulnerable to such policy changes.

Energy markets are closely linked to fertiliser supply chains, especially for nitrogen fertilisers. Since natural gas is both a feedstock and a major energy input for ammonia production, fluctuations in gas prices directly impact fertiliser production costs. When gas prices rise sharply, fertiliser plants may reduce output or temporarily shut down, tightening global supply and increasing prices.

Infrastructure and logistics also play a crucial role. Efficient transport systems—such as ports, railways, and storage facilities—are required to move bulk materials like phosphate rock, potash, and fertilisers. Disruptions due to natural disasters, labour shortages, or technical failures can delay shipments and create supply bottlenecks. In developing regions, limited infrastructure can restrict access to fertilisers even when global supply is sufficient.

Environmental and regulatory pressures are increasingly shaping fertiliser supply chains. Stricter environmental standards for mining, emissions, and waste management can affect production costs and capacity. For example, regulations on carbon emissions may increase the cost of natural gas-based ammonia production, while environmental concerns related to phosphate mining may limit expansion in certain regions.

These supply chain dynamics have direct implications for food security and agricultural productivity. Fertiliser availability influences crop yields, and disruptions can lead to reduced agricultural output and higher food prices. Because fertiliser markets are global, events in one region—such as a geopolitical conflict or export restriction—can have far-reaching effects on farmers and consumers worldwide.

In response to these challenges, there is growing emphasis on resilience and diversification. This includes developing alternative technologies such as green ammonia produced from renewable energy, recycling nutrients from organic waste streams, and improving fertiliser use efficiency. Countries are also seeking to diversify supply sources and invest in domestic production capacity to reduce dependence on imports.

Chapter 5

Ammonia Production (Haber–Bosch Process)

This chapter builds upon the foundational understanding of ammonia production by examining the critical role of energy consumption and the pathways available to improve efficiency within the Haber–Bosch process. As one of the most energy-intensive chemical processes in modern industry, ammonia synthesis sits at the intersection of engineering performance, economic viability, and environmental responsibility. Understanding how energy is consumed, managed, and optimised is therefore essential for both existing operations and future developments.

The discussion that follows explores where energy is used across the ammonia production chain, from hydrogen generation and gas compression to synthesis and heat management, and why these stages demand such high inputs. It then considers the engineering strategies and technological innovations that have been developed to reduce energy intensity, improve process integration, and enhance overall plant performance.

In addition, this chapter introduces emerging approaches aimed at transforming ammonia production in response to global sustainability pressures. These include advancements in catalyst technology, process optimisation, carbon capture, and the transition toward low-carbon and green ammonia systems. Together, these developments highlight a shift from purely scale-driven efficiency to more intelligent, integrated, and environmentally conscious production models.

By the end of this chapter, you will gain a clear understanding of how energy considerations shape the design and operation of ammonia plants, and how ongoing innovation is redefining the future of fertiliser production.

Overview of Ammonia Production (Haber–Bosch Process)

Ammonia production via the Haber–Bosch process is the cornerstone of modern nitrogen fertiliser manufacture. It is the industrial method used to convert atmospheric nitrogen into ammonia (NH_3), a form that plants can use. This process underpins global food production, as ammonia is the starting point for most nitrogen fertilisers, including urea, ammonium nitrate, and ammonium sulphate.

At its core, the Haber–Bosch process combines nitrogen (N_2) from the air with hydrogen (H_2), typically derived from natural gas, to produce ammonia. The overall chemical reaction is:

$$N_2 + 3H_2 \rightleftharpoons 2NH_3$$

This reaction is reversible and exothermic, meaning it releases heat. However, because nitrogen gas is extremely stable due to its strong triple bond, the reaction requires specific industrial conditions to proceed at a useful rate.

Ammonia production involves several integrated steps, beginning with raw material preparation and ending with ammonia synthesis:

1. Hydrogen Production: Hydrogen is most commonly produced from natural gas (methane, CH_4) through a process called steam methane reforming (SMR). In this step, methane reacts with steam at high temperature to produce hydrogen, carbon monoxide, and carbon dioxide. Further reactions (water–gas shift) maximise hydrogen yield, and carbon dioxide is removed.

2. Nitrogen Supply: Nitrogen is obtained from air, typically through air separation. Since air is approximately 78% nitrogen, it provides an abundant and inexpensive source.

3. Gas Purification: Impurities such as sulphur compounds and carbon dioxide are removed because they can poison the catalyst used in ammonia synthesis.

4. Ammonia Synthesis (Haber–Bosch Reactor): Purified nitrogen and hydrogen gases are combined in a ratio of 1:3 and passed over an iron-based catalyst at:

- High pressure (150–300 bar)
- Elevated temperature (400–500°C)

Under these conditions, ammonia is formed.

5. Cooling and Separation: The reaction mixture is cooled, causing ammonia to condense into a liquid. Unreacted nitrogen and hydrogen are recycled back into the reactor to improve efficiency.

Process Conditions and Efficiency

Agricultural Fertiliser Production: Principles, Processes, and Practice

The Haber–Bosch process operates under carefully controlled conditions to achieve an optimal balance between reaction rate and ammonia yield. High pressure is applied to favour the formation of ammonia, as the reaction involves a reduction in the number of gas molecules from reactants to products. According to Le Chatelier's principle, increasing pressure shifts the equilibrium toward ammonia production, improving overall yield.

At the same time, moderate-to-high temperatures are used to ensure that the reaction proceeds at a practical rate. Although lower temperatures would favour higher ammonia yields due to the exothermic nature of the reaction, they would also slow the reaction significantly. Therefore, an elevated temperature is selected as a compromise to maintain both efficiency and productivity.

Catalysts, typically based on iron, play a crucial role in accelerating the reaction without being consumed. They provide a surface that facilitates the breaking of the strong nitrogen bonds and the formation of ammonia, allowing the process to occur at industrially viable conditions.

Because the reaction does not reach completion in a single pass through the reactor, a significant portion of the nitrogen and hydrogen remains unreacted. These gases are continuously separated from the produced ammonia and recycled back into the system. This recycling loop greatly enhances overall efficiency, making the Haber–Bosch process highly effective at large industrial scale.

Role in Fertiliser Production

Ammonia is a key intermediate in fertiliser production rather than simply a final product. While it can be applied directly in some agricultural systems, its primary importance lies in its role as a building block for a wide range of nitrogen-based fertilisers.

It is used to manufacture major fertiliser products such as urea, which is the most widely used nitrogen fertiliser globally, as well as ammonium nitrate and ammonium sulphate. In addition, ammonia is combined with phosphoric acid to produce ammonium phosphates, including monoammonium phosphate (MAP) and diammonium phosphate (DAP), which supply both nitrogen and phosphorus to crops.

Because of this central role, ammonia production underpins the entire nitrogen fertiliser industry. Its availability directly influences the supply of essential nutrients to agricultural systems, making it fundamental to crop productivity and global food production.

Energy and Environmental Considerations

Ammonia production is highly energy-intensive, primarily due to the processes required to generate hydrogen and maintain the operating conditions of the synthesis reaction. A significant portion of the energy demand comes from hydrogen production, which is typically derived from natural gas through steam methane reforming. In addition, the Haber–Bosch process itself requires high pressures and elevated temperatures, further contributing to overall energy consumption.

This energy intensity is closely linked to environmental impact. Ammonia production is a major source of carbon dioxide (CO_2) emissions, particularly from the reforming stage where natural gas is converted into hydrogen. As global attention on climate change increases, the fertiliser industry is under growing pressure to reduce its carbon footprint.

In response, there is increasing interest in more sustainable alternatives. One approach is the development of green ammonia, where hydrogen is produced through water electrolysis using renewable energy sources such as wind or solar power, rather than fossil fuels. Another strategy involves applying carbon capture and storage (CCS) technologies to conventional ammonia plants, allowing carbon emissions to be captured and stored instead of released into the atmosphere. Together, these innovations aim to reduce the environmental impact of ammonia production while maintaining its critical role in global agriculture.

Global Importance

The Haber–Bosch process is one of the most important industrial innovations in history. It enables the large-scale fixation of atmospheric nitrogen, supporting the production of fertilisers that sustain a significant proportion of the world's food supply. Without it, modern agricultural productivity, and global population levels, would be far lower.

Chemistry of Ammonia Synthesis

The chemistry of ammonia synthesis is centred on the combination of nitrogen (N_2) and hydrogen (H_2) to form ammonia (NH_3), a compound that underpins nearly all nitrogen fertilisers. This reaction is carried out industrially using the Haber–Bosch process, one of the most important chemical processes ever developed for large-scale production.

At its simplest, the reaction is:

$$\mathrm{N_2} + 3\mathrm{H_2} \rightleftharpoons 2\mathrm{NH_3}(\Delta H = -92\ \mathrm{kJ/mol})$$

This equation shows that one molecule of nitrogen reacts with three molecules of hydrogen to produce two molecules of ammonia. The reaction is reversible and exothermic, meaning it releases heat.

> This equation shows that one molecule of nitrogen gas (N_2) reacts with three molecules of hydrogen gas (H_2) to produce two molecules of ammonia (NH_3). The numbers in front of each substance (called stoichiometric coefficients) indicate the exact proportions

required for the reaction to occur efficiently. In practical terms, this means that hydrogen must be supplied in three times the amount of nitrogen to achieve the correct balance.

The double arrow (⇌) indicates that the reaction is reversible. This means that ammonia can be formed from nitrogen and hydrogen, but it can also break down back into these gases under certain conditions. As a result, the reaction does not go to completion and instead reaches a state of equilibrium, where the forward and reverse reactions occur at the same rate.

The term $\Delta H = -92$ kJ/mol refers to the enthalpy change of the reaction. The negative sign indicates that the reaction is exothermic, meaning it releases heat as ammonia is formed. Specifically, 92 kilojoules of energy are released for every mole of reaction that occurs (as written in the equation). This release of heat has important implications for process design, as it affects how temperature influences both the rate of reaction and the equilibrium position.

From a molecular perspective, the reaction involves breaking the very strong triple bond in nitrogen (N≡N) and forming new bonds between nitrogen and hydrogen atoms in ammonia. Breaking the nitrogen bond requires significant energy, which is why high temperatures and catalysts are needed. Once the bonds are broken and ammonia forms, energy is released, contributing to the exothermic nature of the reaction.

Molecular Challenges: Breaking the Nitrogen Bond

The key challenge in ammonia synthesis lies in the nature of the nitrogen molecule. Nitrogen exists as N_2, where the two atoms are joined by a triple bond (N≡N). This bond is extremely strong and stable, making nitrogen relatively inert under normal conditions.

To convert nitrogen into ammonia, this triple bond must be broken, a process that requires high energy input. Without assistance, the reaction proceeds extremely slowly, which is why industrial conditions and catalysts are essential.

Reaction Conditions: Temperature, Pressure, and Equilibrium

The ammonia synthesis reaction is governed by the principles of chemical equilibrium and thermodynamics. High pressure, typically in the range of 150–300 bar, is applied because the reaction results in a reduction in the number of gas molecules, from four molecules of reactants to two molecules of product. According to Le Chatelier's principle, increasing pressure shifts the equilibrium toward ammonia formation, thereby improving the overall yield.

At the same time, moderate to high temperatures, generally between 400–500°C, are required to provide sufficient energy to break the strong nitrogen–nitrogen triple bond and to achieve a practical reaction rate. Without this thermal input, the reaction would proceed too slowly to be industrially viable.

However, there is an inherent trade-off between temperature and yield. Lower temperatures favour higher ammonia production because the reaction is exothermic, meaning heat is released and equilibrium shifts toward the product side at lower temperatures. In contrast, higher temperatures increase the rate of reaction, allowing ammonia to form more quickly, but they reduce the equilibrium yield.

As a result, industrial ammonia synthesis operates at an optimised balance between temperature and pressure. These conditions are carefully selected to maximise both reaction speed and ammonia yield, ensuring the process is efficient and economically viable at large scale.

Role of the Catalyst

A catalyst is essential to make ammonia synthesis viable at an industrial scale. The most commonly used catalyst is iron (Fe), which is often enhanced with small amounts of promoter materials such as potassium, aluminium, or calcium oxides. These additives improve the catalyst's activity, stability, and resistance to deactivation during operation.

The catalyst functions by providing an active surface where the reaction can occur more efficiently. Nitrogen and hydrogen gases are first adsorbed onto the surface of the catalyst. This interaction weakens the very strong nitrogen–nitrogen triple bond (N≡N), which is otherwise difficult to break. Once the bond is weakened, nitrogen and hydrogen atoms can combine more readily, forming ammonia molecules that then desorb from the catalyst surface.

Importantly, the catalyst does not alter the equilibrium position of the reaction, meaning it does not change the maximum possible yield of ammonia. Instead, its role is to significantly increase the rate at which the reaction reaches equilibrium. This acceleration is what makes the Haber–Bosch process practical and economically feasible for large-scale ammonia production.

Reaction Mechanism (Simplified)

At the catalyst surface, the reaction proceeds in several steps:

1. Nitrogen molecules adsorb onto the catalyst and dissociate into nitrogen atoms
2. Hydrogen molecules split into hydrogen atoms
3. Nitrogen and hydrogen atoms combine stepwise to form NH, NH_2, and finally NH_3
4. Ammonia molecules desorb from the surface and enter the gas phase

This surface chemistry is critical because the reaction does not occur efficiently in the gas phase alone.

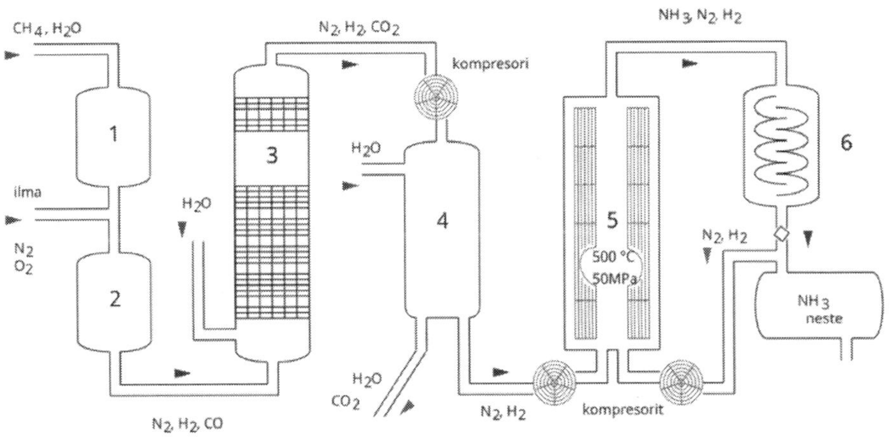

Figure 31: Haber-Bosch. Antimoni, CC BY-SA 3.0, via Wikimedia Commons.

Figure 31 represents the overall industrial ammonia production process, integrating hydrogen generation, gas purification, compression, and the Haber–Bosch synthesis loop. It illustrates how raw feedstocks are progressively transformed into ammonia through a series of tightly controlled and interconnected steps.

The process begins with feed preparation, where natural gas (methane) is used as the primary feedstock, along with steam and air or oxygen. These inputs are conditioned and directed into the hydrogen production section. Prior to entering the main reactors, impurities such as sulphur are removed to prevent catalyst poisoning and ensure efficient downstream operation.

In the hydrogen production stage, methane is converted into hydrogen through a combination of steam methane reforming (SMR) and secondary reforming using air or oxygen. These reactions produce a synthesis gas mixture consisting of hydrogen (H_2), carbon monoxide (CO), and carbon dioxide (CO_2). This stage is highly energy-intensive and forms the foundation of the ammonia production process.

The carbon monoxide generated is then converted in the water–gas shift reaction, where it reacts with steam to produce additional hydrogen and carbon dioxide. This step increases the overall hydrogen yield and prepares the gas mixture for purification.

Following this, the gas undergoes purification to remove carbon dioxide, residual carbon monoxide, and water vapour. This is typically achieved through processes such as amine scrubbing and methanation. The removal of these impurities is critical, as even trace amounts can deactivate the catalyst used in ammonia synthesis.

Nitrogen is then introduced, usually derived from air separation or secondary reforming, to achieve the required stoichiometric ratio of nitrogen to hydrogen (1:3). Maintaining this precise ratio is essential for efficient ammonia synthesis.

The purified gas mixture is subsequently compressed to high pressure, typically between 150 and 300 bar. This compression is necessary because high pressure favours ammonia formation according to chemical equilibrium principles.

The compressed gases enter the ammonia synthesis loop, where nitrogen and hydrogen react over an iron-based catalyst to form ammonia. The reaction is carried out at elevated temperatures of around 400–500°C. Because the reaction is reversible and does not go to completion in a single pass, only a portion of the gas is converted each time it passes through the reactor.

After leaving the reactor, the gas mixture is cooled, allowing ammonia to condense into a liquid form. The remaining unreacted nitrogen and hydrogen gases are separated and recycled back into the system, where they are mixed with fresh feed gas and reintroduced into the reactor. This recycling loop significantly improves overall process efficiency and ammonia yield.

The final product is liquid ammonia, which is stored under pressure or refrigeration. It can be used directly or further processed into fertilisers such as urea or ammonium nitrate.

Overall, the process is continuous and highly integrated. Hydrogen production represents the most energy-intensive stage, while effective purification is essential to protect catalysts and maintain system performance. Gas recycling plays a critical role in maximising efficiency, and the entire system is designed to balance thermodynamic constraints, which influence yield, with kinetic factors, which determine reaction rate.

Continuous Process and Recycling

Because the ammonia synthesis reaction does not go to completion in a single pass, typically achieving only about 10–20% conversion per cycle, industrial systems are designed to operate as a continuous loop. After the reaction takes place, the mixture of gases leaving the reactor contains ammonia along with unreacted nitrogen and hydrogen.

The ammonia is separated from the gas mixture by cooling and condensation, which causes it to liquefy. The remaining unreacted nitrogen and hydrogen gases are then recovered and recycled back into the reactor. These recycled gases are mixed with fresh feed gas and passed through the synthesis loop again.

This continuous recycling process significantly improves overall efficiency and yield, ensuring that a much higher proportion of the input gases is ultimately converted into ammonia.

Source of Reactants

From a feedstock perspective, ammonia synthesis relies on two key inputs: nitrogen and hydrogen. Nitrogen (N_2) is obtained from the air, typically through air separation processes that isolate nitrogen from oxygen and other atmospheric gases. Because air is abundant and readily available, nitrogen supply is generally not a limiting factor in ammonia production.

Hydrogen (H_2), on the other hand, is most commonly produced from natural gas through steam methane reforming. This process converts methane into hydrogen and carbon-based by-products, making it the primary source of hydrogen for conventional ammonia production.

As a result, ammonia synthesis is directly linked to both atmospheric resources and fossil fuel-based hydrogen production. While nitrogen is effectively unlimited, the reliance on natural gas for hydrogen ties ammonia production to energy markets, carbon emissions, and broader considerations around sustainability and resource availability.

Energy and Environmental Considerations

Ammonia synthesis is an energy-intensive process, primarily due to the demands of hydrogen production and the need to maintain high pressure and temperature conditions within the reactor. Producing hydrogen—especially through steam methane reforming—requires significant energy input, while the compression and heating of gases to industrial operating conditions further increase overall energy consumption.

In addition to its energy requirements, ammonia production is a major source of carbon dioxide (CO_2) emissions. Most of these emissions arise from the hydrogen production stage when natural gas is used as the feedstock. As a result, ammonia manufacturing is closely linked to both energy use and greenhouse gas emissions.

To address these challenges, there is growing interest in the development of green ammonia. In this approach, hydrogen is produced through water electrolysis powered by renewable energy sources such as wind or solar. This significantly reduces or eliminates carbon emissions, making ammonia production more sustainable and aligned with global decarbonisation goals.

Importance in Fertiliser Manufacture

Ammonia is not usually applied directly as fertiliser (except in some cases such as anhydrous ammonia). Instead, it serves as the foundation for a wide range of nitrogen fertilisers, including:

- Urea ($CO(NH_2)_2$)
- Ammonium nitrate (NH_4NO_3)
- Ammonium sulphate ($(NH_4)_2SO_4$)

Because of this, the chemistry of ammonia synthesis is central to global agriculture and food production.

The chemistry of ammonia synthesis involves overcoming the stability of nitrogen gas through high pressure, elevated temperature, and catalytic action. By carefully balancing thermodynamics and reaction kinetics, the Haber–Bosch process enables the large-scale production of ammonia, transforming atmospheric nitrogen into a usable form for plants and supporting modern fertiliser systems.

Steam Methane Reforming (SMR)

Steam Methane Reforming (SMR) is the primary industrial method used to produce hydrogen, particularly for ammonia production. It converts methane (CH_4), the main component of natural gas, into hydrogen (H_2) through a series of high-temperature chemical reactions. This process is central to fertiliser manufacture because hydrogen is a key input for the Haber–Bosch synthesis of ammonia.

The overall purpose of SMR is to extract hydrogen from methane, produce a hydrogen-rich gas stream suitable for further processing, and supply hydrogen for ammonia production as well as refining and other chemical industries. The process begins with the primary reforming reaction, which takes place in a reformer furnace at temperatures of approximately 700–1,000°C using a nickel-based catalyst. In this step, methane reacts with steam (H_2O) according to the reaction:

$$CH_4 + H_2O \rightarrow CO + 3H_2$$

This reaction is endothermic, meaning it requires a continuous input of heat. It produces hydrogen along with carbon monoxide (CO), and although it generates most of the hydrogen, the carbon monoxide must be further processed.

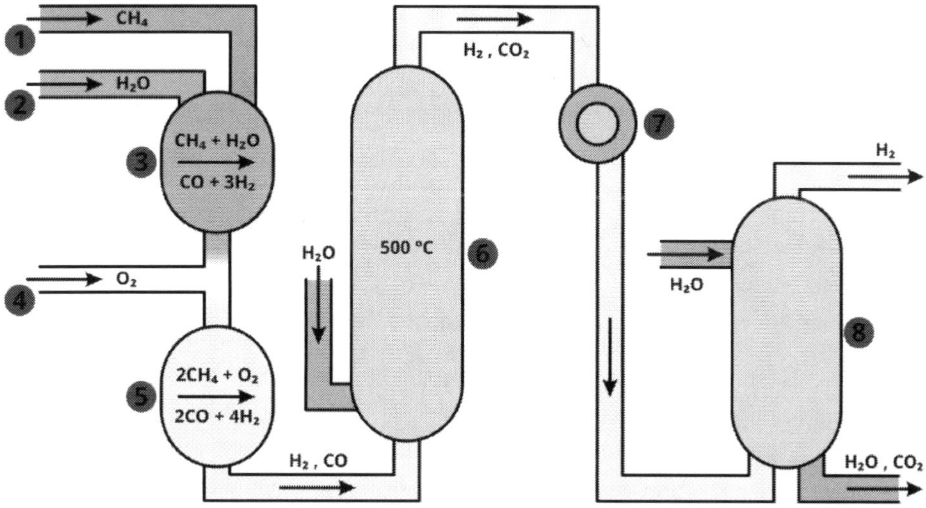

Figure 32: Steam reformer: 1. Adding of methane; 2. Adding of water; 3. primary reformer; 4. Adding of oxygen; 5. secondary reformer; 6. Catalyst; 7. Compressor; 8. Washing. Mrmw, CC0, via Wikimedia Commons.

In many ammonia plants, a secondary reforming stage follows, where methane reacts with oxygen (O_2), typically introduced via air. The reaction can be represented as:

$$2CH_4 + O_2 \rightarrow 2CO + 4H_2$$

This step generates additional hydrogen, provides heat to support the process, and introduces nitrogen (N_2) from the air, which is later used in ammonia synthesis.

The carbon monoxide produced in earlier stages is then converted through the water–gas shift reaction, where it reacts with steam to produce more hydrogen and carbon dioxide:

$$CO + H_2O \rightarrow CO_2 + H_2$$

This reaction increases the total hydrogen yield and produces carbon dioxide (CO_2) as a by-product, which must be removed in subsequent steps.

Following the shift reaction, the gas mixture contains hydrogen (H_2), carbon dioxide (CO_2), residual carbon monoxide (CO), and water vapour. Gas purification processes are used to remove CO_2, commonly through amine absorption, and eliminate trace amounts of CO via methanation. The result is high-purity hydrogen suitable for ammonia synthesis.

The SMR process operates under specific conditions to maximise efficiency and prevent operational issues. Typical operating temperatures range from 700 to 1,000°C, with pressures around 20–30 bar in the reforming stage. A nickel-based catalyst is used to facilitate the reactions, and the steam-to-carbon ratio is carefully controlled to prevent carbon deposition (coking) on the catalyst.

Steam Methane Reforming (SMR) requires a complex, integrated set of industrial equipment designed to operate under high temperatures, moderate pressures, and continuous flow conditions. The hardware is engineered to handle reactive gases, maintain catalyst performance, and maximise hydrogen production efficiency. The process is best understood as a series of interconnected units, each performing a specific function within the overall system.

The process begins with feedstock preparation and desulphurisation. Before methane enters the reformer, it must be purified to remove sulphur compounds that can poison catalysts. This is achieved using gas pre-heaters to raise the temperature of the incoming natural gas, followed by desulphurisation reactors typically packed with zinc oxide (ZnO) to remove hydrogen sulphide (H_2S). Filters and separators are also used to remove particulates and liquids, ensuring a clean feed gas for downstream catalytic reactions.

At the heart of the system is the primary reformer furnace, which provides the high temperatures required for the reforming reaction, typically between 700 and 1,000°C. This unit consists of a fired furnace containing high-alloy steel reformer tubes filled with a nickel-based catalyst. Burners, fuelled by natural gas or process off-gas, supply the necessary heat, while refractory linings insulate the furnace and protect structural components. Within the tubes, methane and steam react to form hydrogen and carbon monoxide, and efficient heat transfer is critical to maintaining reaction performance.

In ammonia production systems, a secondary reformer is often included. This unit introduces air or oxygen into a combustion chamber, where further reactions take place over a catalyst bed. Mixing nozzles ensure proper blending of gases. The secondary reformer increases hydrogen production and introduces nitrogen (N_2) into the process stream, which is required for subsequent ammonia synthesis.

Given the high energy demands of SMR, heat recovery systems are essential. Waste heat boilers capture heat from hot process gases, while heat exchangers transfer energy between streams to improve efficiency. Steam generators produce the steam required for the reforming reaction, significantly enhancing overall energy utilisation.

Following reforming, the gas mixture passes through water–gas shift reactors, where carbon monoxide is converted into additional hydrogen. These typically include a high-temperature shift (HTS) reactor operating at approximately 350–450°C and a low-temperature shift (LTS) reactor operating at around 200–250°C. The reactors contain catalyst beds, commonly iron-based for HTS and copper-based for LTS, which increase hydrogen yield and reduce carbon monoxide levels.

The gas stream is then cooled and conditioned before purification. This involves gas coolers, condensers, and knock-out drums that remove condensed water, preparing the gas for carbon dioxide removal. The CO_2 removal system typically uses absorption towers, such as amine scrubbers, to capture CO_2, followed by regeneration units (strippers) that release CO_2 from the solvent. Pumps and heat exchangers support solvent circulation and regeneration, ensuring efficient operation.

To remove trace amounts of carbon monoxide and carbon dioxide that could damage downstream catalysts, the gas passes through a methanation reactor. In this stage, residual CO and CO_2 are converted to methane using hydrogen, according to the reaction $CO + 3H_2 \rightarrow CH_4 + H_2O$. Nickel-based catalysts are used, ensuring the production of ultra-pure hydrogen suitable for ammonia synthesis.

The purified hydrogen is then compressed using multi-stage compressors, with intercoolers installed between stages to remove heat generated during compression. Pressure control systems ensure that the gas reaches the required conditions for downstream processes such as the Haber–Bosch synthesis loop.

SMR plants rely heavily on instrumentation and control systems to maintain safe and efficient operation. These include temperature and pressure sensors, flow meters, and gas analysers, all integrated into distributed control systems (DCS). Safety systems such as pressure relief valves and emergency shutdown mechanisms are critical due to the hazardous operating conditions.

In addition to the core process equipment, supporting infrastructure is essential for continuous operation. This includes steam generation and water treatment systems, fuel gas systems for burners, flare systems for safe handling of excess gases, extensive piping networks and valves, and cooling water systems.

Overall, Steam Methane Reforming depends on a highly engineered combination of reactors, furnaces, heat exchangers, separation units, and control systems. Each component plays a vital role in converting methane into hydrogen efficiently and safely. The integration of these systems enables SMR plants to operate continuously at industrial scale, forming the backbone of modern ammonia and fertiliser production.

SMR is highly energy-intensive, largely due to the heat required for the endothermic reforming reaction and the overall process conditions. It is also a major source of carbon dioxide emissions, arising both from the chemical reactions and from the combustion of fuel used to supply heat. As a result, SMR significantly contributes to the carbon footprint of ammonia and nitrogen fertiliser production.

The importance of SMR in fertiliser production lies in its role as the primary source of hydrogen for ammonia synthesis. It directly links fertiliser production to natural gas availability and price, and it is a key factor in determining production costs and environmental impact.

Due to environmental concerns, there is increasing interest in alternative hydrogen production methods. These include green hydrogen produced via water electrolysis using renewable energy, carbon capture and storage (CCS) applied to conventional SMR systems, and emerging technologies that aim to reduce emissions and improve sustainability.

Hydrogen Production and Purification

Hydrogen production and purification are critical stages in ammonia manufacturing, as hydrogen is a key reactant in the Haber–Bosch process. These stages involve converting raw feedstocks, most commonly natural gas, into a high-purity hydrogen stream suitable for chemical synthesis. The process is both chemically complex and highly integrated, combining reaction engineering, separation technologies, and process control.

Hydrogen production typically begins with Steam Methane Reforming (SMR), where methane (CH_4) reacts with steam (H_2O) at high temperatures in the presence of a catalyst to produce hydrogen and carbon monoxide. The primary reaction is:

$$CH_4 + H_2O \rightarrow CO + 3H_2$$

This is followed by the water–gas shift reaction, which converts carbon monoxide into additional hydrogen:

$$CO + H_2O \rightarrow CO_2 + H_2$$

The water–gas shift (WGS) reaction is a key step in hydrogen production, particularly in processes such as steam methane reforming. Its primary purpose is to convert carbon monoxide (CO), which is produced during reforming, into additional hydrogen (H_2) while forming carbon dioxide (CO_2). The reaction can be written as:

$$CO + H_2O \rightarrow CO_2 + H_2$$

This reaction is important because it increases the overall yield of hydrogen and reduces the concentration of carbon monoxide, which is undesirable in downstream processes such as ammonia synthesis.

The reaction works by allowing carbon monoxide to react with steam (water vapour) over a catalyst. In this process, the carbon monoxide molecule combines with a water molecule, resulting in the formation of carbon dioxide and hydrogen. Essentially, the oxygen atom from water binds with carbon monoxide to form CO_2, while the remaining hydrogen atoms are released as hydrogen gas. This transformation effectively shifts the composition of the gas mixture toward higher hydrogen content, which is why it is called the "shift" reaction.

From a thermodynamic perspective, the water–gas shift reaction is exothermic, meaning it releases heat. As a result, lower temperatures favour higher equilibrium conversion of carbon monoxide to hydrogen. However, lower temperatures also slow down the reaction rate. To balance these effects, the process is typically carried out in two stages. The first stage is the high-temperature shift (HTS), operating at around 350–450°C, where the reaction proceeds quickly but does not reach maximum conversion. The second stage is the low-temperature shift (LTS), operating at approximately 200–250°C, where slower reaction rates allow for greater conversion efficiency.

Catalysts play a crucial role in enabling the reaction to proceed at practical rates. In the high-temperature stage, iron-based catalysts are commonly used, while in the low-temperature stage, copper-based catalysts are preferred due to their higher activity at lower temperatures. These catalysts provide a surface that facilitates the reaction without being consumed, improving efficiency and control.

The water–gas shift reaction is also important from a process integration perspective. By converting carbon monoxide into carbon dioxide, it simplifies subsequent purification steps. Carbon dioxide can be more easily removed from the gas stream using absorption technologies, whereas carbon monoxide is more difficult to separate and can poison catalysts in later stages, particularly in ammonia synthesis.

Together, these reactions maximise hydrogen yield while converting carbon-containing species into carbon dioxide. In ammonia plants, a secondary reforming step may also be used, where air is introduced to provide nitrogen and generate additional hydrogen.

After hydrogen is produced, the resulting gas mixture contains not only hydrogen but also carbon dioxide (CO_2), residual carbon monoxide (CO), water vapour, and trace impurities. These impurities must be removed because they can poison catalysts used in downstream processes, particularly in ammonia synthesis.

The first major purification step is carbon dioxide removal. This is typically achieved using chemical absorption processes, such as amine scrubbing, where CO_2 is selectively absorbed into a liquid solvent and later removed in a regeneration unit. This step significantly increases hydrogen purity and reduces the carbon content of the gas stream.

Following CO_2 removal, trace amounts of carbon monoxide and carbon dioxide are eliminated through methanation. In this process, these residual gases react with hydrogen over a nickel catalyst to form methane (CH_4) and water:

$$CO + 3H_2 \rightarrow CH_4 + H_2O$$
$$CO_2 + 4H_2 \rightarrow CH_4 + 2H_2O$$

Methanation ensures that even very small concentrations of CO and CO_2 are removed, producing a highly purified hydrogen stream.

Methanation is a gas purification process used to remove trace amounts of carbon monoxide (CO) and carbon dioxide (CO_2) from hydrogen-rich gas streams. It is particularly important in ammonia production because even very small concentrations of these gases can poison the iron-based catalyst used in the Haber–Bosch process. Methanation converts these undesirable components into methane (CH_4), which is inert in the context of ammonia synthesis and does not interfere with the reaction.

The process works through catalytic hydrogenation reactions, where carbon oxides react with hydrogen over a catalyst—typically nickel-based—at elevated temperatures. The key reactions are:

$$CO + 3H_2 \rightarrow CH_4 + H_2O$$
$$CO_2 + 4H_2 \rightarrow CH_4 + 2H_2O$$

In these reactions, hydrogen combines with carbon monoxide or carbon dioxide to form methane and water. Essentially, the oxygen atoms from CO and CO_2 are transferred to hydrogen to form water (H_2O), while the carbon is converted into methane. This effectively removes the reactive carbon oxides from the gas stream.

Methanation is an exothermic process, meaning it releases heat. As a result, temperature control is important to prevent overheating, which could damage the catalyst or reduce efficiency. The reaction is typically carried out at temperatures of around 250–350°C and at moderate pressures, using fixed-bed reactors packed with catalyst material.

Before methanation, most of the CO_2 is usually removed through processes such as amine absorption. Methanation is therefore considered a "polishing" step, designed to eliminate the small residual amounts of CO and CO_2 that remain after primary purification. This ensures that the final gas stream is of very high purity.

The success of methanation depends on maintaining clean operating conditions. Sulphur compounds and other contaminants must be removed upstream, as they can deactivate the nickel catalyst. Proper temperature and flow control are also critical to maintaining catalyst performance and preventing hotspots.

Water vapour is then removed through cooling and condensation, often followed by drying systems such as molecular sieves. The result is hydrogen of very high purity, typically suitable for use in ammonia synthesis.

In some systems, additional purification technologies such as Pressure Swing Adsorption (PSA) may be used. PSA systems separate hydrogen from other gases based on adsorption properties, producing extremely high-purity hydrogen streams.

Pressure Swing Adsorption (PSA) is a gas separation technology used to produce high-purity hydrogen by selectively removing impurities from a gas mixture. It is widely used in hydrogen production systems, particularly after reforming and shift reactions, as a final purification step before ammonia synthesis or other applications.

The process works on the principle that different gases are adsorbed onto solid materials at different rates depending on pressure. In PSA systems, a gas mixture—typically containing hydrogen, carbon dioxide (CO_2), carbon monoxide (CO), methane (CH_4), and nitrogen (N_2)—is passed through a vessel filled with an adsorbent material such as activated carbon, zeolites, or molecular sieves. Under high pressure, impurities such as

CO_2, CO, CH_4, and N_2 are preferentially adsorbed onto the surface of the material, while hydrogen, which is less strongly adsorbed, passes through as the product gas.

Once the adsorbent becomes saturated with impurities, the system undergoes a pressure reduction phase. The pressure is lowered, causing the adsorbed gases to desorb (release) from the material. These impurities are then removed as off-gas. The adsorbent bed is regenerated during this step, allowing it to be reused in the next cycle. This alternating process of high-pressure adsorption and low-pressure desorption is what gives PSA its name.

PSA systems typically consist of multiple adsorption vessels operating in parallel, allowing continuous production. While one vessel is in the adsorption phase, others are undergoing depressurisation, regeneration, or repressurisation. This cyclic operation ensures a steady output of purified hydrogen.

The process can achieve very high hydrogen purity, often exceeding 99.9%, making it suitable for sensitive downstream processes such as ammonia synthesis, fuel cells, and chemical manufacturing. However, PSA does result in some hydrogen loss in the off-gas stream, which is often recycled or used as fuel within the plant.

Operating conditions usually involve moderate to high pressures (depending on the upstream process), and the performance of PSA depends on factors such as adsorbent type, cycle timing, pressure levels, and gas composition. The system requires precise control to optimise recovery and purity.

Hydrogen production and purification are among the most energy-intensive and technically demanding parts of ammonia manufacturing. They determine not only the efficiency of the overall process but also its environmental impact, particularly due to carbon dioxide emissions associated with natural gas reforming.

Hydrogen production and purification for ammonia manufacturing require a highly integrated set of specialised industrial equipment, designed to manage high temperatures, reactive gases, and continuous operation. Each stage of the process, from reforming to final purification, relies on dedicated hardware that ensures efficient conversion, separation, and protection of downstream systems.

The process begins with feed preparation and steam generation. Natural gas must first be treated to remove contaminants such as sulphur, which can poison catalysts. This requires desulphurisation units, typically packed-bed reactors containing zinc oxide (ZnO), along with gas pre-heaters and filtration systems. At the same time, a steam generation system, including boilers and water treatment units, produces high-purity steam required for the reforming reaction.

The core hydrogen production step occurs in the primary reformer, a large fired furnace containing high-alloy reformer tubes filled with a nickel-based catalyst. Burners supply the heat required to drive the endothermic reaction $CH_4 + H_2O \rightarrow CO + 3H_2$. The furnace is lined with refractory materials to withstand extreme temperatures, and precise control systems regulate heat input and gas flow. In ammonia plants, a secondary reformer is often included, consisting of a combustion chamber and catalyst bed where air or oxygen is introduced. This unit not only increases hydrogen production but also supplies nitrogen for ammonia synthesis.

Following reforming, the gas mixture passes through water–gas shift reactors, which are typically arranged in two stages: a high-temperature shift (HTS) reactor and a low-temperature shift (LTS) reactor. These are fixed-bed catalytic reactors containing iron-based and copper-based catalysts, respectively. Heat exchangers and temperature control systems are essential in this section, as the reaction $CO + H_2O \rightarrow CO_2 + H_2$ is exothermic and must be carefully managed to maintain optimal conversion and protect catalyst integrity.

After the shift conversion, the gas stream must be cooled and conditioned before purification. This requires gas coolers, heat exchangers, condensers, and knock-out drums to remove excess water. Effective thermal integration at this stage improves overall plant efficiency and prepares the gas for carbon dioxide removal.

Carbon dioxide removal is typically carried out in an absorption system, most commonly using amine scrubbing. This involves tall absorption columns where CO_2 is selectively absorbed into a liquid solvent, followed by regeneration (stripping) columns where the solvent is heated to release CO_2 for removal. Pumps, heat exchangers, and solvent circulation systems support continuous operation. This stage is critical for significantly increasing hydrogen purity.

To remove trace amounts of carbon monoxide and residual carbon dioxide, the gas is then passed through a methanation reactor. This is a fixed-bed reactor containing a nickel-based catalyst, where CO and CO_2 react with hydrogen according to $CO + 3H_2 \rightarrow CH_4 + H_2O$ and $CO_2 + 4H_2 \rightarrow CH_4 + 2H_2O$. Because methanation is highly exothermic, the reactor requires careful temperature monitoring and control to prevent catalyst damage and ensure consistent performance.

Following methanation, the gas is further dried to remove water vapour. This is typically achieved using condensers followed by adsorption systems such as molecular sieve dryers, which remove residual moisture to produce a dry hydrogen stream suitable for ammonia synthesis.

In some plants, an additional purification step is included using Pressure Swing Adsorption (PSA). PSA units consist of multiple adsorption vessels filled with specialised materials that selectively adsorb impurities under high pressure and release them when pressure is reduced. This allows for the production of extremely high-purity hydrogen.

Compression systems are also essential, as hydrogen must be pressurised for downstream use. Multi-stage compressors with intercooling are used to increase pressure while managing temperature and energy consumption. These systems are supported by extensive instrumentation, including pressure sensors, temperature monitors, flow meters, and gas analysers, all integrated into a distributed control system (DCS) for safe and efficient plant operation.

Supporting infrastructure plays a vital role in ensuring continuous operation. This includes cooling water systems, fuel gas systems for reformer burners, flare systems for safety, and extensive piping networks and valves to manage process flows.

Reactor Design and Catalysts

Reactor design and catalysts are central to the efficiency, safety, and economics of chemical processes such as hydrogen production and ammonia synthesis. Together, they determine how effectively chemical reactions occur, how heat and mass are managed, and how stable and productive the process remains over time.

Reactor design refers to the physical configuration and operating conditions of the vessel in which chemical reactions take place. In industrial systems, reactors must be engineered to withstand high temperatures, high pressures, and continuous operation while ensuring optimal contact between reactants and catalysts. Common reactor types used in fertiliser production include fixed-bed reactors, tubular reactors, and fluidised-bed reactors. For example, in steam methane reforming, long tubular reactors are used inside a furnace, allowing heat to be transferred efficiently to the reacting gases. In the ammonia synthesis loop, fixed-bed reactors are typically used, where gases flow over a packed bed of catalyst.

Figure 33: Severnside fertilizer works. This plant is owned by Terra Nitrogen UK Ltd. and manufactures ammonia and ammonium nitrate. Sharon Loxton/Severnside fertilizer works, CC BY-SA 2.0, via Wikimedia Commons.

A key consideration in reactor design is the management of heat. Some reactions, such as steam methane reforming, are endothermic and require continuous heat input, while others, such as the water–gas shift reaction and ammonia synthesis, are exothermic and release heat. Reactor systems must therefore include heat exchange mechanisms to maintain optimal temperatures, prevent overheating, and maximise efficiency. Poor temperature control can reduce reaction rates, shift equilibrium unfavourably, or damage catalysts.

Another important aspect is mass transfer, which refers to how well reactants are brought into contact with the catalyst surface. Reactor design must ensure efficient gas flow, minimise pressure drop, and avoid channelling or dead zones within the catalyst bed. Proper mixing and flow distribution are essential to achieving uniform reaction conditions and consistent product quality.

Catalysts are materials that accelerate chemical reactions without being consumed in the process. They work by providing an active surface where reactants can adsorb, react, and desorb more easily than in the gas phase. In fertiliser production, catalysts are essential for making reactions economically viable. For example, nickel-based catalysts are used in steam methane reforming to break the strong bonds in methane, while iron-based catalysts are used in the Haber–Bosch process to facilitate the reaction between nitrogen and hydrogen. In water–gas shift reactors, iron and copper-based catalysts are used depending on the operating temperature.

Catalyst performance depends on several factors, including surface area, activity, selectivity, and stability. A high surface area provides more active sites for reactions, while selectivity ensures that the desired reaction pathway is favoured over unwanted side reactions. Stability is critical, as catalysts must operate for long periods without significant loss of activity. Over time, catalysts can degrade due to poisoning (e.g., sulphur contamination), sintering (loss of surface area at high temperatures), or fouling (deposition of carbon or other materials).

To maximise catalyst effectiveness, reactor design must support proper operating conditions. This includes maintaining the correct temperature, pressure, and gas composition, as well as ensuring uniform flow distribution. Catalyst beds are often designed in layers or stages to optimise performance and allow for temperature control. In some systems, multiple reactors are used in series to achieve higher overall conversion.

Catalyst handling and lifecycle management are also important considerations. Catalysts must be loaded carefully to ensure even distribution, and periodic replacement or regeneration may be required. In large industrial plants, catalyst performance is closely monitored using process data and gas analysis to detect signs of deactivation.

Creating a Reactor Design

Designing and operating reactors with appropriate catalysts in processes such as hydrogen production and ammonia synthesis is a structured engineering task that integrates chemical, mechanical, and process design principles. It involves translating reaction chemistry into a practical, safe, and efficient industrial system.

The first step is to clearly define the reaction system. This includes identifying the chemical reactions involved, such as steam methane reforming or ammonia synthesis, and understanding their thermodynamics (whether they are endothermic or exothermic) and kinetics (how fast they proceed). This information determines the required temperature, pressure, and residence time. For example, an endothermic reaction like reforming requires continuous heat input, whereas an exothermic reaction like ammonia synthesis requires heat removal to maintain stability.

Once the reaction requirements are understood, the appropriate reactor type is selected. In fertiliser production, this typically involves choosing between tubular reactors (for high heat transfer, such as in reforming), fixed-bed reactors (for catalytic gas-phase reactions like ammonia synthesis), or fluidised-bed reactors (where improved mixing is required). The selection depends on factors such as heat transfer needs, catalyst type, pressure drop constraints, and scale of operation.

The next stage is detailed reactor design. This includes sizing the reactor based on flow rates and desired conversion, determining the geometry (length, diameter, and internal configuration), and selecting materials that can withstand high temperatures, pressures, and corrosive environments. For example, reformer tubes must be made from high-alloy steels capable of operating at extreme temperatures without failure. The design must also incorporate pressure containment, insulation, and mechanical integrity to meet safety standards.

Heat management is then integrated into the design. For endothermic systems, this involves designing furnaces, burners, and heat transfer surfaces to supply sufficient energy. For exothermic systems, heat exchangers or interstage cooling may be required to remove excess heat and maintain optimal reaction conditions. Effective heat management ensures that the reaction remains within the desired operating window, preventing catalyst damage and improving efficiency.

Mass transfer and flow distribution must also be carefully engineered. This involves designing inlet distributors, ensuring even gas flow across catalyst beds, and avoiding channelling or dead zones. Computational fluid dynamics (CFD) and flow modelling are often used to optimise these aspects. Proper flow design ensures that all reactants have equal access to the catalyst, maximising conversion and maintaining consistent product quality.

Catalyst selection is a critical parallel step. The catalyst must be chosen based on activity (how effectively it accelerates the reaction), selectivity (its ability to favour the desired reaction), and stability under operating conditions. For example, nickel-based catalysts are selected for reforming due to their ability to break methane bonds, while iron-based catalysts are used in ammonia synthesis for nitrogen activation. The physical form of the catalyst—such as pellets, rings, or extrudates—is also chosen to balance surface area and pressure drop.

Once selected, the catalyst must be properly loaded into the reactor. This involves careful placement to ensure uniform packing, often using layered configurations to optimise performance and protect more active catalyst zones. For example, guard beds may be installed upstream to capture impurities before they reach the main catalyst. Correct loading prevents uneven flow and hotspots, which can damage the catalyst and reduce efficiency.

Operational control systems are then implemented to maintain the required conditions. This includes temperature sensors, pressure transmitters, flow meters, and gas analysers connected to a distributed control system (DCS). These systems allow operators to monitor reactor performance in real time, adjust operating parameters, and respond to deviations. Safety systems, such as pressure relief valves and emergency shutdown mechanisms, are also essential.

Over time, catalyst performance must be monitored and managed. This involves analysing process data to detect signs of deactivation, such as reduced conversion efficiency or increased pressure drop. Maintenance strategies may include catalyst regeneration (if possible) or replacement. Preventative measures, such as ensuring feedstock purity and maintaining stable operating conditions, are critical to extending catalyst life.

In practice, reactor design and catalyst management are iterative and integrated processes. Engineers must continuously balance reaction efficiency, energy use, safety, and cost. By carefully aligning reactor configuration with catalyst properties and operating conditions, it is possible to achieve high conversion rates, reliable performance, and long-term operational stability in fertiliser manufacturing systems.

Case Study Example: Optimising Reactor Design and Catalyst Performance in an Ammonia Plant

A hypothetical mid-sized fertiliser manufacturer, Southern AgriChem, set out to develop a new ammonia production facility to supply nitrogen fertilisers to regional agricultural markets. The plant was designed to produce 300 tonnes of ammonia per day using natural gas as the primary feedstock. During early commissioning, the company experienced lower-than-expected hydrogen yields and unstable ammonia synthesis performance, prompting a detailed review of reactor design and catalyst systems.

The first stage of the investigation focused on the steam methane reforming (SMR) unit, which used a tubular reactor design housed within a fired furnace. Engineers identified that while the reactor tubes were correctly specified with high-alloy steel, heat distribution across the furnace was uneven. Some tubes were operating at lower temperatures, reducing methane conversion efficiency in the reaction $CH_4 + H_2O \rightarrow CO + 3H_2$. This led to incomplete reforming and increased carbon monoxide levels downstream.

To address this, the company redesigned the furnace burner configuration and improved heat distribution using advanced control systems. Additional temperature sensors were installed to monitor tube performance in real time. As a result, the reforming reaction achieved more consistent temperatures, significantly improving hydrogen production and reducing variability in the gas stream.

The next issue arose in the water–gas shift (WGS) reactors, where carbon monoxide was converted to hydrogen via $CO + H_2O \rightarrow CO_2 + H_2$. The plant used a two-stage system consisting of a high-temperature shift (HTS) reactor followed by a low-temperature shift (LTS) reactor. However, performance data showed that the HTS reactor was overheating due to insufficient heat removal, causing partial catalyst degradation. This reduced its effectiveness and placed additional load on the LTS reactor.

Engineers implemented interstage cooling and optimised flow distribution to better manage the exothermic heat release. They also replaced the degraded iron-based catalyst in the HTS reactor and upgraded the copper-based catalyst in the LTS reactor to a higher-activity formulation. These changes improved overall hydrogen yield and reduced residual CO levels entering the purification stage.

In the methanation reactor, which removes trace CO and CO_2 via reactions such as $CO + 3H_2 \rightarrow CH_4 + H_2O$, the plant initially experienced hotspot formation within the catalyst bed. This was traced to poor gas distribution caused by uneven catalyst loading. As a result, certain areas of the bed were exposed to higher temperatures, accelerating catalyst sintering and reducing its lifespan.

To resolve this, the catalyst was reloaded using a layered configuration with improved packing techniques and flow distributors installed at the reactor inlet. This ensured even gas flow and eliminated hotspots, restoring stable operation and extending catalyst life.

The most critical improvements were made in the ammonia synthesis reactor, a high-pressure fixed-bed reactor operating at approximately 200 bar and 450°C. Initially, ammonia conversion per pass was below design expectations, and pressure drop across the catalyst bed was increasing over time. Investigation revealed partial fouling and suboptimal catalyst selection for the operating conditions.

The company replaced the original iron-based catalyst with a promoted catalyst formulation that offered higher activity and better resistance to deactivation. In addition, the reactor was reconfigured into a multi-bed system with interstage cooling, allowing better temperature control of the exothermic reaction $N_2 + 3H_2 \rightleftharpoons 2NH_3$. This improved both conversion efficiency and operational stability.

Across the plant, improvements were also made to mass transfer and flow distribution. Computational fluid dynamics (CFD) modelling was used to redesign inlet distributors and optimise gas flow through reactors. This eliminated channelling and ensured uniform contact between reactants and catalysts, further enhancing performance.

From a catalyst management perspective, Southern AgriChem implemented a proactive monitoring system using gas analysis and performance data. This allowed early detection of catalyst deactivation due to poisoning, sintering, or fouling. Scheduled maintenance and regeneration strategies were introduced, reducing unplanned downtime and extending catalyst service life.

As a result of these integrated improvements in reactor design and catalyst selection, the plant achieved a 15% increase in hydrogen production efficiency and a 10% increase in ammonia yield. Energy consumption per tonne of ammonia decreased, and overall operational stability improved significantly.

This case study demonstrates that reactor design and catalysts are not independent elements but are closely interrelated. Effective reactor configuration ensures proper heat and mass transfer, while appropriate catalyst selection and management enable high reaction rates and long-term stability. Together, they directly influence efficiency, safety, and economic performance in fertiliser manufacturing.

Energy Consumption and Efficiency Improvements

Ammonia production today still rides on one big energy ticket: hydrogen. In the classic Haber–Bosch route, you squeeze nitrogen from the air and hydrogen (mostly from natural gas) together at high pressure and high temperature to make ammonia. It's a powerhouse of an industrial process, but its energy bill is steep, and that cost shapes everything from plant design to policy decisions. When you wander through recent discussions, you'll see a simple, stubborn pattern: most of the energy ends up in making the hydrogen, with the rest spread across compressing gases, keeping the synthesis loop hot, and driving the utilities that move heat around the plant. A number of sources repeatedly point to hydrogen production and compression as the dominant energy sinks, while heat recovery and smarter catalysts stand out as the best routes to cut that load [102-109].

In the big picture, the reforming furnace that makes hydrogen from methane (steam reforming) is the energy juggernaut. That endothermic step demands a lot of heat to push the reaction forward, and the downstream shift reaction and purification steps add more energy needs. Techno-economic analyses consistently show hydrogen production can account for roughly three-quarters of the total energy use in a conventional plant, with the remaining share going to compression, air separation for nitrogen, and the high-pressure synthesis loop [102-104, 107, 108]. Even when you account for different feedstocks or process variants, the consensus is clear: attack hydrogen production if you want to drop the plant's energy bill. The synthesis loop itself, while exothermic, still needs energy inputs to maintain the right temperatures and to remove heat so the reactor operates near its best point for ammonia formation [102-104, 106].

Compressing nitrogen and hydrogen to the typical Haber–Bosch pressures (often in the 150–300 bar range) is no minor feat. Multi-stage compressors with intercooling are common, and energy efficiency gains here—reducing pressure drops, improving flow, and tightening integration with other units—translate into real energy savings [102-104, 106]. The synthesis step itself releases heat (it's exothermic), but operators still must supply heat to keep the reacting streams at the right temperature range (roughly 400–500°C) and then rake away excess heat to maintain the right equilibrium and avoid runaway conditions. In practice, this means that even though the chemical reaction is favourable, the plant uses a lot of energy in heat management and maintaining those high operating conditions [102-104, 107].

If you want to shave energy off the overall process, the big wins come from heat integration and smarter catalysts. Modern plants deploy waste heat boilers and heat exchangers to turn hot process streams into steam or heat feedstocks, which cuts the need for fresh energy. This kind of energy recovery is repeatedly highlighted as one of the strongest, most reliable ways to improve thermal efficiency across the whole ammonia chain [102-104, 106, 108]. On the catalyst side, better iron-based catalysts for the Haber–Bosch loop and improved nickel catalysts for reforming can push conversions higher or allow operation at lower temperatures, which directly reduces energy input per tonne of ammonia. Water–gas shift catalysts and configurations that boost hydrogen yield also help, because more hydrogen means less downstream processing energy and less fresh feed gas heating [102-104, 106, 107].

Beyond materials, the way a plant is controlled matters. Process intensification, things like multi-bed synthesis loops with interstage cooling, smarter heat integration networks, and better flow distribution, reduces energy losses in heat transfer and compression. Add in modern digital tools: online monitoring, predictive maintenance, and advanced process control. Taken together, these can push a plant to operate closer to its optimum envelope, reducing wasted energy and downtime [102-104, 107, 108].

There's a growing conversation about greener routes: green ammonia uses hydrogen from water electrolysis powered by renewables, while blue ammonia keeps the Haber–Bosch backbone but adds carbon capture to cut CO_2 from hydrogen production. These options change the energy landscape because the electricity cost and the carbon price become major variables. Several studies show that, in the near term, grey (natural-gas-based) ammonia often wins on price and reliability, but under high renewables or strong carbon pricing, green and blue routes become more attractive. The energy question here isn't only about kilojoules per tonne, it's about the costs and timing of electricity, electrolyzer tech, grid integration, and policy signals like carbon taxes [74, 102, 103, 106, 110, 111].

Green hydrogen, produced by electrolysis with renewable energy, sits at the heart of the green ammonia idea. Modelling and reviews point out that the value of green ammonia depends a lot on electricity costs, electrolyzer performance, and whether the system can ride the renewables curve with storage and flexible operation. In practice, green hydrogen can raise the energy cost per tonne of ammonia unless electricity is cheap and abundant, or the plant can flexibly shift production with the grid and storage. Still, many analyses see long-term decarbonization potential, especially when combined with grid services, storage, and clever siting near abundant renewables [102, 103, 106, 110, 111].

A striking thread in recent work is the idea of decentralizing ammonia production, pushing smaller, modular plants closer to farms or storage hubs, powered by electrolysis and simple ammonia synthesis units. The argument goes that with cheap solar or grid electricity, distributed electrified Haber–Bosch or electrocatalytic routes could cut transport costs and reduce vulnerability to fossil-fuel price swings. The feasibility studies suggest that decentralization could be cost-competitive for a large share of ammonia demand by 2030–2050, depending on energy prices, transport costs, and policy incentives. Still, centralization currently offers cost advantages at scale, so the economics hinge on the energy and policy environment in the coming years [102-104, 106, 107, 110].

In talking about energy in ammonia production, the message is really twofold: first, the energy demand is dominated by hydrogen production and by compression and heat management in the Haber–Bosch loop; second, there is a powerful set of strategies to cut energy use that consistently show up in the literature. Heat integration, smarter catalysts, and tighter process design are the lowest-hanging fruits, while digital optimization and process-control advances help operators squeeze every last drop of efficiency out of the plant. Looking ahead, green and blue ammonia, along with carbon pricing and smarter energy systems, will reshape the economics and energy choices for ammonia plants. Decentralized electrified approaches offer an intriguing vision for localized production that could reduce transport costs and increase resilience, provided electricity prices and capital costs cooperate. As one reader, you can take away a practical core: if you want to cut energy in ammonia, start with the hydrogen production train and the heat-exchange network, then layer in smarter controls and better catalysts, and watch how the economics bend as renewables and policy evolve.

Small Scale Implementation

The Haber–Bosch process can be undertaken on a smaller scale; however, it remains technically complex, capital-intensive, and generally uneconomical when compared with large industrial plants. The process relies on high pressure (typically 150–300 bar), high temperature (around 400–500°C), and specialised catalysts, all of which require robust and carefully engineered systems. Even relatively small-scale facilities must incorporate hydrogen production units (such as steam methane reforming or electrolysis), gas purification systems to remove impurities like sulphur and carbon dioxide, high-pressure reactors and compressors, as well as heat management and recycling loops. These requirements create a minimum viable scale below which production becomes inefficient and cost-prohibitive.

Large-scale ammonia plants dominate the industry primarily due to economies of scale. Facilities producing more than 1,000 tonnes per day benefit from reduced costs per unit of output, improved energy efficiency, and lower capital costs relative to production volume. Continuous operation also enhances process stability and efficiency. For new manufacturers, competing with these large-scale operations using conventional technology is challenging unless there is access to low-cost feedstocks, such as natural gas, and well-developed infrastructure.

Despite these challenges, there is growing interest in smaller and more flexible ammonia production systems. Modular Haber–Bosch units have been developed as compact, skid-mounted systems with lower production capacities, typically ranging from 1 to 50 tonnes per day. While they still operate under high-pressure conditions, they offer greater flexibility and are particularly suited to remote or regional fertiliser production where transport costs are high. In parallel, green ammonia systems are emerging, where hydrogen is produced through electrolysis using renewable energy sources such as solar or wind power. These systems can be integrated with smaller-scale synthesis units and offer a lower carbon footprint, although they are currently more expensive than conventional approaches.

In addition to these developments, alternative technologies such as electrochemical ammonia synthesis, plasma-based processes, and biological nitrogen fixation systems are being explored. However, these methods are still largely in experimental or early commercial stages and are not yet widely competitive with established Haber–Bosch systems.

For small-scale producers, several constraints must be considered. These include the high capital cost of specialised equipment such as pressure vessels and compressors, significant energy requirements, particularly for hydrogen production, technical expertise and strict safety standards, availability of suitable feedstocks, and compliance with regulatory requirements related to pressure systems and emissions.

Small-scale ammonia production can be viable under specific conditions. These include remote agricultural regions where fertiliser transport costs are high, locations with access to low-cost renewable energy for green ammonia production, integrated industrial operations such as mining or large-scale agriculture, and countries seeking greater fertiliser self-sufficiency.

The following is a practical, staged implementation plan for establishing a small-scale ammonia production facility (Haber–Bosch process). It is structured for a new manufacturer and aligns with engineering, regulatory, and commercial realities.

Implementation Plan: Small-Scale Ammonia Production (Haber–Bosch)

1. Define Project Scope and Objectives

The first step is to clearly define the purpose and scale of the plant. Typical small-scale plants range from 1–50 tonnes/day of NH_3.

Key decisions:

- Target production capacity (e.g., 10 t/day)
- End use (fertiliser production, local supply, industrial use)
- Feedstock choice:
  - Natural gas (conventional)
  - Renewable electricity (green ammonia)

Output: Concept definition and business case

2. Feasibility Study

Conduct a detailed techno-economic feasibility study covering:

Technical feasibility

- Hydrogen source (SMR vs electrolysis)
- Nitrogen source (air separation vs PSA)
- Process configuration (batch vs continuous loop)
- Utilities (water, power, cooling)

Economic feasibility

- CAPEX estimate (typically USD $3.5M–$35M/AUD $5M–$50M depending on scale)
- OPEX (energy dominates cost)
- Break-even ammonia price

Site selection

- Proximity to:
  - Feedstock (gas or renewable power)
  - Water supply
  - Transport/logistics

- End users (farms, fertiliser plants)

Output: Feasibility report + go/no-go decision

3. Process Design and Engineering

Develop a basic process flow diagram (PFD) and then a detailed process design.

Core process blocks:

1. **Hydrogen Production**
    - Option A: Steam Methane Reforming (SMR)
    - Option B: Electrolysis (green hydrogen)
2. **Gas Purification**
    - CO_2 removal (amine scrubbing or PSA)
    - Sulphur removal (ZnO beds)
3. **Nitrogen Generation**
    - Air separation unit (ASU) or Pressure Swing Adsorption (PSA)
4. **Ammonia Synthesis Loop**
    - Compressor (to ~150–300 bar)
    - Reactor (iron catalyst)
    - Heat exchangers
    - Condenser (to liquefy NH_3)
    - Gas recycle loop

Simplified Process Flow

Natural Gas / Water → Hydrogen Unit → Gas Purification → Mixing with N_2

↓

Compression

↓

Haber Reactor

↓

Cooling & Condensation

↓

Liquid Ammonia

↑

Gas Recycle Loop

Output: PFD, mass/energy balances, equipment sizing

4. Technology Selection

Choose between **modular vs custom-built systems**:

Option A: Modular (Recommended for small scale)

- Skid-mounted Haber–Bosch units
- Faster deployment (12–24 months)
- Lower engineering complexity

Option B: Custom-built plant

- Higher flexibility
- Longer timeline (2–4 years)

Also select:

- Catalyst supplier (iron-based)
- Reactor type (fixed bed)

- Compression system

Output: Technology package selection

5. Regulatory and Safety Approvals

Ammonia production involves high pressure + hazardous chemicals, so compliance is critical.

Key requirements:

- WHS legislation (pressure vessels, hazardous chemicals)
- Environmental approvals (emissions, water use)
- Dangerous goods storage (ammonia)
- Major hazard facility (MHF) classification (if applicable)

Safety systems:

- Pressure relief systems
- Gas detection (NH_3, H_2)
- Emergency shutdown (ESD)
- Ventilation and containment

Output: Approved permits and safety case

6. Procurement and Construction

Major equipment:

- Reformers or electrolyser units
- Compressors (high-pressure)
- Reactor vessel
- Heat exchangers
- Condensers
- Storage tanks (pressurised NH_3)

Construction phases:

1. Civil works and foundations
2. Equipment installation
3. Piping and instrumentation
4. Electrical and control systems

Output: Installed plant ready for commissioning

7. Commissioning and Start-Up

Commissioning is done in stages:

1. Mechanical completion checks
2. System pressure testing
3. Catalyst loading
4. Initial gas circulation
5. Gradual temperature/pressure ramp-up
6. First ammonia production

Performance is validated against:

- Conversion efficiency
- Energy consumption
- Product purity

Output: Operational plant

8. Operations and Maintenance

Key operational requirements:

- Continuous monitoring of:
  - Pressure and temperature
  - Catalyst performance
  - Gas composition

Maintenance:

- Catalyst replacement (every 5–10 years)
- Compressor servicing
- Heat exchanger cleaning

Staffing:

- Process engineers
- Operators
- Maintenance technicians
- Safety/environment officers

Output: Stable, continuous production

9. Integration with Fertiliser Production

Ammonia can be:

- Stored and sold directly
- Converted onsite into:
  - Urea
  - Ammonium nitrate
  - MAP/DAP

Small plants often benefit from vertical integration.

10. Sustainability and Optimisation

To improve long-term viability:

- Integrate renewable energy
- Implement carbon capture (CCS)
- Optimise heat recovery systems
- Improve nitrogen use efficiency (NUE) downstream

Typical Timeline

Phase	Duration
Feasibility	3–6 months
Design	6–12 months
Approvals	6–12 months
Construction	12–24 months
Commissioning	3–6 months

Key Risks

- Energy cost volatility
- Feedstock supply interruptions
- High capital cost

- Regulatory delays
- Technical complexity

Chapter 6

Nitrogen Fertiliser Production

This chapter builds upon the foundational processes of ammonia and nitric acid production by examining how these key intermediates are transformed into the most widely used nitrogen fertilisers in modern agriculture. As global demand for food production continues to rise, nitrogen fertilisers such as urea, ammonium nitrate, and ammonium sulphate play a critical role in enhancing crop yields and supporting agricultural productivity.

The chapter explores the chemical principles, process steps, and industrial technologies used to convert ammonia into usable fertiliser products. It provides a detailed examination of synthesis pathways, including high-pressure urea production, neutralisation processes, and downstream finishing techniques such as prilling and granulation. In addition to process design, attention is given to equipment selection, process integration, and energy efficiency, highlighting how modern plants optimise performance while managing operational complexity.

Beyond production, the chapter also addresses the practical aspects of product handling, storage, and safety. Given the hazardous nature of certain nitrogen fertilisers—particularly ammonium nitrate—there is a strong emphasis on risk management, regulatory compliance, and lessons learned from major industrial incidents. These considerations reinforce the importance of engineering controls, safe operating practices, and robust oversight in fertiliser manufacturing.

Overall, this chapter provides a comprehensive overview of nitrogen fertiliser production, linking core chemical processes with real-world industrial applications, operational challenges, and safety responsibilities.

Urea Production Process

Urea production is one of the most important processes in nitrogen fertiliser manufacture, as urea is the most widely used nitrogen fertiliser globally due to its high nitrogen content (about 46%) and relatively low cost. The process converts ammonia (NH_3), produced via the Haber–Bosch process, and carbon dioxide (CO_2), typically captured from hydrogen production (e.g., steam methane reforming), into solid urea through a series of chemical reactions and physical processing steps.

The process begins with the synthesis stage, where ammonia and carbon dioxide react under high pressure (typically 150–250 bar) and elevated temperature (around 170–200°C). The first reaction forms ammonium carbamate:

$$2NH_3 + CO_2 \rightarrow NH_2COONH_4$$

This reaction is fast and exothermic, meaning it releases heat. However, ammonium carbamate is not the final product. It must undergo a second reaction, dehydration, to form urea:

$$NH_2COONH_4 \rightarrow (NH_2)_2CO + H_2O$$

This second step is slower and endothermic, requiring heat input. Because the overall conversion to urea is incomplete in a single pass, the process operates as a loop, where unconverted materials are continuously recycled.

Following synthesis, the reaction mixture contains urea, water, and unreacted ammonium carbamate. The next stage is separation and recovery. The mixture is depressurised, causing ammonium carbamate to decompose back into ammonia and carbon dioxide. These gases are captured and recycled back into the reactor, significantly improving efficiency and reducing raw material losses.

The remaining solution, primarily containing urea and water, is then concentrated through evaporation. Water is removed using heat, increasing the urea concentration to form a molten urea solution (typically >99% purity). This molten urea is then processed into solid fertiliser products.

The final stage is solidification, which can be achieved through prilling or granulation. In prilling, molten urea is sprayed from the top of a tall tower, forming droplets that solidify as they fall and cool in air. In granulation, the molten urea is sprayed onto seed particles in a rotating drum or fluidised bed, forming larger, more uniform granules. Granulated urea is generally preferred due to its improved handling properties, reduced dust, and improved storage stability.

From a process integration perspective, urea production is highly efficient because it utilises carbon dioxide generated as a by-product of hydrogen production. This not only improves resource efficiency but also partially offsets CO_2 emissions from ammonia plants.

However, the process is still energy-intensive, particularly in the synthesis and evaporation stages, and careful heat management is required. Modern plants incorporate heat recovery systems and advanced process controls to improve energy efficiency and reduce environmental impact.

Urea Production – Step-by-Step (Process + Equipment)

The process described below is the conventional high-pressure urea production process, most commonly referred to as the **Bosch–Meiser Urea Process (Total Recycle Process).**

The Bosch–Meiser process is a well-established industrial method used for the large-scale production of urea, an essential nitrogen-based chemical widely used in fertilisers. Developed in the early 20th century and patented in 1922, the process is named after German chemists Carl Bosch and Wilhelm Meiser.

The process is based on two main equilibrium reactions, both of which proceed with incomplete conversion and therefore require recycling of unreacted materials. The first step is carbamate formation, a rapid and exothermic reaction in which liquid ammonia reacts with carbon dioxide at elevated temperature and pressure to form ammonium carbamate:

$$2NH_3 + CO_2 \rightleftharpoons NH_2COONH_4 \quad (\Delta H \approx -117 \text{ kJ/mol})$$

The second step is urea formation (dehydration), a slower and endothermic reaction in which ammonium carbamate decomposes to produce urea and water:

$$NH_2COONH_4 \rightleftharpoons CO(NH_2)_2 + H_2O \quad (\Delta H \approx +15.5 \text{ kJ/mol})$$

Overall, the conversion of ammonia and carbon dioxide to urea is exothermic. The heat released during carbamate formation helps drive the subsequent urea formation reaction. However, the operating conditions must balance competing requirements: higher temperatures favour urea formation but reduce the efficiency of carbamate formation, while higher pressures favour carbamate formation.

To optimise both reactions, the process is carried out at moderately high temperatures (typically around 180–200°C) and high pressures (approximately 140–175 bar). These conditions represent a compromise that enables acceptable reaction rates and yields. Carbon dioxide must be compressed to these pressures, whereas ammonia is already available in liquid form from upstream production and can be pumped more efficiently into the system.

Because the conversion to urea is not complete in a single pass and the dehydration step is relatively slow, the process requires a large reaction volume to allow sufficient residence time. As a result, industrial urea reactors are typically large, robust pressure vessels designed to operate continuously under demanding conditions.

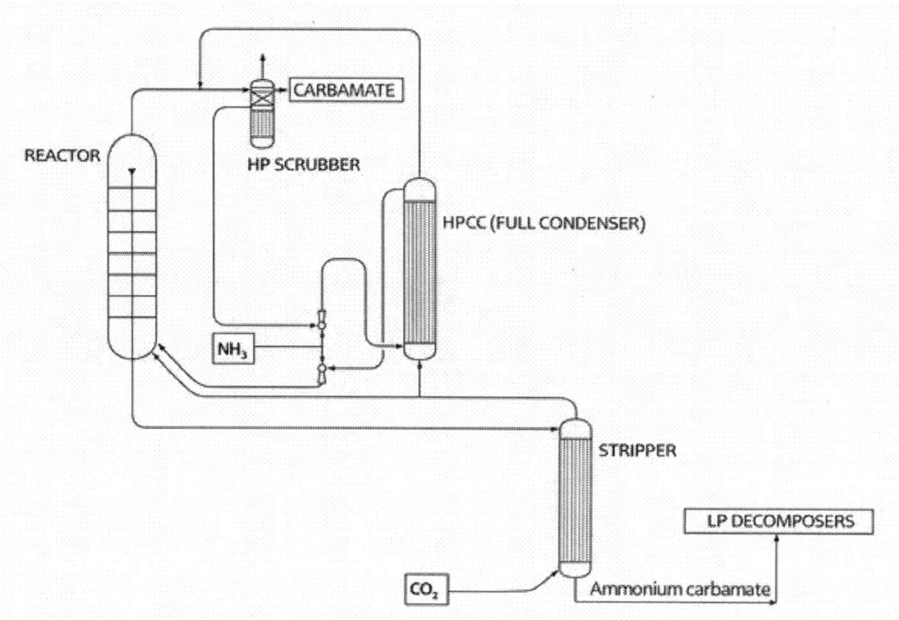

Figure 34: Urea production process, adapted from Fertilizer Industrial Service [112].

Step 1: Feedstock Supply and Preparation

Process

- Supply ammonia (NH_3) from the Haber–Bosch plant
- Supply carbon dioxide (CO_2) from hydrogen production (SMR/WGS)
- Ensure correct NH_3:CO_2 ratio (~3:1) and purity

Key Equipment

- Ammonia storage tanks (pressurised/refrigerated)
- CO_2 compressors
- Feed pumps (high-pressure ammonia pumps)
- Gas filters/dryers

Step 2: High-Pressure Synthesis (Carbamate Formation + Urea Reaction)

Process

- Operates at 150–250 bar and 170–200°C
- Two reactions occur:

Carbamate formation (fast, exothermic):

$$2NH_3 + CO_2 \rightarrow NH_2COONH_4$$

Urea formation (slow, endothermic):

$$NH_2COONH_4 \rightarrow (NH_2)_2CO + H_2O$$

- Only ~50–70% conversion per pass

Key Equipment

- High-pressure urea reactor (autoclave)
- Carbamate condenser (heat recovery)
- High-pressure piping and valves
- Temperature and pressure control systems

Step 3: High-Pressure Recovery (Stripping/Decomposition)

Process

- Unreacted ammonium carbamate is decomposed back into NH_3 and CO_2
- Typically achieved using:
  - Heat
  - Pressure reduction
  - Stripping (steam or CO_2 stripping)

Key Equipment

- High-pressure stripper
- Carbamate decomposer
- Heat exchangers

Step 4: Gas Recovery and Recycle Loop

Process

- Released NH_3 and CO_2 gases are:
    - Separated from liquid urea solution
    - Recombined to form carbamate again
    - Recycled back to reactor
- This creates a closed-loop system for efficiency

Key Equipment

- Carbamate condensers
- Absorbers
- Recycle pumps
- Gas compressors

Step 5: Medium- and Low-Pressure Decomposition

Process

- Further removal of residual NH_3 and CO_2 at lower pressures
- Multi-stage pressure reduction improves recovery

Key Equipment

- Medium-pressure decomposer
- Low-pressure decomposer
- Flash drums
- Gas separators

Step 6: Urea Solution Concentration (Evaporation)

Process

- Urea solution (~70–80%) is concentrated to >99% molten urea
- Water is removed via evaporation

Key Equipment

- Vacuum evaporators
- Falling film evaporators
- Steam heaters

- Vacuum systems (ejectors or pumps)

Step 7: Melt Handling and Conditioning

Process

- Molten urea is:
  - Stabilised
  - Filtered if required
  - Sent to finishing section

Key Equipment

- Melt tanks
- Pumps
- Filters
- Heat exchangers

Step 8: Solidification (Prilling or Granulation)

Option A: Prilling

Process

- Molten urea sprayed from top of tower
- Droplets fall and solidify in air

Equipment

- Prilling tower
- Spray nozzles/bucket system
- Air cooling system

Option B: Granulation (Preferred in modern plants)

Process

- Molten urea sprayed onto seed particles
- Particles grow into uniform granules

Equipment

- Fluidised bed granulator or rotary drum
- Air handling systems
- Recycle conveyors

Step 9: Cooling and Screening

Process

- Solid urea cooled to prevent caking
- Oversize/undersize particles separated

Key Equipment

- Fluidised bed coolers
- Vibrating screens
- Crushers (for oversize recycle)

Step 10: Coating and Storage

Process

- Apply anti-caking agents
- Store and prepare for transport

Key Equipment

- Coating drums
- Storage silos
- Conveyors and bagging systems

Supporting Systems

Supporting systems play a critical role across all stages of urea production, ensuring the process operates efficiently, safely, and reliably. One of the most important aspects is heat integration, where energy generated within the process is recovered and reused. Equipment such as waste heat boilers and heat exchangers capture heat from high-temperature process streams and redirect it to other parts of the plant, reducing overall energy consumption and improving thermal efficiency.

In addition to heat management, a range of utility systems supports continuous operation. Steam systems provide the necessary heat for reactions, stripping, and evaporation processes, while cooling water systems are used to remove excess heat and maintain optimal operating temperatures. Instrument air systems supply clean, compressed air for operating control valves and pneumatic equipment, ensuring stable and responsive process control.

Modern urea plants also rely heavily on advanced control systems to maintain precise operating conditions. A Distributed Control System (DCS) monitors and regulates the entire process in real time, allowing operators to adjust parameters and respond to changes quickly. Supporting this are gas analysers, which continuously measure concentrations of key components such as ammonia (NH_3) and carbon dioxide (CO_2), as well as pressure and temperature sensors that provide critical data for safe and efficient operation. Together, these systems enable high levels of automation, consistency, and process optimisation.

Alternative Production Methods

Commercial urea production remains overwhelmingly based on the ammonia–carbon dioxide chemistry that underpins Bosch–Meiser processes [113-115]. Nevertheless, a spectrum of alternative or modified pathways has been explored, ranging from engineering and process configuration refinements within the NH_3–CO_2 framework to entirely different CO_2 utilization routes and near-term electrification concepts. The literature consistently shows that while genuinely new chemistries exist in niche or early-stage forms (electrochemical, biological/enzyme-inspired, cyanate/isocyanate pathways, and solid-state or plasma concepts), they have not achieved industrial maturity or economic viability at scale for bulk fertilizer production [115-118]. In parallel, there is a robust body of work on integrating CO_2 capture and utilization with urea production, particularly via CCU approaches that reuse CO_2 from point sources such as power plants or industrial flue gases to produce urea or co-produce urea with other products, often within CCS/CCU supply chains [119-123]. This synthesis draws on multiple sources to map the range of alternative pathways, their current state, and the economic and environmental drivers that shape their viability, with careful attention to where chemistry diverges from the conventional Bosch–Meiser route and where it remains a conversion or integration challenge rather than a fundamental chemical shift.

Modified NH3–CO2 processes and engineering innovations

A major strand of alternatives continues to leverage the same underlying NH_3–CO_2 chemistry but adopts different process configurations and internals to reduce energy intensity and improve material recoveries. Notable examples include the Stamicarbon CO_2 stripping concept, which uses CO_2 as the stripping medium to improve carbamate decomposition and recovery, thereby lowering energy consumption relative to traditional exhaust-stream approaches [119]. In parallel, NH_3 stripping approaches, exemplified by Saipem (Snamprogetti) configurations, employ excess ammonia to enhance gas recovery and per-pass conversion, offering improvements in looping performance and process integration [113, 124].

Snamprogetti Urea Process

The Snamprogetti urea process, now part of Saipem technology, is one of the most widely used modern methods for urea production. It is based on the conventional ammonia–carbon dioxide (NH_3–CO_2) chemistry but is distinguished by its use of ammonia stripping to improve process efficiency, reduce energy consumption, and simplify the recovery system. As a refined development of the Bosch–Meiser process, it is extensively adopted in large-scale fertiliser plants around the world.

The core concept of the Snamprogetti process is the use of excess ammonia as the stripping medium in the high-pressure section. Rather than using carbon dioxide or steam to remove unreacted ammonium carbamate, the process uses ammonia itself to promote decomposition and separation. In this stage, ammonia helps decompose ammonium carbamate, strips unreacted carbon dioxide from the liquid phase, and improves gas–liquid contact. This approach significantly enhances overall conversion efficiency and reduces the load on downstream recovery systems.

In operation, ammonia and carbon dioxide are first fed into a high-pressure synthesis reactor, typically operating at pressures of 140–170 bar and temperatures of 170–190°C. Within the reactor, two key reactions occur: the rapid, exothermic formation of ammonium carbamate ($2NH_3 + CO_2 \rightarrow NH_2COONH_4$), followed by the slower, endothermic conversion of carbamate into urea and water ($NH_2COONH_4 \rightarrow (NH_2)_2CO + H_2O$). Because conversion is incomplete in a single pass, the process relies on recycling unreacted materials.

The reactor effluent then flows to the high-pressure ammonia stripper, which is the defining feature of this technology. In this unit, excess ammonia is used to strip carbon dioxide from the solution while simultaneously promoting the decomposition of ammonium carbamate back into ammonia and carbon dioxide. Heat, typically supplied by steam or recovered process energy, supports this step. The result is a reduction in unreacted carbamate in the liquid phase, improved separation efficiency, and lower energy requirements for recycling.

The stripped gases, consisting of ammonia and carbon dioxide, are sent to a high-pressure carbamate condenser, where they recombine to form ammonium carbamate. The heat released during this exothermic reaction is recovered, often for steam generation, improving overall energy efficiency. The condensed carbamate is then recycled back to the reactor, forming a closed-loop system.

Further purification occurs in the medium- and low-pressure recovery sections, where residual ammonia and carbon dioxide are removed from the urea solution through staged pressure reduction and decomposition. This involves equipment such as medium-pressure decomposers, low-pressure flash vessels, and gas separation systems. The recovered gases are absorbed and returned to the process, maximising material utilisation.

The resulting urea solution, typically around 70–80% concentration, is then processed in the evaporation stage, where water is removed under vacuum conditions to produce molten urea with a purity exceeding 99%. This molten product is subsequently sent to the finishing section.

In the final stage, urea is solidified either by granulation or prilling. Granulation, commonly performed in a fluidised bed system, is generally preferred due to the production of larger, more uniform particles with better handling and storage characteristics. Alternatively, prilling involves spraying molten urea into a tower, where droplets solidify as they fall through a cooling air stream.

The Snamprogetti process offers several important advantages. It achieves high efficiency through improved carbamate removal and higher conversion per pass. It reduces energy consumption by lowering recycle loads and incorporating effective heat integration. The process design is relatively simplified, requiring fewer downstream recovery stages compared to older methods. It is also well suited to large-scale production, with capacities exceeding 3,000 tonnes per day, and is regarded as a proven and reliable technology with widespread global application.

More advanced design families, self-stripping or optimised loop systems developed by Toyo Engineering and Casale, target enhanced internal recycling, heat integration, and overall energy efficiency in dedicated reactors and loop configurations [113]. Pool reactors and isobaric process designs have been proposed to maintain uniform pressure across synthesis and recovery stages, reducing pressure losses and corresponding energy demands, with reported benefits in energy intensity and transfer efficiency across the synthesis/recovery cycle [113].

Modified NH_3–CO_2 Processes (Industrial Variants)

These are not fundamentally different chemically, but they represent alternative engineering approaches to the Bosch–Meiser process.

a. CO_2 Stripping Process (e.g., Stamicarbon)

- Uses CO_2 as the stripping medium
- Improves carbamate decomposition and recovery
- Reduces energy consumption

✓ Widely used in modern plants

b. NH_3 Stripping Process (e.g., Saipem/Snamprogetti)

- Uses excess ammonia for stripping
- Enhances efficiency of gas recovery
- Allows higher conversion per pass

c. Self-Stripping / Advanced Loop Designs (e.g., Toyo, Casale)

- Optimised internal recycling
- Improved heat integration
- Lower energy consumption

d. Pool Reactor / Isobaric Processes

- Maintain uniform pressure across synthesis and recovery
- Reduce pressure drops and energy losses

Carbon-source diversification within the NH_3–CO_2 framework

A parallel line of development is to broaden the origin of carbon in the urea synthesis while preserving the core chemistry. CO_2 captured from industrial emissions or power plant flue gases can be integrated into urea production, linking urea synthesis with carbon capture technologies to produce low-carbon fertilisers [120-122]. Biomass-derived CO_2 offers a renewable feedstock pathway that can further lower lifecycle emissions associated with urea when used in conjunction with renewable hydrogen and green ammonia concepts [120-122]. The literature emphasizes that such integrations typically do not alter the stoichiometric chemistry of urea formation but modify feedstock sourcing, purification, and integration with capture technologies to reduce fossil carbon reliance and improve overall carbon footprints [120-122].

Urea Production from Alternative Carbon Sources

a. Urea from Flue Gas CO_2 (Carbon Capture Integration)

- CO_2 sourced from:
  - Power plants
  - Industrial emissions
- Same chemistry, different carbon source

✓ Increasingly important for low-carbon fertiliser production

b. Biomass-Derived CO_2

- CO_2 generated from:
  - Bioenergy plants
  - Fermentation processes

✓ Enables partially renewable urea

Emerging lower-emission and hybrid pathways

Beyond CCU-integrated NH_3–CO_2 routes, several emerging pathways aim to bypass conventional synthesis constraints through alternative energy inputs or catalysis concepts. Electrochemical urea synthesis, pursued in various research programs, envisions direct synthesis of urea from nitrogen and CO_2 using electricity and specialized catalysts, potentially enabling operation at lower pressures than conventional high-pressure reactors; however, the approach remains largely at the laboratory or pilot scale and faces technical barriers to scale-up and economic competitiveness [115, 116]. Green ammonia approaches, hydrogen produced by renewable-energy-powered electrolysis followed by conventional ammonia synthesis and subsequent urea production, offer a near-term route to lower-carbon urea production while retaining the mature downstream chemistry of conventional urea plants [115, 118]. Biological and enzymatic routes to urea production have been explored as potential pathways to lower-temperature, lower-pressure systems, but these remain in early-stage research and are not commercially viable at present [115].

Electrochemical and Green Pathways (Emerging)

a. Electrochemical Urea Synthesis

- Produces urea directly from:
  - Nitrogen (N_2)
  - Carbon dioxide (CO_2)
- Uses electricity instead of high-pressure reactors

Example concept:
$N_2 + CO_2 + H_2O \rightarrow$ Urea (via catalysts)

✓ Still experimental / lab-scale

b. Green Ammonia + CO_2 Route

- Hydrogen from electrolysis (renewables)
- Produces "green ammonia"
- Then conventional urea synthesis

✓ Same process, but low-carbon feedstocks

Alternative chemical pathways and their maturity

Other chemical pathways, including those involving cyanate or isocyanate intermediates, exist in specialized chemical manufacturing but are not suitable for bulk fertiliser production due to cost, safety, and process complexity considerations [115, 116]. Experimental concepts such as solid-state reactions, plasma systems, and microwave-assisted processes have shown theoretical energy, or pressure-reduction potential but have not yet achieved commercial maturity for large-scale urea production [115, 116]. In this context, while these alternative chemistries are scientifically intriguing, the consensus in the literature is that they have not displaced Bosch–Meiser-type processes at industrial scale; instead, they remain niche or early-stage research directions [115, 116].

Biological and Enzymatic Routes (Research Stage)

a. Enzyme-Based Urea Formation

- Mimics natural biochemical pathways
- Uses catalysts similar to urease-related systems

✓ Not commercially viable yet

Table 1: Simple Classification.

Category	Example	Status
Conventional	Bosch–Meiser	Dominant
Improved processes	Stamicarbon, Saipem, Toyo	Commercial
Low-carbon variants	Green ammonia + CO_2	Emerging
Alternative chemistry	Electrochemical urea	Experimental
Biological routes	Enzymatic systems	Research

Scale of Production

Producing urea at a small scale is fundamentally different from industrial practice. The conventional Bosch–Meiser process operates at extremely high pressures (140–250 bar) and elevated temperatures (170–200°C), making it impractical and unsafe outside of large, purpose-built facilities. As a result, small-scale urea production is not simply a scaled-down version of an industrial plant. Instead, it involves a simplified, low-throughput, and carefully controlled system, typically used for pilot operations, research, or demonstration purposes rather than commercial fertiliser manufacturing.

At small scale, several key constraints must be considered. The requirement for high pressure necessitates specialised equipment such as autoclaves, which must be designed to safely withstand demanding operating conditions. Ammonia handling presents additional challenges due to its toxic and corrosive nature, requiring strict safety controls. Heat management is also complex, as the process involves both exothermic and endothermic reactions that must be carefully balanced. Furthermore, recycling unreacted materials—an essential feature of industrial systems—is difficult to implement efficiently at small scale. Consequently, these systems typically operate at lower throughput, use batch or semi-batch reactors, achieve limited recycling efficiency, and accept lower overall conversion yields.

The process begins with a simplified feedstock supply stage. Instead of integrating with a full ammonia production facility, ammonia is supplied from pressurised cylinders or storage vessels, while carbon dioxide is provided from bottled sources or small generators. Basic equipment includes gas cylinders fitted with regulators, metering valves for flow control, and simple gas purification systems such as dryers and filters to ensure feedstock quality.

The core chemical reactions are carried out in a batch high-pressure reactor, usually an autoclave. The process involves charging ammonia into the reactor, introducing carbon dioxide under pressure, and heating the system to approximately 170–190°C. Pressure is maintained—typically at a lower level than industrial systems, such as 80–150 bar—and the mixture is held for a sufficient residence time to allow the reactions to proceed. These reactions include the formation of ammonium carbamate ($2NH_3 + CO_2 \rightarrow NH_2COONH_4$) followed by its conversion into urea and water ($NH_2COONH_4 \rightarrow (NH_2)_2CO + H_2O$). This stage is the most critical and hazardous, requiring robust equipment such as a high-pressure autoclave, heating systems, instrumentation for pressure and temperature monitoring, and safety relief valves.

After the reaction phase, the system is cooled and depressurised in a controlled manner. As pressure is reduced, unreacted ammonia and carbon dioxide are released. In small-scale setups, recycling these gases is typically limited or performed manually rather than through a continuous loop. Equipment used in this stage may include condensers to recover ammonia where possible and gas scrubbing systems to neutralise or safely handle released gases.

The resulting mixture, containing urea, water, and residual ammonium carbamate, then undergoes a simplified separation stage. During depressurisation, ammonium carbamate decomposes back into ammonia and carbon dioxide, which can either be captured in more advanced systems or neutralised in basic setups. Separation equipment may include a flash vessel or separator and an acid-based scrubber for ammonia removal.

The remaining liquid, primarily composed of urea and water, is then concentrated through evaporation. This involves gently heating the solution to remove water, either at atmospheric pressure or under mild vacuum conditions. Equipment for this stage typically includes a heated evaporation vessel, an optional vacuum pump, and a condenser to collect water vapour. The result is a molten or highly concentrated urea solution.

Finally, the urea is solidified using simplified methods, as industrial prilling or granulation systems are not practical at small scale. Instead, the molten urea may be poured into trays or moulds and allowed to cool, or crystallised from solution under controlled conditions. Basic equipment such as cooling trays, pans, or simple crystallisers is sufficient for this stage, producing solid urea suitable for demonstration or limited-use applications.

Safety and Practical Considerations

Safety and practical considerations are critical in small-scale urea production, and this is the area where many attempts fail if systems are not properly engineered and controlled. The process involves several inherent hazards, including the handling of ammonia, which is toxic, corrosive, and poses a serious inhalation risk. In addition, the use of high-pressure equipment introduces the potential for vessel failure if not correctly designed, maintained, and operated. The compression and handling of carbon dioxide also present risks, particularly under pressurised conditions. Furthermore, the combination of exothermic and endothermic reactions creates the possibility of thermal instability or runaway reactions if temperature is not carefully managed.

To operate safely, a number of minimum requirements must be met. These include the use of certified pressure vessels designed to withstand the operating conditions, as well as properly installed pressure relief systems to prevent over-pressurisation. Gas detection systems, particularly for ammonia, are essential to provide early warning of leaks. Adequate ventilation and gas scrubbing systems must be in place to manage any released gases and protect both personnel and the environment. Equally important is the presence of trained and competent operators who understand the process, the associated risks, and the appropriate emergency response procedures.

Without these fundamental safety measures, small-scale urea production becomes extremely hazardous and should not be attempted.

Ammonium Nitrate and Ammonium Sulphate

Both ammonium nitrate and ammonium sulphate are key nitrogen-based fertilisers derived from ammonia, but they differ significantly in their chemistry, production methods, nutrient composition, and agronomic applications.

Ammonium Nitrate (NH_4NO_3)

Ammonium nitrate is produced by reacting ammonia (NH_3) with nitric acid (HNO_3), both of which are derived from ammonia in upstream processes. The nitric acid is typically produced via the Ostwald process, where ammonia is oxidised to form nitrogen oxides and then absorbed in water to produce nitric acid. The neutralisation reaction is straightforward:

$$NH_3 + HNO_3 \rightarrow NH_4NO_3$$

This reaction is highly exothermic and results in an aqueous ammonium nitrate solution. The solution is then concentrated through evaporation and converted into solid fertiliser through prilling or granulation.

From a nutrient perspective, ammonium nitrate is highly valued because it contains both ammonium (NH_4^+) and nitrate (NO_3^-) forms of nitrogen, typically around 33–34% nitrogen content. This dual form provides both immediate and sustained nitrogen availability to plants, nitrate is rapidly absorbed, while ammonium is taken up more slowly or converted in the soil.

In terms of application, ammonium nitrate is widely used in agriculture due to its high efficiency and relatively predictable behaviour in soils. However, it presents significant safety and regulatory challenges. It is classified as an oxidiser and, under certain conditions, can become explosive. As a result, its storage, handling, and transport are strictly regulated in many countries, and its use has declined in some regions in favour of safer alternatives such as urea or ammonium sulphate.

Figure 35: "HD 33,5" ammonium nitrate big bag fertilizer produced by AZF factory in Toulouse, France. Cjp24, CC BY-SA 3.0, via Wikimedia Commons.

Production

Ammonium nitrate production is a well-established industrial process that converts ammonia (NH_3) and nitric acid (HNO_3) into a high-nitrogen fertiliser. The process consists of a sequence of chemical and physical stages, each supported by specialised equipment designed to manage reaction heat, concentration, and product handling efficiently and safely.

The process begins upstream with the production of nitric acid via the Ostwald process. In this stage, ammonia is oxidised in air over a platinum–rhodium catalyst at high temperatures of approximately 800–900°C, according to the reaction $4NH_3 + 5O_2 \rightarrow 4NO + 6H_2O$. The nitric oxide (NO) formed is further oxidised to nitrogen dioxide (NO_2), which is then absorbed in water to produce nitric acid (HNO_3). This stage requires equipment such as ammonia burners fitted with catalyst gauze, oxidation reactors, waste heat boilers for energy recovery, absorption towers for acid formation, and associated compressors and cooling systems.

Once nitric acid is produced, it is reacted with ammonia in the neutralisation stage to form ammonium nitrate. The reaction $NH_3 + HNO_3 \rightarrow NH_4NO_3$ is highly exothermic and produces an aqueous ammonium nitrate solution, typically at a concentration of 60–80%. This step is carried out in a neutraliser reactor, which may operate at atmospheric or elevated pressure. Key equipment includes the neutraliser vessel, ammonia injection systems, acid feed systems, instrumentation for temperature and pH control, and vapour separators to remove steam and non-condensable gases.

The resulting ammonium nitrate solution is then concentrated through evaporation. Water is removed using heat, increasing the concentration to approximately 95–99% to produce a molten ammonium nitrate suitable for further processing. This stage typically employs falling film or vacuum evaporators, along with steam heaters, vacuum systems where required, and condensers to recover water vapour.

Before solidification, the molten ammonium nitrate undergoes melt conditioning. In this stage, the melt may be stabilised and treated with additives to improve properties such as resistance to caking and thermal stability. Equipment used includes melt storage tanks, mixing or conditioning vessels, filtration systems, pumps, and heat exchangers to maintain appropriate temperature and flow conditions.

The conditioned melt is then converted into solid fertiliser through either prilling or granulation. In prilling, molten ammonium nitrate is sprayed from the top of a tall tower, forming droplets that solidify as they fall through a stream of cooling air. In granulation, the melt is sprayed onto seed particles in a rotating drum or fluidised bed, producing larger and more uniform granules. This stage requires prilling towers with spray systems or granulators, along with air handling and cooling systems and, in the case of granulation, recycle conveyors to manage particle size distribution.

Following solidification, the product is cooled and screened. Cooling prevents caking and improves the mechanical strength of the fertiliser. Screening ensures that particles meet the required size specifications, with oversized material crushed and recycled, and undersized material returned to the process. Equipment used includes fluidised bed or rotary coolers, vibrating screens, crushers, and conveying systems.

In the final stage, the product is coated and stored. Anti-caking agents are applied to improve handling and storage characteristics, after which the fertiliser is stored in silos and prepared for distribution. This stage involves coating drums or spray systems, storage silos, bagging or bulk loading systems, and material handling equipment such as conveyors.

Supporting all stages of the process are integrated utility and control systems. Heat integration plays a key role, with waste heat boilers and heat exchangers recovering and reusing energy to improve overall efficiency. Utility systems such as steam for heating, cooling water for temperature control, and instrument air for process equipment operation are essential for continuous operation. Modern plants also rely on advanced control systems, including Distributed Control Systems (DCS), gas analysers for monitoring ammonia and nitrogen oxides, and pressure and temperature sensors to ensure safe, stable, and optimised performance throughout the production process.

Ammonium Sulphate ($(NH_4)_2SO_4$)

Ammonium sulphate is produced by reacting ammonia with sulphuric acid (H_2SO_4):

$$2NH_3 + H_2SO_4 \rightarrow (NH_4)_2SO_4$$

This is also an exothermic neutralisation reaction and is typically carried out in a reactor followed by crystallisation to produce solid fertiliser. In many cases, ammonium sulphate is also recovered as a by-product from industrial processes such as caprolactam production, coke oven gas treatment, or flue gas desulfurisation.

Ammonium sulphate contains approximately 21% nitrogen and 24% sulphur, making it particularly valuable in soils that are deficient in sulphur. Unlike ammonium nitrate, it provides nitrogen only in the ammonium form (NH_4^+), which must be converted in the soil (via nitrification) before becoming nitrate.

Agronomically, ammonium sulphate is especially useful in alkaline soils because it has an acidifying effect, helping to lower soil pH and improve nutrient availability. It is also less prone to volatilisation losses compared to some other nitrogen fertilisers. However, its lower nitrogen content means that larger quantities are required to deliver the same nitrogen input as higher-analysis fertilisers like urea or ammonium nitrate.

From a safety and handling perspective, ammonium sulphate is significantly safer and more stable than ammonium nitrate. It is non-explosive, less reactive, and easier to store and transport, making it a preferred option in situations where safety and regulatory compliance are critical.

Production

Ammonium sulphate ($(NH_4)_2SO_4$) production is a relatively straightforward and widely used process in nitrogen fertiliser manufacturing. It involves the neutralisation of ammonia (NH_3) with sulphuric acid (H_2SO_4), followed by crystallisation and product finishing. Compared to processes such as ammonium nitrate or urea production, it operates at lower pressures and temperatures, making it simpler, safer, and more flexible. In many cases, ammonium sulphate is also produced as a by-product of other industrial operations, which further enhances its economic viability.

The process begins with feedstock supply and preparation. Ammonia is sourced either from an on-site Haber–Bosch plant or delivered as a pressurised or refrigerated liquid, while sulphuric acid is supplied from an acid plant or external supplier. Both reactants must be clean and delivered at controlled flow rates to maintain the correct stoichiometric balance. This stage typically involves ammonia and sulphuric acid storage tanks, metering pumps, flow control systems, and corrosion-resistant piping, valves, and safety equipment.

The next stage is neutralisation, where ammonia reacts with sulphuric acid in a reactor to form ammonium sulphate according to the reaction $2NH_3 + H_2SO_4 \rightarrow (NH_4)_2SO_4$. This reaction is exothermic and occurs at moderate temperatures, producing an aqueous ammonium sulphate solution. The process is carried out in a neutraliser reactor, which may be a tank or pipe reactor, supported by mixing systems, ammonia injection devices, temperature and pH monitoring instrumentation, and vent systems to safely handle vapours.

Following neutralisation, the solution undergoes crystallisation. The ammonium sulphate solution is concentrated and cooled to promote the formation of crystals as the solution becomes supersaturated. This can be achieved through evaporative crystallisation, where water is removed, or cooling crystallisation, where the temperature is reduced. Equipment used in this stage includes crystallisers, such as forced circulation or vacuum types, evaporators if additional concentration is required, heat exchangers for temperature control, and slurry handling systems.

The crystal slurry is then subjected to solid–liquid separation. In this stage, solid ammonium sulphate crystals are separated from the remaining liquid, known as the mother liquor. The mother liquor is often recycled back into the process to improve efficiency and reduce waste. This separation is typically achieved using centrifuges, although filter presses or vacuum filters may also be used, along with pumps and slurry transfer systems.

After separation, the crystals are dried to remove residual moisture and improve storage stability. Drying ensures that the final product is free-flowing and less prone to caking during storage and handling. This stage commonly uses rotary dryers or fluidised bed dryers, supported by hot air generation systems and dust collection equipment such as cyclones.

Once dried, the product is screened and sized to meet required specifications. Vibrating screens are used to separate particles by size, with oversized material crushed and recycled, and undersized material returned to the crystallisation stage. Conveyors and crushers or mills support this process, ensuring consistent product quality.

In the final stage, the product is coated and stored. Anti-caking agents may be applied to improve handling characteristics and long-term storage stability. The finished ammonium sulphate is then stored in silos and prepared for distribution, either in bulk or packaged form. This stage involves coating drums or spray systems, storage silos, bagging or bulk loading equipment, and conveying systems.

Supporting all stages of the process are essential utility and control systems. Heat integration, using heat exchangers and evaporators, helps optimise energy efficiency. Utility systems such as steam for heating, cooling water for temperature control, and compressed air for instrumentation are critical for continuous operation. Modern facilities also rely on advanced control systems, including Distributed Control Systems (DCS), along with pH monitoring, temperature sensors, and flow measurement devices to ensure safe, stable, and efficient plant performance.

Comparative Perspective in Nitrogen Fertiliser Production

From an industrial and production standpoint, both fertilisers are closely linked to ammonia as the primary nitrogen source, but they differ in their integration within the broader fertiliser production system:

- Ammonium nitrate is closely tied to nitric acid production and is often integrated into facilities that produce explosives or specialised fertilisers. It offers high nitrogen efficiency but requires strict safety controls.
- Ammonium sulphate is more flexible in production, often generated as a by-product in other industries, and provides the added benefit of sulphur, making it valuable in balanced fertilisation programs.

Ammonium nitrate is a high-analysis, fast-acting nitrogen fertiliser with regulatory challenges, while ammonium sulphate is a lower-analysis, safer fertiliser that also supplies essential sulphur and improves soil conditions. Both play important roles in modern fertiliser systems, with their use determined by agronomic needs, safety considerations, and regulatory frameworks.

Ammonium sulphate can be produced at small scale relatively safely and practically, whereas ammonium nitrate production at small scale, although technically possible, is strongly discouraged due to significant safety and regulatory risks.

Ammonium sulphate is one of the most suitable nitrogen fertilisers for small-scale production because the process is simple, operates at low pressure, and involves relatively stable chemistry. The reaction occurs at atmospheric pressure and moderate temperatures, typically below 100°C, making it far less demanding than many other fertiliser processes. The chemistry itself is straightforward, based on the neutralisation reaction $2NH_3 + H_2SO_4 \rightarrow (NH_4)_2SO_4$, and does not require high-pressure reactors or complex recycle systems.

A typical small-scale approach involves a controlled neutralisation process, where ammonia—either as a gas or in solution—is reacted with diluted sulphuric acid in a stirred tank reactor. The resulting solution is then subjected to crystallisation, either by cooling or by concentrating the solution to form solid crystals. These crystals are subsequently separated and dried to produce the final fertiliser product. The equipment required for such a system is relatively simple and includes an acid-resistant reaction vessel (such as polypropylene or stainless steel), an agitator or mixer, pH and temperature monitoring instruments, a basic crystalliser or cooling tank, filtration or centrifugation equipment, and a drying system such as a tray dryer or air drying unit.

While there are risks associated with ammonia exposure and the handling of corrosive acids, these are generally manageable with appropriate safety measures. The use of personal protective equipment (PPE), adequate ventilation, and proper process controls allows ammonium sulphate production to be carried out safely at pilot or small industrial scale.

In contrast, ammonium nitrate presents a much higher risk profile despite its simple reaction chemistry ($NH_3 + HNO_3 \rightarrow NH_4NO_3$). It is a strong oxidising substance with the potential for explosive decomposition under certain conditions. The material is highly sensitive to temperature, contamination, and confinement, which makes process control critical. Key stages such as neutralisation, concentration, and solidification must be tightly controlled to prevent thermal runaway or instability.

Additional complications arise from the production of nitric acid, which typically involves the Ostwald process. This requires very high temperatures (800–900°C), specialised catalysts, and complex equipment, making it impractical for small-scale operations. Furthermore, concentrating ammonium nitrate to a molten state introduces additional hazards due to its thermal sensitivity.

From a regulatory perspective, ammonium nitrate is strictly controlled in many countries, including Australia. Licensing is often required for its manufacture, storage, and transport, reflecting its classification as a hazardous and potentially explosive material. Any attempt to produce ammonium nitrate at small scale would require advanced engineering controls, including a carefully designed neutraliser with effective heat removal, precise temperature control systems, explosion protection measures, gas scrubbing systems, and full compliance with dangerous goods regulations.

In practice, these requirements make small-scale ammonium nitrate production rarely justified or permitted, particularly when safer and more practical alternatives such as ammonium sulphate are available.

Table 2: Direct Small Scale Production Comparison.

Factor	Ammonium Sulphate	Ammonium Nitrate
Pressure	Low	Low–moderate
Temperature	Moderate	Moderate–high (risk sensitive)
Reaction risk	Low–moderate	High
Explosion risk	Very low	Significant
Process complexity	Simple	Complex control required
Small-scale viability	✓ Yes	⚠ Generally no

Nitric Acid Production (Ostwald Process)

Nitric acid production is a critical step in nitrogen fertiliser manufacturing, particularly for products such as ammonium nitrate. The most widely used industrial method is the Ostwald process, a continuous catalytic process that converts ammonia into nitric acid through a series of oxidation and absorption reactions. This process links ammonia production (via Haber–Bosch) to downstream fertiliser and chemical manufacturing.

Overview of the Ostwald Process

The Ostwald process involves three main stages:

1. Catalytic oxidation of ammonia to nitric oxide (NO)
2. Oxidation of nitric oxide to nitrogen dioxide (NO_2)
3. Absorption of nitrogen dioxide in water to form nitric acid (HNO_3)

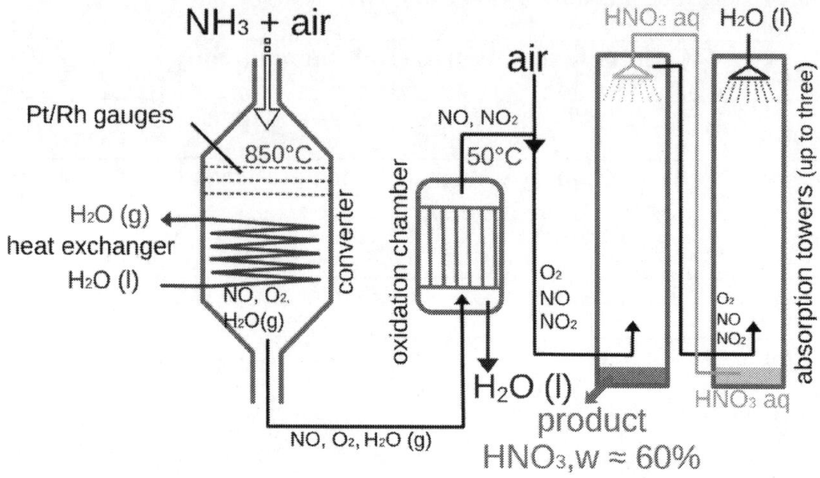

Figure 36: Ostwald Process Schematic. Andi schmitt, CC BY-SA 4.0, via Wikimedia Commons.

Figure 36 represents a simplified flow of the Ostwald process for nitric acid production, illustrating the key stages from high-temperature ammonia oxidation through to absorption and final product formation. It provides a clear visual overview of how ammonia is progressively converted into nitric acid through a series of controlled chemical and physical transformations.

The process begins with high-temperature ammonia oxidation in the first vessel, which operates at approximately 850°C. In this stage, ammonia (NH_3) reacts with oxygen from air over a platinum–rhodium catalyst to produce nitric oxide (NO) and water, according to the reaction $4NH_3 + 5O_2 \rightarrow 4NO + 6H_2O$. The diagram indicates the presence of a catalyst bed within the reactor, typically shown as internal lines. Water is produced in both vapour and liquid forms, and the reaction is highly exothermic, generating significant heat that must be managed. This stage is the core of the process, where ammonia is initially converted into nitric oxide.

Following this, the process moves to a cooling and secondary oxidation stage, represented by the smaller unit labelled at approximately 50°C. Here, the hot gases from the reactor are cooled, which facilitates further reaction. As the temperature decreases, nitric oxide reacts with oxygen to form nitrogen dioxide (NO_2), according to the reaction $2NO + O_2 \rightarrow 2NO_2$. The cooling also allows water vapour to condense into liquid form. This stage is critical for preparing the gas mixture for efficient absorption, as it both reduces the temperature and converts NO into the more reactive NO_2 required for nitric acid formation.

The next section of the diagram shows the absorption system, consisting of two vertical columns where nitric acid is produced. In the first absorption column, nitrogen dioxide is absorbed into water in a counter-current flow arrangement, where water is sprayed from the top while gases rise from the bottom. The reaction $3NO_2 + H_2O \rightarrow 2HNO_3 + NO$ occurs, forming nitric acid while regenerating some nitric oxide. This regenerated NO is not wasted but continues through the system for re-oxidation, improving overall process efficiency.

The second absorption column further enhances recovery and conversion. Any remaining nitrogen oxides are absorbed to produce additional nitric acid, ensuring maximum utilisation of reactants. The diagram indicates that the final product collected is an aqueous nitric acid solution with a concentration of approximately 60% by weight, which is typical for many industrial applications, including ammonium nitrate production.

Each stage is carefully controlled to maximise yield, recover energy, and minimise emissions.

Stage 1: Ammonia Oxidation (Primary Reaction)

The process begins with the oxidation of ammonia using air over a platinum–rhodium catalyst at high temperatures (approximately 800–900°C):

$$4NH_3 + 5O_2 \rightarrow 4NO + 6H_2O$$

This reaction is highly exothermic, releasing significant heat. The catalyst is typically in the form of fine gauze, which provides a large surface area for the reaction. High conversion efficiency (often >90%) is achieved under carefully controlled conditions.

Key equipment

- Ammonia–air mixer
- Burner/reactor with platinum–rhodium catalyst gauze
- Waste heat boiler (to recover heat as steam)

Stage 2: Oxidation of Nitric Oxide

The nitric oxide (NO) produced in the first stage is further oxidised in the presence of oxygen to form nitrogen dioxide:

$$2NO + O_2 \rightarrow 2NO_2$$

This reaction occurs as the gas mixture cools and moves through the system. It is relatively fast at lower temperatures and does not require a catalyst.

Key equipment

- Gas cooling systems
- Oxidation chambers or ducting

Stage 3: Absorption (Nitric Acid Formation)

Nitrogen dioxide is absorbed in water to form nitric acid:

$$3NO_2 + H_2O \rightarrow 2HNO_3 + NO$$

Some nitric oxide (NO) is regenerated in this step and is recycled back into the oxidation stage, improving overall efficiency.

The absorption typically takes place in a packed or tray column, where gas flows upward and water flows downward (counter-current flow), maximising contact between phases.

Key equipment

- Absorption tower (packed column or plate column)
- Water feed system
- Gas recirculation systems
- Acid collection tanks

Concentration and Product Handling

The nitric acid produced is typically 50–70% concentration. For certain applications, it may be further concentrated using additional processes such as distillation or dehydration.

Key equipment

- Acid storage tanks (corrosion-resistant materials)
- Concentration units (if required)
- Pumps and transfer systems

Energy Integration and Efficiency

The Ostwald process is highly integrated from an energy management perspective, with significant emphasis placed on recovering and reusing heat generated during operation. One of the key features of the process is the exothermic oxidation of ammonia, which releases a substantial amount of heat during the initial reaction stage.

Rather than allowing this heat to dissipate, it is captured using waste heat boilers. These systems convert the thermal energy into useful steam, which can then be utilised in other parts of the plant. This may include driving turbines, supporting heating requirements, or supplying energy to downstream processes.

As a result, the effective recovery and reuse of heat significantly improves the overall energy efficiency of the plant. This integrated approach not only reduces the need for external energy inputs but also lowers operating costs, making the process more economically and environmentally sustainable.

Environmental and Safety Considerations

Nitric acid production involves the handling of hazardous gases and operation at high temperatures, which necessitates strict control measures to ensure safe and reliable performance. One of the primary concerns is the management of nitrogen oxides (NO_x), which are harmful pollutants. To address this, plants incorporate tail gas treatment systems that capture and reduce emissions before release to the atmosphere.

Another important aspect is catalyst management. The process relies on platinum-based catalysts, which are both expensive and critical to efficiency. Modern systems are designed to recover and recycle platinum losses, improving both economic performance and sustainability.

Due to the highly corrosive nature of nitric acid and associated process streams, equipment must be constructed from corrosion-resistant materials. This ensures long-term durability and reduces the risk of leaks or equipment failure. In addition, continuous temperature and pressure monitoring systems are essential for maintaining stable operating conditions and preventing unsafe deviations.

To further minimise environmental impact, modern nitric acid plants incorporate advanced emission control technologies such as tail gas scrubbers or catalytic reduction systems. These systems significantly reduce NO_x emissions, ensuring compliance with environmental regulations and improving overall process sustainability.

Role in Fertiliser Production

Nitric acid is a critical intermediate in the production of nitrogen-based fertilisers, serving as a key link in the conversion of ammonia into more widely used agricultural products. Its primary application is in the manufacture of ammonium nitrate (NH_4NO_3), one of the most common high-nitrogen fertilisers used globally. In addition, nitric acid is used to produce a range of other nitrate-based fertilisers that support plant growth and soil fertility.

As a result, the Ostwald process plays a vital role in the overall fertiliser production chain. By converting ammonia into nitric acid, it effectively bridges upstream ammonia synthesis with downstream fertiliser manufacturing, enabling the efficient production of essential nitrogen fertilisers.

Granulation and Prilling Techniques

Granulation and prilling are two key techniques used in fertiliser production to convert molten or concentrated solutions into solid, free-flowing particles. These processes are essential for producing fertilisers such as urea and ammonium nitrate in forms that are easy to handle, store, transport, and apply in the field. While both methods achieve solidification, they differ in particle formation, equipment, and final product characteristics.

Prilling Technique

Prilling is a relatively simple and widely used method in which molten fertiliser is converted into small, spherical particles known as prills. In this process, the molten material is fed to the top of a tall prilling tower and sprayed through nozzles or a rotating bucket. As the droplets fall through a counter-current stream of air, they cool and solidify before reaching the bottom of the tower.

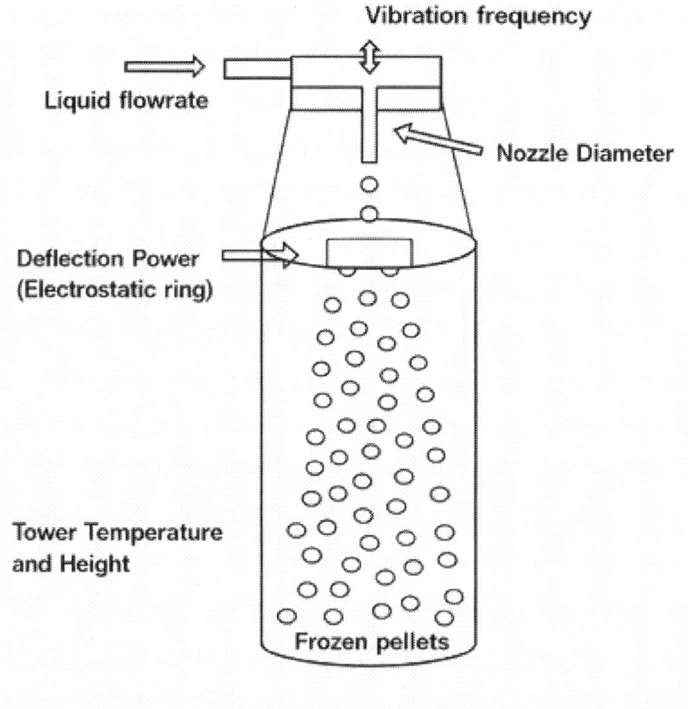

Figure 37: Prilling illustration indicating main process parameters. Adapted from Clenet, et al. (2019) [125].

The key principle behind prilling is droplet formation and air cooling. The size of the prills depends on factors such as nozzle design, melt viscosity, and airflow conditions. The resulting particles are typically small, smooth, and spherical.

Prilling equipment includes a prilling tower, spray nozzles or rotary distributors, air handling systems, and product collection conveyors. While the process is relatively low-cost and straightforward to operate, prilled products tend to have lower mechanical strength, making them more susceptible to breakage and dust formation during handling.

<u>Prilling Process</u>

The process begins with melt preparation and feed. A molten fertiliser, such as urea or ammonium nitrate, is prepared and maintained at the correct temperature and viscosity to enable effective droplet formation. The melt must remain above its melting point and be free of impurities that could block spray systems or affect product quality. This stage typically involves insulated and heated melt storage tanks, heat exchangers to regulate temperature, pumps to transfer the molten material, and filtration systems to remove contaminants.

Once prepared, the molten fertiliser is transported to the top of a prilling tower for distribution. At this point, the liquid is evenly dispersed and atomised into droplets using either fixed spray nozzles or a rotating bucket (centrifugal distributor). The goal is to produce droplets of consistent size, which directly determines the uniformity of the final prills. This stage requires feed piping, distribution manifolds, and flow control systems to ensure stable and even delivery.

As the molten material exits the distributor, it breaks into droplets due to gravity and, in the case of rotating systems, centrifugal forces. These droplets then begin to fall through the height of the tower. The size of the droplets, and therefore the resulting prills, is influenced by several factors, including nozzle design, rotational speed, melt viscosity, and flow rate. Controlled droplet formation is critical, as it directly affects particle size distribution and overall product consistency.

During the falling phase, the droplets are cooled and solidified by a stream of air flowing through the tower, typically in a counter-current direction, where air moves upward while the particles fall downward. This airflow removes heat from the droplets, allowing them to solidify into spherical prills before reaching the bottom. The cooling system consists of the prilling tower itself, along with air intake and exhaust systems, fans or blowers to regulate airflow, and air distribution systems to ensure uniform cooling throughout the tower.

By the time the particles reach the base of the tower, they have solidified and are collected as finished prills. These are discharged into collection hoppers and transferred via conveyors for further handling. During this stage, some fine particles or dust may be generated, so dust collection systems such as cyclones or filters are used to maintain product quality and control emissions.

Following collection, the prills may undergo additional cooling, screening, and conditioning. Cooling ensures the product reaches a stable temperature for storage, while screening separates particles into the desired size range. Oversized particles may be crushed and recycled, and undersized material can be returned to the process. In some cases, coating agents are applied to reduce caking and improve handling characteristics. Equipment used in this stage includes coolers, vibrating screens, crushers, and coating drums.

Overall, prilling is an efficient and widely used method for producing fertiliser particles. It combines relatively simple equipment with continuous operation, but requires precise control of process conditions to produce uniform, high-quality prills suitable for storage, transport, and agricultural application.

Granulation Technique

Granulation is a more advanced technique that produces larger, stronger, and more uniform particles. Instead of forming droplets in air, granulation involves building up particles by layering material onto small "seed" particles. This is typically carried out in a rotating drum or a fluidised bed, where molten fertiliser or concentrated solution is sprayed onto the seeds.

As the particles move and tumble within the granulator, successive layers of material adhere and solidify, causing the granules to grow in size. Airflow and temperature are carefully controlled to ensure proper drying and solidification. The result is a product with improved physical properties, including higher strength, lower dust generation, and better resistance to caking.

Granulation systems typically include granulators (rotary drum or fluidised bed), spray systems, air handling units, recycle conveyors (to return undersized particles as seeds), and screening equipment. Although more complex and capital-intensive than prilling, granulation produces a higher-quality fertiliser that is preferred in modern agricultural applications.

The main difference between prilling and granulation lies in how particles are formed. Prilling relies on droplets solidifying during free fall, while granulation builds particles through layering and growth. As a result, prills are generally smaller and less robust, whereas granules are larger, stronger, and more uniform.

From an operational perspective, prilling is simpler and less expensive but produces a lower-quality product. Granulation, on the other hand, requires more sophisticated equipment and control but delivers superior handling, storage, and application characteristics.

Granulation Process

Granulation is a controlled particle-growth process used to produce fertiliser granules that are larger, stronger, and more uniform than prilled products. Rather than forming droplets in free fall, granulation builds particles by depositing successive layers of material onto small "seed" particles within a controlled environment. This produces dense, mechanically robust granules with superior handling, storage, and application characteristics.

The process begins with feed preparation and conditioning. The feed material may be a molten fertiliser, such as urea or ammonium nitrate, or a concentrated solution or slurry. It must be carefully conditioned to achieve the correct temperature, viscosity, and composition to ensure effective adhesion to seed particles. This stage typically involves melt tanks or slurry preparation vessels, agitators and mixers to maintain uniformity, heat exchangers for temperature control, and pumps and feed lines to deliver the material to the granulator.

Granulation relies on the presence of seed particles to initiate growth. These seeds are usually recycled undersized particles from the process or deliberately generated nuclei. The recycle loop is a critical component, as it continuously supplies particles onto which additional layers can be deposited. Equipment used in this stage includes recycle conveyors or elevators, seed bins or surge hoppers, and feed distribution systems to regulate particle flow.

The core of the process occurs in the granulation stage, where particle growth takes place. This is typically carried out in a rotary drum granulator or a fluidised bed system. Seed particles are introduced into the granulator, and the molten or liquid fertiliser is sprayed onto them through nozzles or injection systems. As the particles tumble within a rotating drum or remain suspended in a fluidised air stream, successive layers of material adhere and solidify, causing the granules to increase in size. The growth process is carefully controlled by balancing factors such as spray rate, particle residence time, moisture content, and temperature. Key equipment includes the granulator itself, spray systems, drive systems for rotating drums, and internal lifters or air distribution plates.

Once granules are formed, they undergo drying and solidification. Moisture must be removed to harden the particles and prevent agglomeration. Drying is achieved using heated air, which reduces moisture content and stabilises the granule structure. This stage typically uses rotary dryers or fluidised bed dryers, along with hot air generation systems, air handling and ducting, and cyclones or dust collectors for particulate control.

After drying, the granules are cooled to improve their mechanical strength and prevent caking during storage. Cooling is typically achieved using fluidised bed coolers or rotary coolers, supported by air fans and cooling systems that regulate temperature and airflow.

The cooled product is then screened and classified to achieve the desired particle size distribution. Vibrating screens or sieves separate the granules into size fractions. Oversized particles are crushed and recycled back into the process, while undersized particles are returned as seed material for further growth. This stage ensures consistent product quality and efficient material utilisation.

Following size classification, the granules may undergo coating and finishing. Anti-caking agents or conditioning additives are applied to improve flowability, reduce moisture absorption, and enhance storage stability. This is typically carried out in coating drums or spray systems, supported by additive dosing equipment.

Finally, the finished granulated product is transferred to storage and handling systems. Granules are stored in silos or bins and prepared for transport using conveyors, bagging systems, or bulk loading equipment. Proper handling at this stage ensures product integrity and minimises degradation.

Throughout the process, precise control of key operating parameters is essential. Temperature must be carefully managed during melting, drying, and cooling stages. Moisture content plays a critical role in particle growth and drying efficiency. Airflow is important for both drying and fluidisation, while the recycle ratio influences particle size distribution. Residence time within the granulator determines the extent of particle growth and final granule size. Together, these controls ensure the production of high-quality, uniform fertiliser granules.

Storage, Handling, and Safety Risks

Storage, handling, and safety risks in nitrogen fertiliser production are critical considerations due to the chemical properties of materials such as ammonia, urea, ammonium nitrate, and ammonium sulphate. These substances can present hazards including toxicity, corrosivity, fire risk, explosion potential, and environmental harm if not properly managed. Effective risk control requires a combination of engineering systems, operational procedures, and regulatory compliance.

Storage Risks and Requirements

Storage conditions for fertilisers are highly dependent on the chemical type, but a unifying requirement across all fertilisers is rigorous control of temperature, contamination risk, and containment design to prevent releases and propagation of hazards. Ammonia, when stored as a pressurised or refrigerated liquid, poses specific risks of toxic release and vapour cloud formation, necessitating storage tanks that can withstand pressure and incorporate safety relief systems, with siting in well-ventilated areas to mitigate accumulation of hazardous vapours [126-128]. Solid fertilisers such as urea and ammonium sulphate are generally more stable yet are susceptible to moisture uptake, which can lead to caking and chemical degradation; therefore, these materials are stored in silos or bulk storage sheds under controlled humidity and temperature conditions to minimize hydration and clumping [126, 128, 129].

In contrast, ammonium nitrate (AN) carries a markedly higher safety risk due to its strong oxidising properties, requiring storage away from heat sources, contaminants, and combustible materials because contamination or confinement can trigger decomposition or detonation; reviews and incident analyses repeatedly highlight AN's tendency to undergo exothermic decomposition and potential detonation under adverse storage conditions, particularly when heat, confinement, or contaminants are present [126, 128, 130-132]. Across these categories, a central storage risk framework emerges: chemical degradation or self-heating for nitrates, moisture uptake and caking for soluble and hygroscopic fertilisers, contamination with incompatible materials, fire or explosive hazards for AN-containing products, and structural or containment failures of storage systems under environmental and thermal stress [126-128, 130, 132, 133] .

Ammonia storage systems must account for the potential formation of toxic vapours and pressure build-up, which informs design choices such as pressure-rated tanks and integrated relief devices to manage overpressure events. The requirement for ventilation is underscored by the need to prevent vapour accumulation that could lead to exposure or toxic release scenarios, a priority echoed in analyses of ammonia handling and the safety implications of pressurised liquid storage in industrial settings [126-128, 130, 133]. For solid fertilisers, moisture control remains the principal operational driver. Urea and ammonium sulphate's hygroscopic tendencies necessitate humidity-controlled environments and silo or bulk shed storage, with temperature regulation to preserve physical form and chemical integrity, thus reducing the risk of caking, clumping, and degradation that can compromise stability and handling safety [126, 128, 129, 133].

Ammonium nitrate represents the most complex storage hazard among fertilisers due to its oxidising nature and historical association with catastrophic explosions when stored improperly. AN can be kept in safe storage under many circumstances, but accidents have repeatedly occurred where heat input, confinement, humidity, contamination with organic or combustible substances, or poor ventilation precipitated self-heating, decomposition, and explosive events. Consequently, storage recommendations emphasize keeping AN away from heat sources and contaminants, ensuring adequate ventilation, prohibition of contact with fuels or organic materials, and robust separation from combustibles; post-incident analyses and safety guidelines frequently reference the Beirut and West Texas events to illustrate the consequences of storage mismanagement and the importance of preventing ignition, contamination, and heat build-up in AN storage facilities [126, 128, 130-132, 134-136].

Beirut Explosion (2020)

The Beirut explosion occurred on 4 August 2020 at the Port of Beirut in Beirut. Approximately 2,750 tonnes of ammonium nitrate had been stored in a warehouse for several years without adequate safety controls.

Figure 38: Damage after the Beirut explosion. Mehr News Agency, CC BY 4.0, via Wikimedia Commons.

The material was originally confiscated from a ship and left in storage under poor conditions. Over time, the stockpile became contaminated and degraded, and it was stored near combustible materials such as fireworks. A fire broke out in the warehouse, which then escalated into a catastrophic explosion when the ammonium nitrate detonated.

Figure 39: Aftermath of the 2020 Beirut explosion. Freimut Bahlo, CC BY-SA 4.0, via Wikimedia Commons.

The blast caused over 200 deaths, thousands of injuries, and widespread destruction, severely damaging the port and surrounding city. The explosion generated a shockwave felt across large distances and is considered one of the largest non-nuclear explosions in history.

Key contributing factors:

- Long-term storage of a large quantity of ammonium nitrate
- Poor storage conditions and lack of segregation
- Presence of combustible materials (fireworks)
- Lack of regulatory oversight and risk management

Key lessons:

- Strict limits on storage quantities
- Separation from incompatible materials
- Continuous inspection and hazard assessment

- Clear regulatory accountability

West Texas Explosion (2013)

The West Fertilizer Company explosion occurred on 17 April 2013 in West, Texas. The incident involved a fertiliser storage and distribution facility that stored ammonium nitrate.

A fire broke out at the facility, and emergency responders attended the scene. Shortly after, the ammonium nitrate detonated, resulting in a massive explosion equivalent to a small earthquake.

Figure 40: Aerial photo of the west explosion site taken several days after blast (4/22/2013). Shane.torgerson, CC BY-SA 3.0, via Wikimedia Commons.

The explosion caused 15 deaths (including first responders), over 160 injuries, and extensive property damage, destroying nearby homes, a school, and a nursing home.

Key contributing factors:

- Inadequate storage conditions for ammonium nitrate
- Lack of proper hazard classification and reporting

- Fire exposure leading to detonation
- Insufficient emergency planning and awareness

Key lessons:

- Proper hazard communication and chemical inventory reporting
- Fire protection systems and separation distances
- Emergency response planning for high-risk materials
- Land-use planning around hazardous facilities

Both disasters highlight the inherent risks associated with ammonium nitrate when it is stored improperly, becomes contaminated, or is exposed to fire. Although ammonium nitrate is widely used as a fertiliser and is generally stable under normal conditions, it can become highly dangerous if key safety controls are not maintained.

A number of common risk factors were evident in both events. Large quantities of ammonium nitrate were stored in a single location, significantly increasing the potential impact of any incident. Contamination with fuels or incompatible substances further heightened the risk of decomposition or detonation. Exposure to heat or fire acted as the primary trigger in both cases, while poor ventilation or confinement allowed pressure and heat to build. In addition, weak regulatory oversight and inadequate enforcement of safety standards contributed to the escalation of these hazards.

To manage these risks effectively, several critical control measures must be implemented. Storage volumes and durations should be strictly limited to reduce the consequences of any incident. Ammonium nitrate must be segregated from combustible materials and other incompatible substances to prevent contamination. Proper ventilation and temperature control are essential to avoid heat buildup and decomposition. Fire detection and suppression systems should be installed to identify and respond to incidents at an early stage. Finally, strong regulatory compliance, regular auditing, and clear accountability are necessary to ensure that safety standards are consistently maintained.

In addition, expert reviews point to the critical role of external fires as a predominant driver of AN accidents, reinforcing the need for vigilant fire protection, intrastorage separation, and emergency response planning designed to prevent ignition and limit rapid heat accumulation in AN stockpiles [135-139].

Key storage risks common to all fertilisers include chemical degradation or structural breakdown due to temperature excursions, moisture ingress and moisture-induced degradation (especially for hygroscopic solids), contamination from incompatible materials, fire and explosion hazards (with AN presenting the clearest risk profile), and the risk of structural failure of storage infrastructure under thermal or mechanical stress. The literature consistently links these risks to practical countermeasures: for ammonia, pressure-rated containment with relief systems and ventilation; for solid fertilisers, humidity and temperature control to prevent moisture-driven degradation and caking; and for AN, stringent separation from heat and contaminants, robust ventilation, and explicit avoidance of contact with combustible materials, along with fire safety and emergency planning. Across sources, the overarching conclusion is that storage safety hinges on matching containment design and environmental controls to the specific chemical risk profile of each fertiliser, plus robust risk management practices informed by historical accidents and regulatory guidance [126-128, 130-132, 134-136, 139].

Handling Risks and Controls

Handling operations involve transferring, conveying, and packaging fertilisers, often in large volumes. These activities can generate dust, expose workers to hazardous chemicals, and create risks of spills or leaks.

Ammonia handling is particularly hazardous due to its toxicity and corrosive nature. Exposure can occur through inhalation or skin contact, requiring the use of personal protective equipment (PPE) such as respirators, gloves, and eye protection. Properly sealed transfer systems and gas detection equipment are essential.

For solid fertilisers, mechanical handling systems such as conveyors, elevators, and loaders must be designed to minimise dust generation and product degradation. Dust from fertilisers can pose respiratory hazards and, in some cases, contribute to explosion risks in confined spaces.

Operational risks include manual handling injuries, chemical exposure, dust inhalation, and equipment-related hazards such as entanglement or mechanical failure.

Safety Risks in Production and Processing

Nitrogen fertiliser production involves high temperatures, high pressures, and reactive chemicals, all of which introduce significant safety risks. Processes such as ammonia synthesis, nitric acid production, and urea formation require strict control to prevent incidents.

Ammonium nitrate is particularly hazardous because it can undergo thermal decomposition and, under certain conditions, detonate. Risks increase if the material is contaminated with organic substances, exposed to high temperatures, or confined in enclosed spaces. Historical incidents have demonstrated the importance of strict storage segregation and temperature monitoring.

Ammonia presents risks of toxic release, which can form hazardous vapour clouds affecting both workers and surrounding communities. Nitric acid production involves nitrogen oxides (NO_x), which are toxic and environmentally harmful.

Key safety risks include:

- Toxic gas release (NH_3, NO_x)
- Fire and explosion hazards (especially ammonium nitrate)
- High-pressure system failure (reactors, pipelines)
- Thermal runaway reactions
- Corrosion leading to equipment failure

Control Measures and Risk Mitigation

Effective risk management in nitrogen fertiliser production relies on multiple layers of control:

Engineering Controls

- Pressure relief valves and rupture discs
- Gas detection systems (NH_3, NO_x monitoring)
- Ventilation and scrubbing systems
- Corrosion-resistant materials and coatings
- Explosion-proof equipment and plant design

Operational Controls

- Standard operating procedures (SOPs)
- Regular inspection and maintenance programs
- Controlled handling and storage practices
- Emergency shutdown systems

Personal Protective Equipment (PPE)

- Respirators and gas masks
- Chemical-resistant gloves and clothing
- Eye and face protection

Emergency and Environmental Controls

- Spill containment systems
- Fire suppression systems
- Emergency response plans
- Tail gas treatment and emissions control systems

Regulatory and Compliance Considerations

Regulatory and compliance frameworks governing nitrogen fertiliser production and storage are extensive and highly structured across most jurisdictions, reflecting the hazardous properties of materials such as ammonia and ammonium nitrate. These frameworks typically integrate dangerous goods legislation, workplace health and safety (WHS/OHS) requirements, environmental protection laws, and, in many cases, security and anti-terrorism controls. The level of regulatory oversight increases significantly for substances such as ammonium nitrate due to their classification as both hazardous chemicals and security-sensitive materials.

In Australia, nitrogen fertilisers are regulated under a combination of national model laws and state or territory legislation. The Safe Work Australia framework, including the Model Work Health and Safety (WHS) Regulations, governs hazardous chemicals, requiring risk assessments, safety data sheets (SDS), labelling, and the implementation of control measures in accordance with the hierarchy of control. Ammonium nitrate is specifically addressed under the Australian Dangerous Goods (ADG) Code, where it is classified as an oxidising substance (Class 5.1). Additional controls apply under state-based dangerous goods legislation, such as licensing requirements for storage and handling, security risk assessments, and mandatory separation distances.

A critical element in Australia is the regulation of security-sensitive ammonium nitrate (SSAN). Jurisdictions such as Queensland, Western Australia, and New South Wales require licensing for manufacture, storage, transport, and use. Operators must implement strict inventory tracking, personnel background checks, secure storage (including access control systems), and incident reporting protocols. For example, in Western Australia, the Department of Mines, Industry Regulation and Safety enforces detailed SSAN security plans that include site security, transport chain-of-custody controls, and reconciliation of stock movements.

In the United States, regulation is similarly multi-layered. The Occupational Safety and Health Administration (OSHA) enforces standards related to hazardous chemicals, including Process Safety Management (PSM) for highly hazardous substances such as ammonia. The Environmental Protection Agency (EPA) administers the Risk Management Program (RMP), requiring facilities to assess worst-case release scenarios and implement emergency response planning. Ammonium nitrate storage is also regulated under the Department of Homeland Security (DHS) Chemical Facility Anti-Terrorism Standards (CFATS), which impose security requirements for facilities handling chemicals of interest. Following the West Texas explosion, additional scrutiny has been placed on reporting, emergency planning, and land-use controls around fertiliser storage sites.

Within the European Union, regulatory control is driven by major directives such as the Seveso III Directive, which applies to sites storing significant quantities of hazardous chemicals, including ammonium nitrate. Facilities exceeding threshold quantities must implement Major Accident Prevention Policies (MAPP), safety management systems, and detailed safety reports, as well as coordinate with local authorities on emergency planning and land-use restrictions. In parallel, the REACH Regulation governs the registration, evaluation, and safe use of chemical substances, while the Classification, Labelling and Packaging (CLP) Regulation ensures consistent hazard communication.

Globally, dangerous goods transport is harmonised through frameworks such as the United Nations Model Regulations, which underpin modal codes including road, rail, sea (IMDG Code), and air (ICAO/IATA). These systems classify ammonium nitrate under UN numbers (e.g., UN 1942, UN 2067) and prescribe packaging, segregation, labelling, and documentation requirements to minimise risks during transport.

Environmental compliance is another critical dimension. Facilities must manage emissions such as ammonia (NH_3) and nitrogen oxides (NO_x), wastewater discharges, and solid waste. In Australia, this is typically enforced through state environmental protection authorities (e.g., EPA Victoria), requiring environmental licences, monitoring programs, and reporting. Similar regimes exist globally, such as the EU Industrial Emissions Directive and U.S. Clean Air Act requirements, which mandate emission limits, pollution control technologies, and continuous monitoring.

Across all jurisdictions, a common theme is the requirement for comprehensive risk management systems. These include hazard identification (e.g., HAZOP studies), process safety controls, maintenance and inspection regimes, emergency response planning, and workforce training. Documentation and auditability are essential, with regulators requiring evidence of compliance through records, incident reporting, and periodic inspections.

Nitrogen fertiliser production and storage operate within a tightly controlled regulatory environment worldwide. Ammonium nitrate, in particular, is subject to enhanced controls due to its dual-use nature and potential for catastrophic incidents. Compliance requires not only adherence to legal requirements but also the implementation of robust safety, security, and environmental management systems that are continuously monitored and improved.

Storage, handling, and safety risks in nitrogen fertiliser production are closely linked to the chemical and physical properties of the materials involved. Ammonia introduces toxicity and pressure-related hazards, ammonium nitrate presents significant explosion risks, and even relatively stable products like urea require careful handling to maintain quality and safety. Effective risk management depends on robust engineering design, strict operational controls, appropriate PPE, and compliance with regulatory frameworks to ensure safe and efficient operations.

Structuring a Urea Manufacturing Operation

A urea project usually begins with front-end project definition. This includes confirming feedstock availability, especially ammonia and carbon dioxide, defining the target product form (prill or granule), estimating plant capacity, and determining whether the plant will be integrated with an ammonia facility or supplied externally. At this stage, the project team also assesses utilities such as steam, cooling water, electricity, instrument air, wastewater treatment, and storage infrastructure.

The next stage is technology selection and licensing. Commercial urea plants are typically built using licensed process technologies rather than bespoke chemistry developed in-house. The project sponsor engages a reputable licensor and engineering firm to select the synthesis and recovery configuration, define the process guarantees, and develop the design basis. This includes choosing the finishing method, such as granulation or prilling, and establishing the required control philosophy, emissions approach, and product quality targets.

This is followed by engineering design, typically progressing through concept design, front-end engineering design, and detailed design. At a high level, the plant is organised into feed preparation, synthesis, recovery, concentration, finishing, storage, and utility/support systems. Engineering work focuses on metallurgy, corrosion control, pressure containment, heat integration, instrumented safeguards, relief systems, hazardous area classification, ventilation, and emergency isolation. Formal process safety studies such as HAZID, HAZOP, layers of protection analysis, and consequence modelling are central to this stage.

A parallel stream is regulatory and compliance approval. A legitimate operation requires environmental approvals, dangerous goods and hazardous chemical approvals, land-use compatibility review, building and pressure equipment compliance, emissions and wastewater controls, emergency response planning, and worker safety systems. Security, inventory control, and transport compliance also become important where ammonia and other hazardous materials are involved.

The project then moves into procurement and construction. This includes long-lead equipment acquisition, construction of civil foundations, installation of process equipment, piping, electrical systems, automation systems, fire protection, gas detection, and environmental control infrastructure. Commissioning planning begins well before mechanical completion and includes operating procedures, maintenance systems, spare parts strategy, training, and emergency drills.

After construction, the plant enters commissioning and start-up under tightly controlled procedures led by experienced process, mechanical, electrical, and safety specialists. This phase validates equipment integrity, control systems, utility readiness, emissions controls, safety interlocks, and product quality. Only after extensive testing and staged introduction of hazardous materials does the plant move into commercial operation.

Once operational, the site requires robust operations, maintenance, and safety management systems. These include competency-based staffing, preventive maintenance, corrosion monitoring, turnaround planning, incident investigation, management of change, contractor control, product quality assurance, and environmental reporting. Continuous monitoring of process conditions, emissions, equipment condition, and storage integrity is essential.

For a legitimate industrial urea project, the first thing to understand is that "cost" depends heavily on the project boundary. A quoted number might mean only the EPC process plant, or it might include offsites, storage, utilities, water supply, conveyor/loadout, site works, owner's costs, financing, and contingency. Recent Australian project data shows exactly how wide that gap can be: the Perdaman/Ceres project is cited at about A$6.0 billion total financing for a plant producing about 2.3 million tonnes per year of urea, while the EPC contract publicly disclosed for a 2.14 million tonnes per year version of the same project was about US$2.7 billion [140, 141]. Using recent exchange rates, that is roughly US$4.21 billion / €3.64 billion for the A$6.0 billion figure, and roughly A$3.84 billion / €2.34 billion for the US$2.7 billion EPC figure.

In scale terms, urea projects usually fall into three broad bands. Pilot or demonstration units are generally below full commercial scale and are mainly used for process validation or research. Regional commercial plants are often in the few hundred thousand tonnes per year range. World-scale integrated plants are commonly above one million tonnes per year, and very large projects can exceed two million tonnes per year. Recent examples illustrate this spread: a published green-urea techno-economic study modelled a 220,000 t/y plant, while the Perdaman/Ceres project is a 2.14–2.3 Mt/y world-scale development [140, 141].

A typical commercial integrated urea project therefore is not just a "urea reactor." It is usually a site with several major plant blocks. At minimum, you should expect a feedstock and storage section for ammonia and carbon dioxide, a high-pressure synthesis and recovery section for urea formation and carbamate recycle, a concentration section to produce molten urea, a finishing section such as granulation or prilling, and a large package of utilities and offsites. Saipem's description of the Ceres project, for example, explicitly identifies ammonia production, urea and granulation, utilities, storage, water supply, and port/export infrastructure as major elements of scope.

At the equipment level, the core process equipment for a conventional large-scale urea plant typically includes ammonia storage and transfer systems, CO_2 compression and drying, a high-pressure urea reactor, stripper, carbamate condenser, decomposers and absorbers for recycle, vacuum or falling-film evaporators, molten urea tanks, and then either a granulation plant or prilling tower for finishing. Around that core sits a substantial balance-of-plant package: steam and condensate systems, cooling water, power distribution, instrument air, distributed control and safety-instrumented systems, wastewater treatment, flare/vent treatment, bulk storage, bagging or loadout, and often port, rail, or truck export infrastructure for large projects. On projects of the Perdaman/Ceres type, the offsites are material enough to move total project cost by billions, which is why EPC-only and "all-in" numbers differ so much.

For a rough cost frame, a modern large conventional project can be thought of in two ways. On an EPC-only basis, the Perdaman EPC contract implies a capital intensity of roughly US$1,262 per annual tonne of urea capacity. On a broader fully financed/project basis, the same development implies something closer to A$2,609 per annual tonne when owner's scope and infrastructure are included. Those two numbers are not contradictory; they show why investors must define scope precisely before comparing projects.

For a mid-sized regional project, a published 2024 green-urea study estimated total capital investment at US$510.79 million for a 220,000 t/y plant. Converted at current reference rates, that is about A$727 million or €442 million. That figure should be treated carefully, because it is a modelled low-carbon case with electrolysis and carbon capture assumptions, not a like-for-like conventional natural-gas-fed EPC quote. Still, it is useful as an order-of-magnitude indicator that smaller plants do not simply scale down proportionally; their capital intensity is usually higher per tonne than very large plants.

In practical project development terms, the biggest cost drivers are usually capacity, whether ammonia is made onsite, site remoteness, utilities and water availability, export/logistics infrastructure, finishing choice (granulation vs prilling), materials of construction for high-pressure carbamate service, environmental controls, and whether the project is brownfield or greenfield. A brownfield add-on beside an existing ammonia plant with existing utilities, roads, and storage will usually be far cheaper than a remote greenfield integrated complex with its own water supply, port conveyor, and export systems. The Ceres project illustrates this clearly because its published scope includes not only process units but also supporting infrastructure, port linkage, and water supply.

If you need a realistic budgeting rule, it is safer to estimate urea projects in budget classes rather than one fixed number. A pilot or demonstration plant is usually a special-purpose engineering project with very high cost per tonne and is not a good benchmark for commercial supply. A regional commercial plant in the few hundred thousand tonnes per year range will generally need hundreds of millions of dollars. A world-scale integrated export project can easily run into the multi-billion-dollar range once ammonia, utilities, offsites, and logistics are included. Recent public examples support that framing.

For currency reference, recent quoted rates are about 1 EUR = 1.1555 USD = 1.6351 AUD, and about 1 AUD = 0.7025 USD = 0.6068 EUR. Those are suitable for order-of-magnitude conversion, but any live budget should lock a date and methodology because FX can materially affect multi-billion-dollar projects.

A useful way to understand urea project costs is to break total capital expenditure (CAPEX) into plant areas, rather than treating the plant as a single number. This helps identify where money is actually spent and how design choices (e.g. granulation vs prilling, integration with ammonia) affect overall cost.

Below is a conceptual CAPEX breakdown based on typical large-scale, modern urea plants. These are indicative ranges (not exact values), but they align with industry project structures.

<u>Conceptual CAPEX Breakdown – Urea Plant (by Area)</u>

Agricultural Fertiliser Production: Principles, Processes, and Practice

1. Feedstock Handling & Preparation (≈ 5–10%)

This section covers the infrastructure required to receive, store, and prepare ammonia (NH_3) and carbon dioxide (CO_2).

It includes ammonia storage tanks (refrigerated or pressurised), transfer pumps, and safety systems, as well as CO_2 compression, drying, and purification systems. While relatively small compared to the rest of the plant, this area is critical due to safety requirements and material handling complexity.

Cost drivers:

- Storage capacity (buffer vs continuous supply)
- CO_2 compression requirements
- Safety systems (containment, detection, relief)

2. High-Pressure Synthesis & Recovery Section (≈ 30–40%)

This is the **core and most expensive part of the plant.**

It includes the urea reactor, stripper, carbamate condenser, and high-pressure loop. These units operate at very high pressure (140–250 bar) and temperature, requiring specialised metallurgy (corrosion-resistant alloys) and heavy-duty equipment.

It also includes the recycle loop, where unreacted NH_3 and CO_2 are recovered and returned to the reactor.

Key equipment:

- High-pressure reactor
- Stripper (NH_3 or CO_2 stripping)
- Carbamate condenser
- High-pressure piping and valves

Why it's expensive:

- Extreme operating conditions
- Special materials (e.g. duplex stainless steels)
- Thick-walled pressure vessels
- Complex fabrication

3. Medium/Low Pressure Recovery (≈ 10–15%)

After the high-pressure section, remaining unreacted materials are recovered at lower pressures.

This includes decomposers, flash vessels, absorbers, and associated heat exchange systems. While less expensive than the high-pressure section, this area still contributes significantly due to multiple stages and recycle integration.

Key equipment:

- Decomposers
- Flash drums
- Absorbers
- Gas-liquid separators

4. Evaporation & Concentration (≈ 10–15%)

This section removes water from the urea solution to produce molten urea (>99%).

It typically uses vacuum evaporators or falling film evaporators, along with steam systems and condensers.

Key equipment:

- Evaporators
- Vacuum systems
- Steam heaters
- Condensers

Cost drivers:

- Energy integration design
- Materials (corrosion/urea service)
- Scale of evaporation duty

5. Finishing Section (Prilling or Granulation) (≈ 10–20%)

This is where molten urea becomes a solid product.

Prilling (lower cost option)

- Simpler system
- Lower CAPEX (~10–12%)

Granulation (higher cost, preferred)

- More complex system
- Higher CAPEX (~15–20%)
- Better product quality

Key equipment:

- Prilling tower or granulator
- Air handling systems
- Dryers, coolers
- Recycle conveyors

Cost drivers:

- Technology choice (granulation is more expensive)
- Product quality requirements
- Environmental controls (dust, emissions)

6. Storage, Handling & Loadout (≈ 5–10%)

This includes infrastructure for storing and shipping finished urea.

Key equipment:

- Storage silos or bulk sheds
- Conveyors
- Bagging systems
- Rail/truck/port loading systems

Costs can increase significantly if export infrastructure (e.g. port conveyors) is required.

7. Utilities & Offsites (≈ 20–35%)

This is often underestimated but can be a major cost component.

It includes all supporting infrastructure required to run the plant:

Utilities:

- Steam generation systems
- Cooling water systems
- Power supply and distribution
- Instrument air systems

Offsites:

- Water supply and treatment
- Wastewater treatment
- Flare systems
- Fire protection systems
- Buildings, roads, and site infrastructure

In large greenfield projects, this category can rival or exceed the process plant cost.

Table 3: Typical Overall Distribution (Summary).

Plant Area	Typical % of Total CAPEX
Feedstock Handling	5–10%
High-Pressure Synthesis	30–40%
Medium/Low Pressure Recovery	10–15%
Evaporation	10–15%
Finishing (Prilling/Granulation)	10–20%
Storage & Loadout	5–10%
Utilities & Offsites	20–35%

Example (World-Scale Plant ~US$3 Billion EPC)

Approximate breakdown:

- **High-pressure section:** ~US$1.0–1.2B
- **Utilities/offsites:** ~US$0.8–1.0B
- **Finishing:** ~US$0.3–0.5B
- **Evaporation + recovery:** ~US$0.5–0.7B
- **Others (storage/feed):** remainder

Converted (approx):

- **USD:** \$3.0B
- **AUD:** ~\$4.3B
- **EUR:** ~\$2.6B

Chapter 7

Phosphate Fertiliser Production

This chapter shifts the focus from core process chemistry and production pathways to one of the most critical, and increasingly scrutinised, aspects of phosphate fertiliser manufacture: the role of impurities and their environmental implications. While earlier chapters have explored how phosphate rock is converted into plant-available nutrients, it is equally important to understand that this transformation is not clean or selective. Natural phosphate feedstocks contain a wide range of impurities, and these materials influence not only process performance, but also waste generation, emissions, and long-term environmental risk.

In particular, the wet-process phosphoric acid (WPA) route, which underpins most global phosphate fertiliser production, generates large volumes of by-products and waste streams. Among these, phosphogypsum stands out as the most significant environmental footprint of the industry, produced in quantities of several tonnes for every tonne of phosphoric acid generated . Alongside this, fluorine compounds, heavy metals such as cadmium, and process-related emissions present ongoing technical, regulatory, and sustainability challenges.

This chapter examines how impurities originate in phosphate rock, how they behave during processing, and how they impact both plant operation and environmental performance. It explores the formation, composition, and management of phosphogypsum, the control of gaseous emissions such as fluorides, and the growing regulatory pressures associated with contaminants like cadmium. In doing so, it highlights the strong link between feedstock quality, process design, and environmental outcomes.

Understanding these issues is essential not only for safe and compliant plant operation, but also for the future of the fertiliser industry. As environmental standards tighten and resource efficiency becomes more important, the management of impurities and by-products is increasingly central to sustainable fertiliser production and the transition toward circular and lower-impact industrial systems.

Mining and Beneficiation of Phosphate Rock

Phosphate rock is the primary raw material used to produce a range of phosphorus-based fertilisers, which are essential for plant growth, particularly for root development, energy transfer (ATP), and flowering. However, raw phosphate rock is generally not directly usable by plants in most soils, so it is first chemically processed into more soluble and plant-available forms.

Mining and beneficiation transform low-grade phosphate ore into a high-quality concentrate suitable for fertiliser production. While mining extracts the resource, beneficiation is the critical step that determines product quality, process efficiency, and overall economic viability of phosphate fertiliser production.

Mining of Phosphate Rock

Phosphate rock is predominantly mined using surface (open-pit) methods, as most deposits occur close to the earth's surface. This approach is generally more economical and efficient than underground mining, which is only used in limited cases where deposits are deeper or geological conditions require it. Surface mining allows for large-scale extraction and straightforward access to the ore body.

Figure 41: Phosphorite mine near Oron, Negev. Wilson44691, CC0, via Wikimedia Commons.

The process begins with the removal of overburden, which consists of topsoil and non-ore materials covering the phosphate deposit. This layer is carefully stripped away to expose the phosphate-bearing rock beneath. In well-managed operations, the removed overburden is stockpiled and later used for land rehabilitation, helping to restore the site after mining activities are completed.

Once the ore is exposed, extraction can commence. The method of extraction depends on the hardness and composition of the deposit. In softer formations, large machines such as draglines, bucket-wheel excavators, and hydraulic excavators can remove the ore directly without the need for blasting. In harder rock deposits, drilling and controlled blasting may be required to fragment the material before it can be excavated and transported.

Figure 42: Train loaded with phosphate rock; mine dumps and workings in background. Metlaoui, Tunisia. Dennis G. Jarvis, CC BY-SA 2.0, via Wikimedia Commons.

In many modern phosphate mining operations, particularly in regions such as Florida and Morocco, the extracted ore is mixed with water to form a slurry. This slurry is then transported via pipelines to the beneficiation plant, where it undergoes further processing. This method reduces the need for truck haulage and improves efficiency over long distances.

The equipment used in phosphate mining reflects the scale and nature of the operation. Typical equipment includes draglines and excavators for material removal, drilling and blasting systems where necessary, slurry pumps and pipelines for transport, and haul trucks and conveyors for moving materials within the site.

Several key factors influence the effectiveness and cost of phosphate mining. These include variability in ore grade, which affects processing efficiency and product quality, and the strip ratio, which determines how much waste must be removed relative to the amount of ore extracted. Environmental considerations are also critical, particularly in relation to land disturbance, water use, and rehabilitation practices, all of which must be carefully managed to ensure sustainable operations.

Beneficiation of Phosphate Rock

Beneficiation of phosphate rock is the process of upgrading raw ore by increasing the concentration of phosphate minerals, typically expressed as P_2O_5, while removing unwanted impurities such as silica (sand), clays, carbonates, and organic matter. This step is essential because naturally occurring phosphate rock is often of low grade, typically containing around 5–15% P_2O_5. Through beneficiation, the material is upgraded to a commercially viable product, usually in the range of 28–35% P_2O_5, suitable for downstream fertiliser production.

The beneficiation process begins with crushing and screening, where the mined ore is reduced in size to liberate phosphate minerals from the surrounding waste material. Primary crushing is typically carried out using jaw or gyratory crushers, followed by secondary crushing and screening to achieve the desired particle size. Vibrating screens and conveyor systems are used to classify and transport the material through this stage.

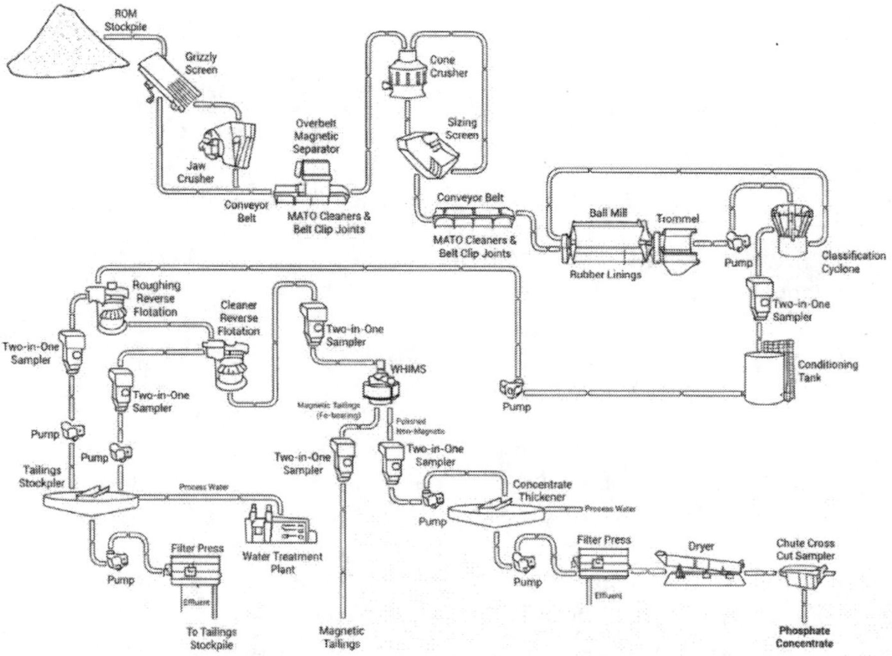

Figure 43: Typical Phosphates Benefication Flow. Adapted from Multotec [142].

Following size reduction, the ore undergoes washing and desliming to remove clay and fine particles that can interfere with later separation processes. Water is used to wash the ore, separating fine clay fractions and creating a slurry. This slurry is then classified based on particle size using equipment such as scrubbers, log washers, hydrocyclones, and screens. This step is particularly important, as the presence of clays can significantly reduce the efficiency of subsequent flotation processes.

The most critical stage in phosphate beneficiation is flotation, which is the primary method used to upgrade the ore. In this process, chemical reagents such as collectors and frothers are added to the slurry, and air is introduced to create bubbles. Phosphate particles attach to these bubbles and rise to the surface, while unwanted materials sink. Depending on the ore characteristics, either direct flotation (where phosphate is floated) or reverse flotation (where impurities are removed) may be used. This stage relies on specialised equipment including flotation cells, agitators, air injection systems, and reagent dosing systems.

Once the phosphate concentrate has been separated, the slurry undergoes thickening and filtration to remove excess water. In thickeners, solid particles settle out, allowing water to be recovered and recycled within the process. The concentrated material is then further dewatered using vacuum or pressure filtration systems, producing a moist phosphate concentrate suitable for drying.

The final stage of beneficiation involves drying and product preparation. The filtered phosphate rock is dried, typically in rotary dryers, to reduce moisture content and improve handling, storage, and transport characteristics. Cyclones and dust collection systems are used to manage particulates, and the final product may be sized or blended before being stored in silos or prepared for shipment.

Overall, beneficiation transforms low-grade phosphate ore into a high-quality concentrate, enabling efficient and economically viable fertiliser production while ensuring that impurities are effectively removed.

Phosphate Rock Deposits

Global phosphate rock (PR) deposits exhibit a highly uneven geographic distribution, with a small set of regions containing the bulk of known reserves and production capacity. This unevenness is a fundamental driver of fertiliser supply, since geology sets where deposits occur while economics and geopolitics determine which deposits are actively mined and processed for fertiliser manufacture [143-145]. The contemporary picture typically emphasizes North Africa as the dominant reserve holder, while production is concentrated in other regions due to infrastructure, investment, and policy choices that shape mining activity and downstream processing [83, 143, 146]. The following synthesis integrates key geological, production, and policy dimensions drawn from the provided sources to reflect the current understanding of global phosphate resource geography and its implications for fertiliser supply chains [83, 143, 145, 146].

Morocco and Western Sahara collectively hold the world's largest concentration of phosphate rock reserves, commonly cited as comprising upwards of 70% of global known reserves. These deposits are predominantly sedimentary phosphorites, which are generally easier to mine and process relative to igneous varieties, and are spearheaded by major operations such as Khouribga and Bou Craa that are largely state-influenced via the OCP Group, underscoring the region's strategic importance for long-term global supply stability [83, 143, 146]. The geographic concentration in North Africa contrasts with production patterns in other major mining regions; while reserves are dominated by Morocco/Western Sahara, actual production has diversified through large-scale operations in other countries, with China currently the world's largest producer and significant output also emanating from the United States, Russia, and Saudi Arabia, reflecting the historical disconnect that can exist between reserve heft and immediate production volumes [143-145].

China, although possessing smaller overall reserves than Morocco, is the world's largest producer of phosphate rock. Its major mining regions include Yunnan, Guizhou, and Hubei, where both sedimentary and igneous deposits are exploited. A defining characteristic of China's phosphate sector is the predominance of domestic consumption to support a large agricultural sector, with comparatively modest export volumes relative to its production capacity. This pattern illustrates how a country's internal demand, coupled with policy and capacity to process phosphate rock into fertilisers, shapes its role in global supply despite reserve size [143-145].

Historically, the United States has been a major phosphate producer, featuring key mining districts in Florida, Idaho, and, to a lesser degree, North Carolina. The majority of US deposits are sedimentary and of relatively high quality, supported by advanced beneficiation technologies and stringent environmental controls. Although current production levels have declined from historical peaks, the United States remains an important contributor to global supply, illustrating how high-grade sedimentary PR can sustain significant regional output even in periods of global market flux [143-145].

In Russia and parts of Central Asia, phosphate deposits are predominantly igneous in origin, with notable occurrences in the Kola Peninsula and in Kazakhstan and Uzbekistan. Igneous phosphate rocks generally present more challenging mining and processing requirements due to rock hardness, and operations often align with downstream chemical industries to optimise value chains. This geology-driven difference in ore type has implications for energy use, processing technology, and regional integration with fertiliser production, contributing to the global pattern where geological abundance does not always align with immediate production emphasis [143-145].

The Middle East, including countries such as Saudi Arabia and Jordan, hosts significant phosphate mining and fertiliser production activities, frequently organized as large-scale, integrated projects with an emphasis on export markets. These operations contribute to global fertiliser supply chains, highlighting how regional scale and export orientation interact with global demand dynamics and pricing pressures. Africa beyond Morocco also hosts phosphate activities of regional significance, with countries such as Senegal, Togo, and

South Africa contributing regional deposits and capacity that complement domestic fertiliser industries where feasible [143-145].

Beyond the leading producers, other countries such as Brazil and India are expanding domestic production capacity and relying on imports to meet fertiliser needs, while Australia maintains smaller deposits and historical operations (e.g., Christmas Island, parts of Queensland). The distribution of these additional producers underscores the role of national resource endowments, policy choices, and investment climate in shaping alternate supply sources and diversification opportunities within the global PR market [143-145]. The distinction between reserves and production remains critical; reserves are concentrated in Morocco/Western Sahara, while production is distributed across China, the United States, Russia, and Saudi Arabia, among others, reinforcing the geographic mismatch often observed between resource endowments and actual mining activity [143-145].

From a geological perspective, phosphate rocks fall into two principal categories: sedimentary deposits, formed in ancient marine environments and generally easier to mine with open-pit methods, and igneous deposits, composed mainly of apatite, formed through magmatic processes and typically harder to process. Sedimentary phosphate deposits are widespread and include examples in Morocco, Florida, and the Middle East, while igneous deposits are famously represented by the Kola Peninsula and other alkaline complexes in Russia and Brazil. This dichotomy has practical consequences for mining, beneficiation, and processing strategies across regions, influencing energy intensity, capital costs, and environmental controls required to convert PR into fertiliser feedstocks [83, 143, 144].

Phosphate rock is a critical global resource because phosphorus is indispensable for fertiliser production and agricultural productivity, with no viable substitute for phosphorus in agriculture. The supply of PR is highly geographically concentrated, which creates geopolitical and supply chain risks, particularly given the dominance of a single region in reserves and the scale of export controls that can accompany strategic minerals. This concentration has spurred interest in phosphorus recovery and recycling within a circular economy framework to bolster resilience, reduce vulnerability to supply disruptions, and diversify supply sources. The interplay of geology, geopolitics, and policy thus remains central to forecasting future fertiliser supply and informing strategic discussions on resource management and recycling of phosphorus across the value chain [143, 145, 146].

Overall, the global phosphate rock landscape is defined by a striking concentration of known reserves in North Africa (notably Morocco and Western Sahara), with a production profile that places major current output in China, the United States, Russia, and Saudi Arabia, among others. This disparity between reserve location and production geography reflects not only ore genesis but also mining, processing infrastructure, investment priorities, and policy environments that determine where mining and fertiliser production occur. The sedimentary–igneous classification of deposits captures the core geological determinants of mining difficulty and processing requirements, while the political economy surrounding OCP's management of Moroccan reserves and Western Sahara's status illustrates the broader geopolitical significance of PR as a globally sensitive commodity. Taken together, these sources underscore the central challenge: ensuring a stable phosphorus supply for fertilisers in a world where geological abundance does not automatically translate into predictable, geographically diversified, or geopolitically neutral production and trade pathways [83, 143-146].

Types of Fertilisers Made from Phosphate Rock

Phosphate rock is the primary raw material used to produce a range of phosphorus-based fertilisers, but it is rarely applied directly in its raw form. Instead, it is typically processed into more soluble and plant-available products. The most important intermediate in this process is phosphoric acid (H_3PO_4), which is produced by reacting phosphate rock with sulphuric acid in what is known as the wet process. Although phosphoric acid itself is not commonly used directly as a fertiliser, it serves as the key building block for manufacturing high-analysis phosphate fertilisers.

One of the earliest and most widely used fertiliser types derived from phosphate rock is superphosphate. Single superphosphate (SSP) is produced by reacting phosphate rock with sulphuric acid. It contains approximately 16–20% phosphorus (as P_2O_5) and also includes gypsum (calcium sulphate) as a by-product. While SSP has a relatively low nutrient concentration compared to other fertilisers, it provides both phosphorus and sulphur, making it particularly useful in soils that are deficient in sulphur. Triple superphosphate (TSP), on the other hand, is produced by reacting phosphate rock with phosphoric acid. This results in a much higher phosphorus content, typically around 44–48% P_2O_5. TSP does not contain gypsum, making it more concentrated and efficient to transport and apply.

Another major group of fertilisers derived from phosphate rock is ammonium phosphates. These are produced by reacting phosphoric acid with ammonia. Diammonium phosphate (DAP) is one of the most widely used fertilisers globally and contains both nitrogen (approximately 18%) and phosphorus (around 46% P_2O_5). It is highly water-soluble and provides a balanced supply of key nutrients for crop growth. Monoammonium phosphate (MAP) is produced using a different ammonia-to-acid ratio, resulting in a product with slightly lower nitrogen content (around 10–12%) but higher phosphorus content (approximately 48–52% P_2O_5). MAP is slightly acidic, which makes it particularly suitable for use in alkaline soils.

Phosphate rock can also be used to produce nitrophosphate fertilisers through a process involving nitric acid. In this route, phosphate rock reacts with nitric acid to form phosphoric compounds and calcium nitrate. The resulting products include NP (nitrogen and phosphorus) fertilisers and NPK blends, which contain nitrogen, phosphorus, and potassium. These fertilisers are designed to provide multiple nutrients in a single product and can be tailored to suit specific crop and soil requirements.

In some limited cases, finely ground phosphate rock may be applied directly to soil as a fertiliser. This approach is generally only effective in acidic soils, where the rock can slowly dissolve and release phosphorus over time. However, due to its low solubility and slow nutrient availability, this method is less efficient and is mainly used in specific agricultural settings or in developing regions.

Overall, phosphate rock is processed through several key pathways to produce usable fertilisers. The sulphuric acid route leads to products such as single superphosphate, while the phosphoric acid route produces higher-grade fertilisers including triple superphosphate, diammonium phosphate, and monoammonium phosphate. The nitric acid route is used to manufacture nitrophosphate and NPK fertilisers.

Wet-Process Phosphoric Acid Production

Wet-process phosphoric acid production is the primary industrial method used to convert phosphate rock into phosphoric acid (H_3PO_4), which serves as a key intermediate in the manufacture of fertilisers such as DAP, MAP, and TSP. This process is widely adopted across the fertiliser industry because it is economical, scalable, and well suited to large-volume production. However, compared to thermal phosphoric acid processes, it produces acid of lower purity, which is generally acceptable for fertiliser applications.

The process is based on the chemical reaction between phosphate rock, primarily composed of fluorapatite, and sulphuric acid (H_2SO_4). This reaction produces phosphoric acid as the desired liquid product and gypsum (calcium sulphate dihydrate) as a solid by-product. The overall reaction can be represented as:

$$Ca_5(PO_4)_3F + 5H_2SO_4 + 10H_2O \rightarrow 3H_3PO_4 + 5CaSO_4{\cdot}2H_2O + HF$$

In addition to phosphoric acid and gypsum, fluorine-containing compounds such as hydrogen fluoride (HF) are released and must be carefully managed to prevent environmental emissions.

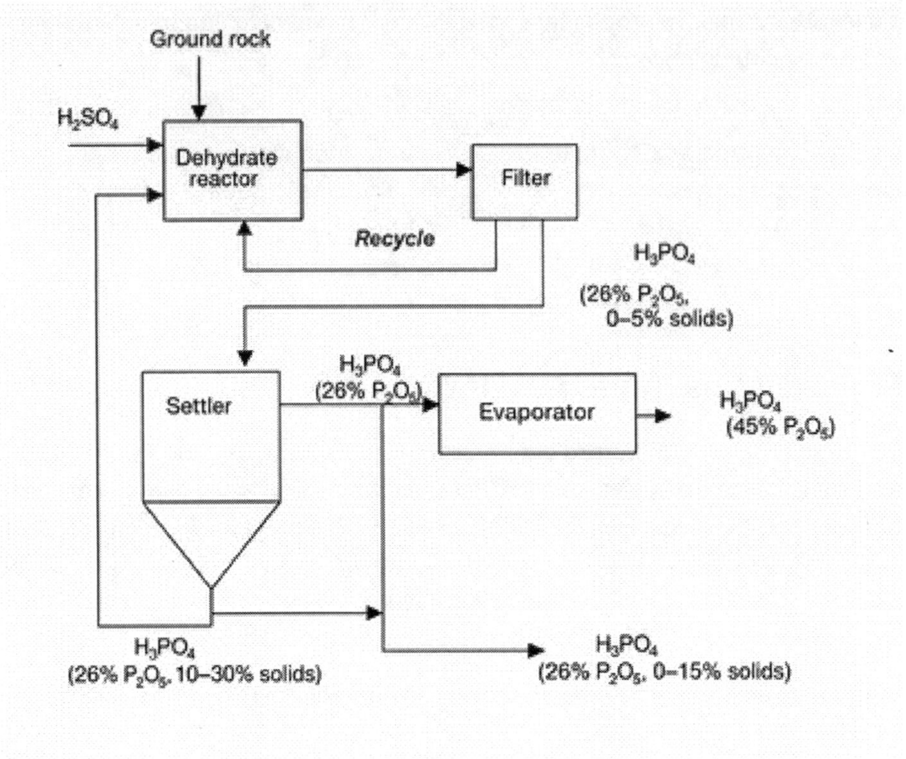

Figure 44: Schematic flow diagram of a wet process plant (phosphoric acid process). Adapted from Valco [147].

The process begins with rock preparation and grinding, where phosphate rock is dried if necessary and then finely ground to increase its surface area and improve reaction efficiency. This is typically carried out using crushers and grinding mills such as ball mills or rod mills, along with drying systems where required.

The ground phosphate rock is then introduced into the acidulation stage, where it reacts with sulphuric acid in a series of reactor tanks. These reactors operate at temperatures of approximately 70–80°C and are equipped with agitation systems to maintain proper mixing. The reaction is continuous and highly exothermic, requiring careful temperature control. Due to the corrosive nature of the process, equipment is constructed from corrosion-resistant materials such as rubber-lined steel or specialised alloys.

The reaction of ground phosphate rock with sulphuric acid—commonly referred to as the acidulation stage—is the core step in wet-process phosphoric acid production. It is a continuous, controlled process in which solid phosphate rock is chemically converted into a slurry containing phosphoric acid and gypsum. This stage must be carefully managed due to its highly exothermic nature and the corrosive environment.

The process begins with the controlled feeding of finely ground phosphate rock into the reactor system. The rock is typically introduced as a dry powder or as a slurry, depending on plant design. At the same time, sulphuric acid is metered into the reactor using an acid dosing system. Maintaining the correct acid-to-rock ratio is critical to ensure complete reaction while minimising excess acid and inefficiencies.

Once both reactants enter the reactor, intensive mixing is applied using agitators. These agitators ensure uniform contact between the phosphate particles and sulphuric acid, promoting efficient reaction and preventing settling of solids. The reactor is usually part of a series of interconnected tanks, allowing the material to move progressively through different stages and ensuring sufficient residence time for the reaction to proceed toward completion.

As the reaction takes place, phosphate minerals react with sulphuric acid to form phosphoric acid in solution and gypsum crystals as a solid phase. The reaction releases a significant amount of heat, making it highly exothermic. To maintain optimal reaction conditions, the temperature is controlled within a typical range of 70–80°C. This is achieved through a combination of heat dissipation, controlled feed rates, and sometimes cooling systems. Proper temperature control is essential, as it influences reaction efficiency, gypsum crystal formation, and downstream filtration performance.

During the reaction, the mixture transitions into a thick slurry composed of liquid phosphoric acid and suspended gypsum crystals. The characteristics of this slurry—particularly the size and structure of the gypsum crystals—are influenced by operating conditions such as temperature, mixing intensity, and residence time. These factors are carefully controlled to produce crystals that can be easily separated in later filtration stages.

The entire system is constructed using corrosion-resistant materials, such as rubber-lined steel or specialised alloys, due to the aggressive chemical environment created by sulphuric acid, phosphoric acid, and fluoride compounds. Reliable materials selection is critical to ensure equipment longevity and safe operation.

As the reaction progresses, gypsum crystals form within the slurry. The size and quality of these crystals are important because they directly affect the efficiency of downstream filtration. Process conditions such as temperature, residence time, and impurity levels are carefully controlled to promote the formation of filterable gypsum crystals.

The crystallisation of gypsum is a critical stage within the wet-process phosphoric acid system, occurring simultaneously with and immediately after the acidulation reaction. As phosphate rock reacts with sulphuric acid, calcium released from the rock combines with sulphate ions to form calcium sulphate (gypsum), which begins to precipitate out of solution as solid crystals within the reaction slurry. The objective of this stage is not only to form gypsum, but to control how it crystallises so that it can be efficiently separated in downstream filtration.

The process begins as the reaction mixture becomes supersaturated with respect to calcium sulphate. Under these conditions, nucleation occurs, where small gypsum crystals initially form. These nuclei then grow as more calcium and sulphate ions deposit onto their surfaces. This transformation from dissolved species to solid crystals occurs within the agitated reactor system, typically across multiple tanks in series, allowing progressive development of crystal size and structure.

A key step in this stage is controlled crystal growth, rather than excessive nucleation. If too many small crystals form, the resulting slurry becomes difficult to filter because fine particles can block filter media and reduce filtration rates. Therefore, operating conditions are adjusted to encourage fewer, larger, and well-formed crystals. These larger crystals settle and filter more easily, improving overall plant efficiency.

Temperature plays a central role in controlling gypsum crystallisation. The process is typically maintained within a range of approximately 70–80°C, which supports stable crystal growth while maintaining the correct phase (usually gypsum dihydrate). If the temperature is too high or fluctuates significantly, it can lead to the formation of undesirable crystal forms or excessively fine particles, both of which negatively impact filtration performance.

Residence time is another critical factor. The slurry must remain in the reactor system long enough to allow crystals to grow to an adequate size. This is achieved by using multiple reactors in series, effectively increasing the time the material spends under controlled conditions. Insufficient residence time leads to incomplete crystal growth and poor filtration characteristics, while excessive time can reduce process throughput.

Impurity levels in the phosphate rock and process stream also influence crystallisation behaviour. Elements such as silica, organic matter, and fluorides can interfere with crystal formation, either by inhibiting growth or by promoting the formation of irregular or fragile crystals. Managing these impurities—through raw material selection and process control—is essential to maintaining consistent crystal quality.

The end result of this stage is a slurry containing well-formed gypsum crystals suspended in phosphoric acid, with properties optimised for efficient solid–liquid separation. Proper control of crystallisation directly impacts filtration efficiency, acid recovery, and overall plant performance, making it one of the most important aspects of wet-process phosphoric acid production.

The resulting slurry, containing both phosphoric acid and gypsum, is then subjected to solid–liquid separation. This is typically achieved using vacuum filtration systems such as belt filters or rotary drum filters. The phosphoric acid passes through as filtrate, while the gypsum is retained as a solid. Multiple washing stages are used to recover as much phosphoric acid as possible from the gypsum before it is discharged.

The solid–liquid separation stage is a critical step in wet-process phosphoric acid production, where the reaction slurry—containing liquid phosphoric acid and solid gypsum crystals—is divided into its two components. The primary objective is to recover as much phosphoric acid as possible while efficiently removing the gypsum for disposal or further handling.

The process begins with the transfer of the reaction slurry from the reactor system to the filtration unit. This slurry is typically hot, acidic, and contains suspended gypsum crystals of varying sizes. It is evenly distributed across the filtration surface, which may be part of a vacuum belt filter or a rotary drum filter, depending on the plant design. Uniform distribution is important to ensure consistent filtration and avoid uneven cake formation.

Once the slurry is applied, a vacuum is created beneath the filter medium, drawing the liquid phase (phosphoric acid) through the filter cloth while retaining the solid gypsum on the surface. As the liquid passes through, it is collected as filtrate in tanks below the filter. This filtrate contains the bulk of the phosphoric acid and is sent forward for further processing, such as concentration.

As filtration progresses, a filter cake of gypsum builds up on the filter surface. This cake still contains residual phosphoric acid trapped within its pores, so additional steps are required to maximise recovery. To achieve this, the process incorporates multiple washing stages, where water or weak acid is sprayed over the gypsum cake. These wash liquids displace the remaining phosphoric acid from the solid material, allowing it to be recovered and returned to the process.

The washing process is typically carried out in counter-current stages, meaning that the cleanest wash water contacts the most washed gypsum, while progressively more contaminated wash liquids are reused upstream. This arrangement improves efficiency and reduces overall water consumption while maximising acid recovery.

After washing, the gypsum cake is further dewatered under vacuum to reduce its moisture content. The partially dried gypsum is then discharged from the filter, usually via mechanical scrapers or belt discharge systems, and transported to disposal areas such as phosphogypsum stacks or, in some cases, for limited reuse applications.

Throughout this stage, careful control of operating conditions is essential. Factors such as vacuum pressure, slurry flow rate, cake thickness, and washing efficiency all influence filtration performance and acid recovery. Proper operation ensures that the majority of phosphoric acid is recovered as filtrate, while the gypsum is removed as a manageable solid by-product.

The filtered phosphoric acid, which typically contains around 25–35% P_2O_5, is then concentrated through evaporation to reach the desired strength of approximately 50–55% P_2O_5 for fertiliser production. This is achieved using evaporators, often operating under vacuum, along with heat exchangers and steam systems to remove water efficiently.

The acid concentration stage is the final major step in wet-process phosphoric acid production, where the relatively dilute phosphoric acid obtained from filtration is concentrated to a strength suitable for fertiliser manufacture. The filtered acid typically contains around 25–35% P_2O_5, which is too low for efficient downstream processing. To increase its concentration to approximately 50–55% P_2O_5, excess water must be removed through controlled evaporation.

The process begins with the transfer of filtered phosphoric acid to the evaporation system, where it is fed into one or more evaporator units. These systems are commonly designed to operate under vacuum conditions, which lowers the boiling point of the liquid. Operating under vacuum allows water to be removed at lower temperatures, reducing the risk of scaling, decomposition, or excessive corrosion, and improving overall energy efficiency.

Once inside the evaporator, the acid is heated indirectly using steam, typically through heat exchangers. The heat causes water within the acid to vaporise, while the phosphoric acid becomes progressively more concentrated. The evaporated water is removed as vapour and directed to condensers, where it is cooled and collected, often for reuse within the plant.

As evaporation continues, the acid concentration gradually increases. The process is carefully controlled to maintain stable operating conditions, as factors such as temperature, pressure, and flow rate directly affect evaporation efficiency and product quality. Multiple-effect evaporators are often used in large plants, where vapour from one stage is reused as the heating medium for the next, significantly improving energy efficiency.

Throughout the process, heat exchangers play a critical role in transferring thermal energy from the steam to the phosphoric acid without direct contact. Steam systems provide the necessary thermal input, and their integration with the rest of the plant—particularly heat recovery systems—helps optimise overall energy consumption.

As the target concentration of approximately 50–55% P_2O_5 is reached, the concentrated acid is withdrawn from the evaporator system and sent to storage or directly to downstream fertiliser production units, such as those producing DAP, MAP, or TSP.

During the process, fluorine compounds released in the reaction are captured and treated using gas scrubbing systems. These systems prevent harmful emissions and often produce by-products such as fluorosilicic acid (H_2SiF_6), which can be used in other industrial applications.

Fluorine recovery and gas treatment is an essential environmental control step in wet-process phosphoric acid production. During the acidulation reaction, fluorine present in the phosphate rock is released in gaseous forms, primarily as hydrogen fluoride (HF) and silicon tetrafluoride (SiF_4). These gases are hazardous and corrosive, and if not properly managed, can cause serious environmental and health impacts. The purpose of this stage is to capture these gases, neutralise or convert them into stable compounds, and prevent their release to the atmosphere.

The process begins with the collection of off-gases from key parts of the plant, particularly the reactors, evaporators, and filtration units. These gases contain fluorine compounds along with water vapour and other minor constituents. The gas streams are directed through ducting systems to dedicated gas treatment units, ensuring that emissions are contained and controlled at the source.

The primary treatment step involves wet scrubbing, where the fluorine-containing gases are brought into contact with a liquid—usually water or a weak acidic solution—inside scrubbers. As the gases pass through the scrubber, they dissolve into the liquid phase. Hydrogen fluoride readily dissolves in water, while silicon tetrafluoride reacts with water to form fluorosilicic acid (H_2SiF_6). This conversion is beneficial because it transforms hazardous gaseous fluorine into a stable liquid compound that can be handled more safely.

Scrubbers are typically designed as packed towers or spray towers, where large contact areas between gas and liquid phases are created to maximise absorption efficiency. In some systems, multiple scrubbing stages are used to ensure that fluorine removal is as complete as possible. The liquid used in the scrubbers is continuously circulated, and its composition is monitored to maintain effective absorption conditions.

The resulting liquid stream, now containing dissolved fluorine compounds, is collected as fluorosilicic acid, which is often recovered as a by-product. This material has commercial value and can be used in applications such as water fluoridation, aluminium processing, or chemical manufacturing, depending on purity and local regulations.

After scrubbing, the cleaned gas stream is typically passed through additional gas treatment or polishing systems to remove any remaining traces of contaminants before being safely discharged to the atmosphere. These systems may include mist eliminators or secondary scrubbers to ensure compliance with environmental standards.

Throughout this process, careful monitoring and control are required. Parameters such as gas flow rate, liquid circulation rate, temperature, and pH are managed to maintain high removal efficiency. The equipment used—including scrubbers, pumps, piping, and ducts—is constructed from corrosion-resistant materials due to the aggressive nature of fluorine compounds and acidic conditions.

A significant by-product of the wet process is phosphogypsum ($CaSO_4{\cdot}2H_2O$), which is generated in large volumes. This material contains impurities such as fluorides and trace radionuclides and is typically stored in large stacks. Managing phosphogypsum presents environmental and logistical challenges, as reuse options, such as in construction or soil conditioning, are limited.

There are several variations of the wet process, primarily distinguished by the form in which gypsum is produced. The dihydrate process is the most common, while the hemihydrate process produces more concentrated acid. Some plants use a combined hemihydrate–dihydrate (HDH) process to improve efficiency and recovery.

The overall process requires a range of specialised equipment, including crushers and grinding mills, reactor vessels with agitators, vacuum filtration systems, evaporators, gas scrubbers, and extensive piping and pumping systems. Supporting infrastructure such as storage tanks and process control systems is also essential.

In terms of advantages, the wet process is cost-effective, well suited to large-scale operation, and easily integrated with downstream fertiliser production. However, it has limitations, including the production of lower-purity acid, the generation of large quantities of gypsum waste, and the need for corrosion-resistant materials to withstand the aggressive process environment.

Phosphogypsum

Phosphogypsum (PG) is the calcium sulphate by-product generated during wet-process phosphoric acid production and is produced in very large quantities, presenting both a major disposal challenge and a potential resource for the fertiliser and construction industries. The sheer global volume of PG, estimated in the literature to exceed hundreds of millions of tonnes annually, underpins the motivation to identify and scale reuse pathways, while regulatory, environmental, and radiological concerns constrain reuse options [148-152]. PG contains impurities such as fluorides, residual acids, and trace radionuclides (notably Ra-226), which complicate processing and necessitate treatment, washing, or stabilization steps to meet safety requirements for specific end-uses; regulators frequently impose dose, leaching, and material-compatibility limits before PG can be employed in construction, agriculture, or consumer products [28, 150, 151]. Because of these constraints, much PG remains stored in large stockpiles or disposed of in landfills or tailings facilities, highlighting the ongoing gap between potential valorisation and actual utilisation [28, 148, 153-155].

In the construction sector, PG is frequently considered as a substitute for natural gypsum in plaster products, cement additives that modify setting, and other building materials, and it can be used as a road-base stabiliser or filler in concrete and plasterboard formulations. The appeal lies in the chemical similarity of PG to natural gypsum, which makes substitution technically feasible; however, the presence of impurities and radionuclides requires rigorous regulatory approvals and material characterisation to ensure compliance with safety standards and performance criteria, leading to regional variation in uptake and often limiting PG's share in construction materials [28, 156, 157]. Several reviews and studies detail how PG incorporation can reduce the need for virgin gypsum and enable substitution in cementitious systems when processing steps—such as washing to reduce soluble impurities and stabilization to immobilise contaminants—are implemented and ensure that leaching risk remains within permitted limits; complementary life-cycle analyses and regulatory guidance further shape feasible PG use in construction [28, 158, 159]. In practice, the commercial deployment of PG in road base and cementitious products frequently hinges on achieving acceptable ettringite formation and controlling strength development, which can be influenced by interactions between PG impurities and cement hydration, underscoring the need for careful mix design and quality control [28, 156, 158].

In agriculture, PG functions as a soil amendment akin to natural gypsum, contributing calcium and sulphur, improving soil structure, reducing compaction, enhancing water infiltration, and aiding in reclamation of sodic or saline soils. While these benefits are well established in the broader gypsum literature, the agricultural use of PG is tightly regulated due to potential contaminants, including radionuclides and trace metals; trials and regulatory frameworks emphasise that contaminant concentrations must remain within acceptable limits, and many agricultural applications are restricted to controlled environments or carefully monitored settings [28, 155, 157]. Recent reviews highlight PG's potential to improve soil physicochemical properties and plant growth when applied in combination with other amendments, yet they also stress the necessity of assessing heavy metal and radionuclide content and ensuring compatibility with crop uptake and soil health objectives [28, 155, 160]. The use of PG in land reclamation and soil amendment thus occupies a regulatory and technical niche where site-specific assessments govern viability, and where co-application with organic or inorganic amendments can mitigate some risks while enhancing agronomic benefits [28, 155, 160].

Beyond direct substitution roles, PG serves as a feedstock or additive in several industrial processes. In fertilizer production, PG can contribute to the manufacture of ammonium sulphate fertiliser or function as a calcium and sulphate source in various formulations, albeit with constraints related to contaminants and market acceptability; in some cases, PG has been explored as a raw material for sulfuric-acid production through thermal decomposition or as a filler in paints and polymers, with commercial viability largely dependent on impurity profiles, regulatory acceptance, and end-user performance requirements [28, 149, 153, 158]. The exploration of PG as a component in composite materials and polymer fillers is supported by studies that demonstrate the stabilising and pozzolanic contributions of PG when used in conjunction with cementitious matrices, although leaching of contaminants and long-term immobilisation are critical considerations that must be addressed in product design and environmental risk assessments [28, 158].

In civil engineering and infrastructure contexts, PG has been proposed for use in embankments, fills, and soil-stabilisation applications for road construction. These applications can harness the large volumes of PG, but they require careful design strategies such as encapsulation, blending with other materials, and robust containment to manage contaminant mobility and prevent environmental releases. The need for encapsulation or blending strategies is repeatedly emphasised in the literature as essential for reducing leaching and ensuring structural performance, with regulatory and environmental risk management playing central roles in determining whether such applications can be scaled up [28, 153, 157, 161]. In addition, environmental considerations in civil applications extend to mine reclamation and site rehabilitation contexts, where PG-inclusive materials may contribute to capping, backfilling, or lightweight fill schemes, provided that stored contaminants remain controlled and monitored over time [28, 157, 161, 162].

In environmental applications, PG has been investigated for pollutant capture and remediation, including wastewater treatment through heavy-metal or phosphate removal and land reclamation projects where PG-containing materials can assist in stabilising contaminated soils or providing structural barriers. The adsorption and immobilisation characteristics of PG and related calcium sulphate-based materials underpin these roles, and the literature emphasizes the importance of contaminant control, long-term stability, and performance under field conditions. Reviews and primary studies alike discuss the use of PG to immobilise metals and phosphates, with attention to potential radiological hazards and the need for site-specific risk assessments to ensure environmental protection while delivering remediation outcomes [28, 155, 160, 163]. In parallel with remediation, PG-derived materials are explored in the area of rare earth element (REE) recovery and advanced ceramics, where the presence of REEs in PG provides a potential feedstock for value-added processing, though economic and technical barriers remain for large-scale deployment [151, 158, 160, 164].

Emerging and advanced applications advocate value recovery from PG through processes such as REE extraction, carbonation to form calcium carbonate or ammonium sulphate fertilisers, syntheses of calcium sulphate whiskers for reinforcement, and the development of PG-based geopolymers and other cementitious systems. The carbonation route and co-precipitation strategies have been studied as avenues to convert PG into marketable products while immobilising impurities; several studies highlight the potential to produce $CaCO_3$ with REE enrichments and to leverage CaSO4 hydration chemistry in engineered materials, though techno-economic barriers and energy intensity remain key challenges for scaling these pathways to commercial reality [28, 151, 155, 158, 160, 165]. Reviews also point to synergies with other industrial wastes (e.g., slag, fly ash) to enhance immobilisation and strength in PG-containing binders, offering potential routes to more sustainable construction materials and reduced environmental footprint, provided that long-term leaching is demonstrated to meet regulatory limits [28, 157, 158, 164].

Despite the breadth of potential end-uses, the current reality remains that the majority of PG is still stored or disposed, driven by the combination of very large generation rates, impurity burdens, and regulatory constraints that govern leachability, radioactivity, and product performance. Analyses of stockpile volumes, regulatory limits, and environmental risks consistently indicate that substantial volumes of PG will require ongoing management, monitoring, and risk mitigation in the foreseeable future. Several studies underscore the imperative of controlled storage, containment, and monitoring programs to prevent leaching and structural instability, reinforcing that safe and compliant reuse will depend on integrated waste management, regulatory alignment, and continued research into stabilization, remediation, and valorisation technologies [28, 148, 150, 151, 153-155, 161].

PG presents a spectrum of reuse opportunities spanning construction, agriculture, industry, civil engineering, and environmental applications, alongside substantial regulatory and environmental hurdles that shape the pace and scale of adoption. The literature consistently demonstrates that when PG is properly processed, treated, and characterised, it can substitute or complement traditional materials and processes, contributing to resource efficiency and circular economy objectives, while ensuring radiological and chemical safety through rigorous assessment, encapsulation, and monitoring strategies. The balance between opportunity and risk hinges on site-specific evaluation, adherence to regulatory limits (especially Ra-226 equivalents and leachate concentrations), and robust design approaches that mitigate contaminant mobility, a nuance repeatedly highlighted across the reviewed sources [28, 148, 150, 151, 153, 155-158, 160].

Process Variations

In wet-process phosphoric acid production, the form in which calcium sulphate (gypsum) crystallises determines both the process configuration and the quality/concentration of the acid produced. The three main industrial variants are the dihydrate (DH), hemihydrate (HH), and hemihydrate–dihydrate (HDH) processes. Each differs in operating conditions, crystal form, equipment design, and overall efficiency.

1. Dihydrate Process (DH) – Most Common

Process Overview

The dihydrate process produces gypsum in the form of calcium sulphate dihydrate ($CaSO_4 \cdot 2H_2O$). It is the most widely used method globally due to its simplicity, reliability, and tolerance to a wide range of phosphate rock qualities.

Key Reaction

Phosphate rock reacts with sulphuric acid to form phosphoric acid and gypsum (dihydrate crystals).

Operating Conditions

- Temperature: ~70–80°C
- Moderate acid concentration
- Atmospheric or slightly elevated pressure

Process Steps

The ground phosphate rock is reacted with sulphuric acid in a series of agitated reactors, forming a slurry of phosphoric acid and gypsum crystals. The conditions are controlled to promote the formation of stable dihydrate crystals, which are relatively large and easier to filter.

The slurry is then sent to filtration, where gypsum is separated from the acid. Multiple washing stages are used to recover phosphoric acid from the gypsum cake. The resulting acid is typically weak, with a concentration of around 25–35% P_2O_5, and must be further concentrated by evaporation.

Equipment

- Reactor tanks (series configuration)
- Agitators
- Vacuum filters (belt or drum)
- Wash systems
- Evaporators (for concentration)

Advantages

- Simple and well-proven technology
- Handles lower-quality or variable phosphate rock
- Produces easily filterable gypsum

Limitations

- Produces lower concentration acid
- Requires significant evaporation (higher energy use downstream)
- Generates large volumes of phosphogypsum

2. Hemihydrate Process (HH) – Higher Concentration Acid

Process Overview

The hemihydrate process produces calcium sulphate in the form of hemihydrate ($CaSO_4 \cdot 0.5H_2O$), rather than dihydrate. This allows for the production of more concentrated phosphoric acid directly from the reactor.

Operating Conditions

- Temperature: ~90–110°C
- Higher acid concentration
- More controlled environment

Process Steps

Phosphate rock reacts with sulphuric acid under higher temperature and acid strength conditions. These conditions favour the formation of hemihydrate crystals instead of dihydrate. The reaction produces a slurry of concentrated phosphoric acid and hemihydrate solids.

Because less water is incorporated into the crystal structure, more water remains in the liquid phase, resulting in a higher acid concentration, typically around 40–50% P_2O_5 directly after filtration.

The slurry is filtered to separate the hemihydrate solids from the acid. However, hemihydrate crystals are generally more difficult to filter and can be less stable than dihydrate crystals, requiring tighter process control.

Equipment

- High-temperature reactor systems
- Specialised filtration units
- Corrosion-resistant materials (more severe conditions)

Advantages

- Produces higher concentration acid (less evaporation required)
- Lower energy consumption in concentration stage
- Potentially reduced gypsum volume

Limitations

- More complex and sensitive process control
- Poorer filtration characteristics
- Less tolerant of impurities in phosphate rock

3. Hemihydrate–Dihydrate Process (HDH) – Combined Approach

Process Overview

The HDH process combines the advantages of both the hemihydrate and dihydrate processes. It is designed to maximise phosphoric acid recovery while improving efficiency and product quality.

Concept

- First stage: hemihydrate formation (high acid concentration)
- Second stage: conversion to dihydrate (improved filtration and recovery)

Process Steps

In the first stage, phosphate rock is reacted with sulphuric acid under hemihydrate conditions (higher temperature and acid concentration), producing a slurry of concentrated phosphoric acid and hemihydrate crystals.

The hemihydrate solids are then transferred to a second stage, where they are recrystallised into dihydrate crystals by adding water or dilute acid under controlled conditions. During this conversion, additional phosphoric acid trapped in the crystals is released into the liquid phase.

The final dihydrate crystals are then filtered and washed, allowing for high recovery of phosphoric acid and improved filtration performance compared to the pure hemihydrate process.

Equipment

- Two-stage reactor system (HH + DH)
- Conversion vessels
- Advanced filtration systems
- Integrated washing circuits

Advantages

- Higher acid recovery efficiency
- Produces relatively high concentration acid
- Improved filtration compared to HH alone
- Better utilisation of phosphate rock

Limitations

- More complex plant design
- Higher capital cost
- More sophisticated control requirements

Table 4: Wet-process phosphoric acid production key comparison.

Feature	Dihydrate (DH)	Hemihydrate (HH)	HDH
Crystal form	$CaSO_4 \cdot 2H_2O$	$CaSO_4 \cdot 0.5H_2O$	Both (converted)
Temperature	70–80°C	90–110°C	Dual-stage
Acid concentration	Low (25–35%)	Medium–High (40–50%)	Medium–High

Feature	Dihydrate (DH)	Hemihydrate (HH)	HDH
Filtration	Easy	Difficult	Improved
Energy use	Higher (evaporation)	Lower	Optimised
Complexity	Low	Medium–High	High

The following provides three case study–style scenarios that illustrate when each wet-process phosphoric acid route (DH, HH, HDH) would be selected in industry. Each scenario links technical, economic, and operational drivers to process choice.

Case Study 1: Dihydrate (DH) Process – Developing Region, Variable Ore Quality

Scenario

A fertiliser company is establishing a new phosphoric acid plant in a developing country where phosphate deposits are low-grade and variable in composition, with relatively high levels of clay and impurities. The project has limited capital funding, and the operator prioritises robustness and reliability over maximum efficiency.

Process Selection Rationale

The dihydrate (DH) process is selected because it is the most forgiving and well-proven technology. It can handle fluctuations in ore quality without major operational instability, and it produces large, stable gypsum crystals that are easy to filter—even when impurities are present.

Operational Context

The plant operates at standard conditions (~70–80°C), using multiple agitated reactors and conventional vacuum filtration. Although the resulting phosphoric acid is relatively dilute (25–35% P_2O_5), the facility includes a dedicated evaporation unit to increase concentration for downstream fertiliser production.

Outcome

- Reliable long-term operation with minimal process upsets
- Lower technical risk and easier operator training
- Higher downstream energy consumption due to evaporation
- Large phosphogypsum stacks due to high by-product volume

Why DH was chosen:

Robustness, simplicity, and ability to process lower-quality, inconsistent phosphate rock.

Case Study 2: Hemihydrate (HH) Process – Energy-Constrained, High-Quality Ore

Scenario

A large, modern fertiliser complex is built in a region where energy costs are high, but high-grade phosphate rock is readily available. The facility is designed to produce high-analysis fertilisers (DAP/MAP) and aims to minimise energy consumption, particularly in evaporation.

Process Selection Rationale

The hemihydrate (HH) process is selected because it produces higher concentration phosphoric acid (40–50% P_2O_5) directly from the reactor. This significantly reduces the need for energy-intensive evaporation.

Operational Context

The plant operates at higher temperatures (90–110°C) and uses specialised reactors and filtration systems. Tight process control is required to manage crystal formation and maintain stable operation. Because the ore is high quality and relatively low in impurities, the process performs efficiently despite its sensitivity.

Outcome

- Reduced steam and energy consumption
- Smaller evaporation section required
- More complex operation and tighter control requirements
- Filtration challenges due to finer, less stable crystals

Why HH was chosen:

To reduce energy costs and maximise efficiency where high-quality feedstock supports stable operation.

Case Study 3: Hemihydrate–Dihydrate (HDH) Process – World-Scale, Efficiency-Driven Plant

Scenario

A multinational fertiliser company develops a world-scale integrated complex in a resource-rich region (e.g. Middle East or North Africa). The project has strong financial backing and aims to maximise resource efficiency, acid recovery, and long-term profitability, while maintaining high production capacity.

Process Selection Rationale

The HDH process is selected because it combines the benefits of both HH and DH processes. It enables:

- High acid concentration (like HH)
- Efficient filtration and recovery (like DH)

Operational Context

The plant uses a two-stage reactor system. In the first stage, hemihydrate crystals form under high-temperature conditions, producing concentrated acid. In the second stage, these crystals are converted into dihydrate form, releasing additional phosphoric acid and improving filtration characteristics.

Advanced control systems and integrated washing circuits are used to maximise recovery and maintain consistent product quality.

Outcome

- Higher overall phosphoric acid recovery
- Reduced losses in gypsum waste
- Lower energy consumption compared to DH
- More complex plant design and higher capital cost

Why HDH was chosen:

To achieve maximum efficiency, recovery, and performance in a large-scale, capital-intensive operation.

The dihydrate process remains the industry standard due to its simplicity and robustness, but the hemihydrate and HDH processes offer efficiency advantages, particularly in reducing energy consumption and improving acid concentration. Modern plants increasingly adopt HDH or advanced variants to balance performance, recovery, and operational reliability.

Superphosphate (SSP, TSP) Manufacturing

Superphosphate fertilisers are produced by chemically treating phosphate rock to convert insoluble phosphorus into forms that are readily available for plant uptake. The two principal types are Single Superphosphate (SSP) and Triple Superphosphate (TSP). Although both are derived from the same raw material, they differ in the acid used, processing approach, and final nutrient concentration, which in turn influences their applications and economic value.

Single Superphosphate (SSP) is manufactured by reacting finely ground phosphate rock with sulphuric acid (H_2SO_4).

Phosphate rock (fluorapatite) reacts with sulphuric acid:

$$Ca_3(PO_4)_2 + 2H_2SO_4 + 4H_2O \rightarrow Ca(H_2PO_4)_2 + 2CaSO_4 \cdot 2H_2O$$

This reaction converts insoluble calcium phosphate into monocalcium phosphate, a water-soluble compound that plants can readily absorb. At the same time, gypsum (calcium sulphate) is formed and remains in the final product, contributing an additional source of sulphur. The overall reaction produces a mixture of monocalcium phosphate and gypsum in a moist, semi-solid form.

The SSP production process begins with rock preparation, where phosphate rock is crushed and ground to increase surface area and improve reaction efficiency. The ground rock is then mixed with sulphuric acid in an acidulation unit, typically a mixer or reactor, where an exothermic reaction produces a thick paste. This material is transferred to a curing area, often referred to as a den, where it solidifies and continues reacting over a period of days or weeks. This curing stage is critical to ensure complete chemical conversion. After curing, the material may be granulated to improve handling and application characteristics, followed by drying and screening to achieve the desired particle size. Oversized or undersized material is recycled within the process.

The equipment used in SSP production typically includes crushers, grinding mills, acidulation mixers or reactors, curing dens, optional granulators, dryers, screening units, and conveying and storage systems. The final product generally contains around 16–20% P_2O_5 and includes gypsum, making it suitable for soils that require both phosphorus and sulphur. The process is relatively simple and low cost, but it produces a lower nutrient concentration product and requires extended curing time, which limits throughput and increases handling requirements.

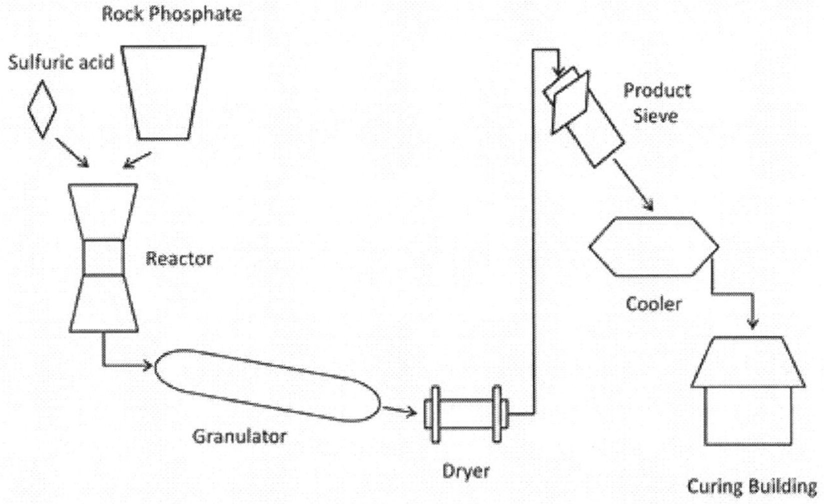

Figure 45:Diagram of the main operations units involved in superphosphate manufacture and further granulation. Adapted from Erro et al. (2016) [166].

Triple Superphosphate (TSP), by contrast, is produced by reacting phosphate rock with phosphoric acid (H_3PO_4) rather than sulphuric acid.

Key reaction:

$$Ca_3(PO_4)_2 + 4H_3PO_4 \rightarrow 3Ca(H_2PO_4)_2$$

This eliminates the formation of gypsum and results in a much higher concentration phosphorus fertiliser. The reaction produces primarily monocalcium phosphate, yielding a product with significantly higher nutrient density.

The TSP manufacturing process also begins with rock preparation, where phosphate rock is crushed and finely ground. The ground rock is then reacted with phosphoric acid in a reactor, producing a slurry of monocalcium phosphate. This slurry is fed into a granulator, typically a rotary drum, where granules are formed and grown using recycled seed particles. The granules are subsequently dried using hot air to remove moisture and improve strength, then cooled and screened to achieve the required size distribution. Off-size material is recycled back into the granulation stage, and the final product may be coated with anti-caking agents before storage and distribution.

The equipment used in TSP production includes crushers, grinding mills, acidulation reactors, granulators, dryers, coolers, screening systems, recycle conveyors, and coating drums. The final product typically contains around 44–48% P_2O_5 and does not contain gypsum, making it a high-analysis fertiliser with excellent transport and application efficiency. However, the process requires phosphoric acid as a feedstock, which adds complexity and cost, and is usually integrated with upstream phosphoric acid production facilities.

In comparison, SSP is a lower-cost, simpler fertiliser that provides both phosphorus and sulphur, but with lower nutrient concentration and slower production rates due to curing requirements. TSP, on the other hand, is a more concentrated and efficient fertiliser, produced via a continuous process, but requires more complex infrastructure and higher capital investment.

In terms of plant integration, SSP facilities are often standalone operations located close to phosphate rock sources, reflecting their simpler design and lower capital requirements. TSP plants, however, are typically integrated into larger fertiliser complexes that include wet-process phosphoric acid production, enabling efficient use of intermediates and supporting large-scale, high-efficiency fertiliser manufacturing.

Small Scale Production

Both Single Superphosphate (SSP) and Triple Superphosphate (TSP) can be produced on a smaller scale, but there are significant differences in feasibility, complexity, safety, and regulatory requirements. In practice, SSP is far more suitable for small-scale or decentralised production, while TSP is generally not viable at small scale except in very controlled or specialised settings.

Single Superphosphate (SSP) is one of the few fertilisers that can be realistically produced at a small to medium scale, typically in the range of approximately 5,000 to 50,000 tonnes per year. Historically, many countries have operated small regional SSP plants to support local agricultural demand, particularly where decentralised fertiliser supply chains are beneficial. Its suitability for small-scale production stems from the relative simplicity of the process and the availability of key inputs.

The SSP process works effectively at smaller scale because it uses sulphuric acid, which is widely available and easier to handle compared to more complex reagents. The reaction operates at moderate temperatures and pressures and does not require highly sophisticated or tightly controlled equipment. In addition, the process is relatively tolerant of variations in phosphate rock quality, allowing operators to use locally available raw materials without extensive pre-treatment.

A typical small-scale SSP plant includes basic equipment such as a rock crushing and grinding unit, an acidulation mixer or simple reactor (such as a pug mill), and a den or curing area where the material is allowed to solidify and complete the chemical reaction. Depending on product requirements, optional granulation equipment may be included to improve handling characteristics, along with basic drying and screening systems. These facilities are often designed for batch or semi-continuous operation and may involve manual or semi-automated handling systems, contributing to lower overall capital costs compared to more complex fertiliser processes.

Despite its advantages, SSP production at small scale does present several constraints. The handling of sulphuric acid introduces significant safety risks due to its corrosive nature, requiring appropriate materials, training, and safety systems. The curing stage, which can take days to weeks, limits throughput and requires sufficient space for storage during processing. Dust generation and acidic emissions must also be managed to meet environmental and workplace safety requirements. Additionally, the final product has a relatively low nutrient concentration, which can increase transportation costs per unit of nutrient.

Overall, SSP is well suited to small-scale production and is commonly manufactured in developing regions, by agricultural cooperatives, and within local fertiliser supply chains where simplicity, flexibility, and lower capital investment are key considerations.

In contrast, Triple Superphosphate (TSP) is generally not practical for small-scale production. The primary limitation is that TSP relies on phosphoric acid as a feedstock, which itself is produced through a large-scale and complex wet-process operation. This creates an inherent dependency on upstream infrastructure that is not feasible to replicate at small scale.

The TSP process is also significantly more complex, involving controlled acidulation, continuous granulation systems, and integrated drying, cooling, and recycling circuits. These operations require precise control systems, well-integrated plant design, and skilled operators to maintain product quality and process stability. As a result, even a relatively small TSP facility requires substantial capital investment in equipment such as rotary drum granulators, dryers, coolers, and air handling systems, making the economics unfavourable at reduced scale.

Environmental and safety requirements further complicate small-scale TSP production. Handling phosphoric acid introduces additional hazards, while upstream phosphoric acid production generates fluoride emissions that must be carefully managed. Dust control and off-gas treatment systems are also essential to ensure compliance with environmental regulations, adding to both capital and operating costs.

In limited cases, TSP production may be undertaken at a smaller scale if phosphoric acid is sourced from a large external supplier and the operation is designed as a modular, controlled system producing specialised or niche products. However, such arrangements are uncommon and typically not economically competitive compared to large integrated plants.

SSP is a practical and widely used option for small-scale fertiliser production due to its simplicity and flexibility, whereas TSP is inherently a large-scale, integrated industrial process that is not well suited to standalone small-scale operations.

Table 5: Key Comparison – Small Scale Feasibility.

Factor	SSP	TSP
Small-scale feasibility	✓ Yes	✗ Generally no
Process complexity	Low	Moderate–high
Acid required	Sulphuric acid	Phosphoric acid
Capital cost	Low–moderate	High
Integration needed	Minimal	High (with acid plant)
Typical scale	Small to large	Medium to very large

Small-scale Single Superphosphate (SSP) Plant Implementation Example

<u>Process Steps (Small-Scale SSP Production)</u>

A small SSP plant follows the same core chemistry as large plants, but with simpler, often semi-continuous operation.

Step 1: Raw Material Preparation

Phosphate rock is crushed and ground to a fine powder to maximise reaction efficiency. The finer the particle size, the better the acid reaction.

Output: Fine phosphate rock powder

Step 2: Acidulation (Reaction)

The ground rock is mixed with sulphuric acid in a mixer or pug mill reactor. This is an exothermic reaction producing a paste of:

- Monocalcium phosphate (fertiliser)
- Gypsum (by-product)

This stage is the core chemical conversion step

Step 3: Denning / Curing

The paste is discharged into a curing area (den) where it:

- Solidifies
- Completes chemical reaction over days to weeks

This step replaces complex reactors used in other fertilisers

Step 4: Granulation (Optional but Recommended)

The cured material is crushed and fed into a granulator to form uniform particles for easier handling and spreading.

Step 5: Drying and Screening

Granules are:

- Dried to reduce moisture
- Screened for size
- Oversize and fines recycled

Step 6: Storage and Bagging

Final SSP is stored in silos or sheds and then:

- Bagged (25–50 kg) or
- Sold in bulk

<u>Equipment Requirements (Small-Scale Plant)</u>

A typical small SSP plant (5,000–50,000 t/year) uses relatively simple equipment:

Core Equipment

- Crushers and grinding mills
- Acidulation mixer (pug mill or reactor)
- Acid storage tanks (sulphuric acid)
- Den/curing bays

Optional / Upgraded Equipment

- Granulator (disc or rotary drum)
- Dryer (rotary)
- Cooler (optional)
- Vibrating screens
- Conveyors and elevators
- Dust collection systems

Typical SSP granulation systems include mixers, granulators, dryers, coolers, and screening systems

Plant Setup (Layout & Infrastructure)

A small SSP plant is relatively simple and can be set up on 1–3 hectares.

Key Areas

Raw material storage

- Phosphate rock stockpile
- Sulphuric acid tanks

Reaction section

- Mixer/reactor
- Acid dosing system

Curing area (den)

- Covered bays or sheds

Granulation & finishing

- Granulator, dryer, screens

Utilities

- Water supply
- Power (electric motors, fans)
- Basic steam/heat (if drying used)

Storage & dispatch

- Warehouse or silos
- Bagging unit

Capital Cost (CAPEX) – Small Scale SSP

Typical Cost Range

- Small SSP plant: USD $100,000 – $5,000,000 depending on capacity and level of automation

Converted to Major Currencies (Approx.)

Table 6: Typical Costs.

Scale	USD	AUD	EUR
Very small (basic, no granulation)	$100k – $300k	$150k – $450k	€90k – €280k
Small (5k–20k t/year)	$300k – $1.5M	$450k – $2.3M	€280k – €1.4M
Medium-small (20k–50k t/year)	$1.5M – $5M	$2.3M – $7.5M	€1.4M – €4.6M

Costs increase significantly if:

- Granulation is included
- Environmental controls are upgraded
- Automation is added

Operating Costs (OPEX)

Main Cost Drivers

- Phosphate rock: 70–80% of OPEX
- Sulphuric acid
- Labour
- Utilities (power, water)

Typical production cost:

- ~USD $100–250 per tonne (varies widely)

Safety and Compliance Requirements

Even at small scale, SSP plants must manage:

Key Risks

- Sulphuric acid (highly corrosive)
- Dust (respiratory hazard)
- Acid fumes (SO_2, mist)

Required Controls

- Acid-resistant materials (rubber-lined steel)
- Ventilation and scrubbers
- PPE and training
- Spill containment systems

Practical Setup Example

Example: Regional SSP Plant

- Capacity: 20,000 t/year
- CAPEX:
    - ~USD $800,000 – $2M
    - ~AUD $1.2M – $3M
    - ~€750k – €1.8M

Configuration

- Simple crusher + grinder
- Pug mill reactor
- Open curing shed
- Disc granulator + small dryer
- Manual bagging

Suitable for:

- Agricultural cooperatives
- Regional fertiliser supply
- Developing markets

A small-scale SSP plant is one of the most achievable fertiliser manufacturing projects because:

- Simple chemistry and process
- Moderate capital investment
- Flexible operation
- Can use local raw materials

However, success depends on:

- Safe acid handling
- Managing curing time
- Controlling dust and emissions
- Securing reliable phosphate rock supply

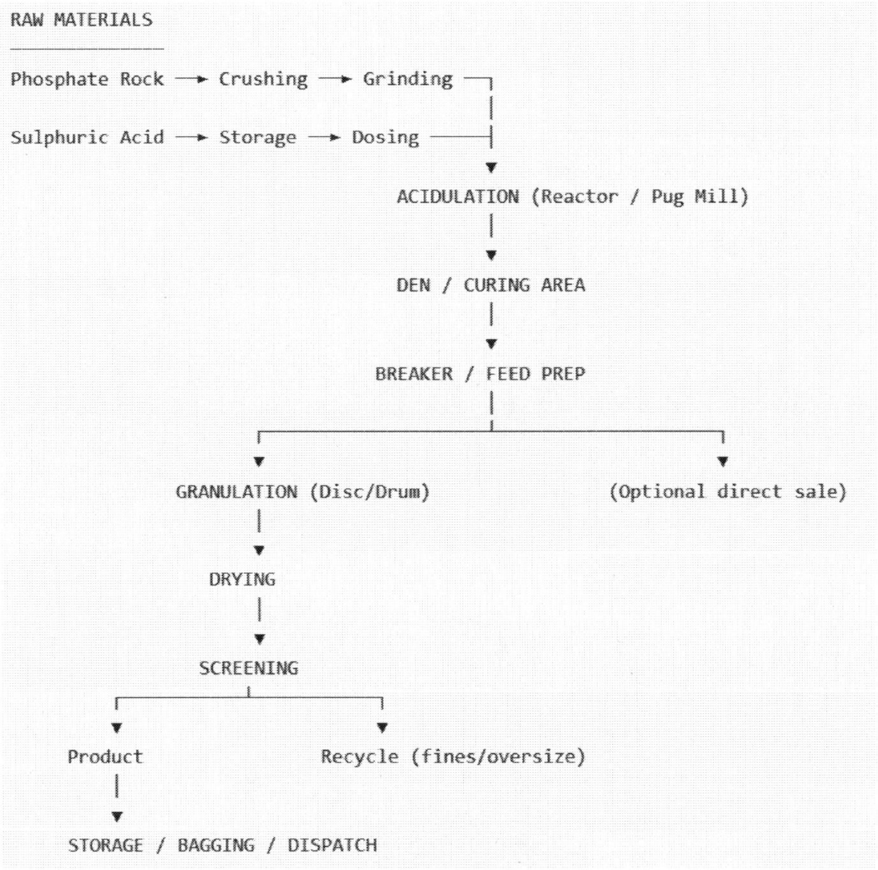

Figure 46: Conceptual Plant Layout.

Plant Layout by Physical Area

A typical site (1–3 hectares) is divided into functional zones:

A. Raw Materials Area

- Phosphate rock stockpile
- Acid storage tanks (bunded)
- Truck unloading

B. Reaction Section

- Crusher + grinding mill
- Acidulation unit (pug mill/reactor)
- Acid dosing system

C. Curing (Den Area)

- Large covered bays or sheds
- Front-end loader access
- Drainage + containment

D. Finishing Section (Optional but Recommended)

- Granulator (disc or drum)
- Dryer
- Screens
- Recycle conveyors

E. Utilities & Environmental Controls

- Power distribution
- Water supply
- Dust collection systems
- Acid mist scrubbers

F. Storage & Loadout

- Product shed or silos
- Bagging line
- Truck loading

DAP and MAP Production

Diammonium Phosphate (DAP) and Monoammonium Phosphate (MAP) are among the most widely used phosphate fertilisers in modern agriculture. Both are ammonium phosphate fertilisers, produced by reacting phosphoric acid (H_3PO_4) with ammonia (NH_3). They provide essential nutrients, phosphorus (P) and nitrogen (N), in highly soluble forms that are readily available to plants. The key difference between DAP and MAP lies in the ammonia-to-acid ratio, which determines their chemical composition, nutrient content, and behaviour in soil.

The production of DAP and MAP begins with wet-process phosphoric acid, typically containing around 50–55% P_2O_5. This acid is produced upstream from phosphate rock and serves as the primary phosphorus source. Ammonia, usually supplied as a liquid or gas, is then introduced into the process. The reaction between ammonia and phosphoric acid is highly exothermic and must be carefully controlled to maintain the desired product composition and prevent excessive temperatures.

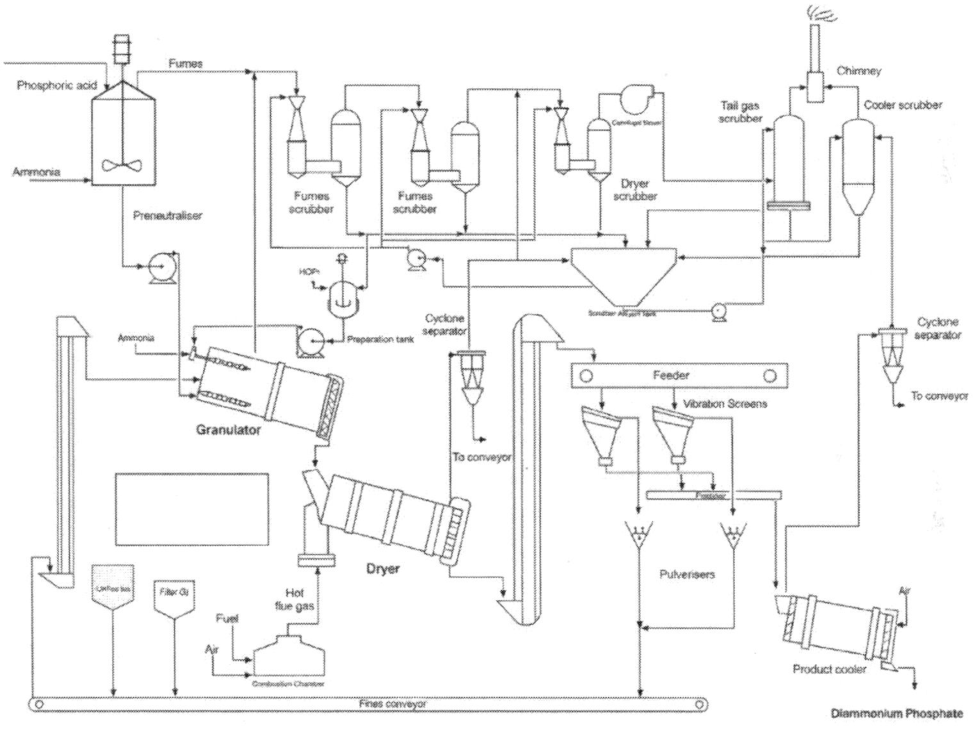

Figure 47: Process flow diagram of DAP fertilizer manufacturing. Adapted from Engineers Guide [167].

Figure 47 shows a typical industrial Diammonium Phosphate (DAP) granulation plant, integrating reaction, granulation, drying, gas scrubbing, and product finishing into a continuous process. Below is a step-by-step explanation aligned to the flow shown.

1. Pre-neutralisation (Reaction Stage)

The process begins in the preneutraliser, where phosphoric acid and ammonia are introduced and partially reacted. This vessel is agitated to ensure proper mixing and control of the highly exothermic reaction.

- The goal here is partial neutralisation (not full DAP formation yet)
- Produces a slurry of ammonium phosphate
- Helps stabilise the reaction before granulation

Fumes generated (containing ammonia and fluorides) are extracted and sent to fume scrubbers for treatment.

2. Slurry Conditioning and Feed to Granulator

The slurry from the preneutraliser is transferred to a preparation tank, where it may be adjusted for:

- Temperature
- Composition
- Flow consistency

A cyclone separator may be used to remove entrained solids or droplets from gas streams.

The conditioned slurry is then pumped into the granulator.

3. Granulation (Core Particle Formation)

In the granulator (rotary drum):

- The slurry is sprayed onto recycled seed particles
- Additional ammonia may be injected to complete neutralisation to DAP
- Particles grow through layering and agglomeration

This is the key step where:

Liquid slurry → solid granules

Off-gases from the granulator are directed to scrubbers to remove ammonia and dust.

4. Drying

The wet granules are discharged into a rotary dryer, where:

- Hot gases from a combustion chamber remove moisture

- Fuel + air generate hot flue gas

This step:

- Hardens the granules
- Reduces moisture for stability

Dust and vapours from the dryer are treated in a dryer scrubber system.

5. Product Separation and Recycling

After drying, the material is conveyed to:

- Cyclone separators (remove dust from air streams)
- Screens (vibrating screens) for size classification

The material is split into:

- Product-sized granules → continue forward
- Oversize material → sent to pulverisers (crushers)
- Fines (undersize) → recycled back to granulator

This recycle loop is essential for efficient granulation.

6. Cooling

The correctly sized granules are sent to a product cooler, where:

- Air is used to reduce temperature
- Prevents caking and degradation

Dust-laden air from cooling is cleaned via cyclones or scrubbers.

7. Final Product Handling

The cooled DAP product is:

- Conveyed to storage or bagging
- Prepared for transport

At this stage, the fertiliser has:

- Correct size distribution
- Adequate strength

- Stable moisture content

8. Gas Handling and Environmental Control

Throughout the process, multiple gas streams are generated (from reactor, granulator, dryer, cooler). These are treated via:

Fume Scrubbers

- Capture ammonia and acid vapours
- Prevent atmospheric release

Dryer Scrubber

- Removes dust and residual gases from drying stage

Tail Gas Scrubber

- Final cleaning before discharge

Cooler Scrubber

- Treats cooling air emissions

Cleaned gas is released via the chimney (stack).

9. Key Supporting Systems

Combustion Chamber

- Generates hot gas for drying
- Uses fuel + air

Cyclone Separators

- Remove particulate matter from gas streams

Conveyors

- Move material between stages
- Enable recycle loops

Overall Process Flow (Simplified)

1. Phosphoric acid + ammonia → **preneutraliser**
2. Slurry → **granulator (particle formation)**
3. Granules → **dryer (moisture removal)**
4. Product → **screening (size control)**
5. Recycle → back to granulator
6. Product → **cooler → storage**
7. Off-gases → **scrubbers → chimney**

In the case of DAP production, a higher ratio of ammonia is used. Phosphoric acid reacts with ammonia to form diammonium phosphate according to the reaction:

$$2NH_3 + H_3PO_4 \rightarrow (NH_4)_2HPO_4$$

This results in a fertiliser with a typical nutrient content of approximately 18% nitrogen and 46% P_2O_5. DAP is slightly alkaline in reaction and is widely used because of its high nutrient concentration and versatility across a range of crops and soils.

For MAP production, a lower ammonia-to-acid ratio is applied. The reaction is:

$$NH_3 + H_3PO_4 \rightarrow NH_4H_2PO_4$$

This produces monoammonium phosphate, with a typical composition of around 10–12% nitrogen and 48–52% P_2O_5. MAP is slightly acidic, which makes it particularly suitable for alkaline soils, as it can help improve phosphorus availability under such conditions.

Industrial production of both DAP and MAP typically follows a similar process configuration. Phosphoric acid and ammonia are fed into a reactor or ammoniator, where the neutralisation reaction occurs. The resulting slurry of ammonium phosphate is then directed to a granulation unit, commonly a rotary drum granulator. In this unit, the slurry is sprayed onto recycled seed particles, allowing granules to grow to the desired size through layering and agglomeration.

Following granulation, the material is passed through a rotary dryer, where moisture is removed using hot air. This step is critical to achieve the required product strength and stability. The dried granules are then cooled in a cooler to prevent caking and to stabilise the product for storage. After cooling, the fertiliser is screened to ensure a consistent particle size distribution. Oversized granules are crushed and recycled, while undersized particles are returned to the granulator as seed material.

The final stage involves coating and storage. Anti-caking agents are often applied to improve flowability and prevent moisture absorption during storage and transport. The finished product is then stored in bulk or bagged for distribution.

The equipment used in DAP and MAP production includes ammoniators or reactors, granulators, rotary dryers, coolers, screening systems, crushers, conveyors, and gas handling systems. Because ammonia is used, the plant must also include ammonia storage and handling systems, along with appropriate safety controls.

From an operational perspective, DAP and MAP plants are typically continuous, large-scale operations integrated with phosphoric acid production facilities. They require precise control of temperature, pH, and reactant ratios to ensure consistent product quality. Environmental controls are also critical, particularly for managing ammonia emissions, dust, and fluoride-containing gases.

Small Scale Diammonium Phosphate (DAP) and Monoammonium Phosphate (MAP) Production

DAP and MAP can be produced on a smaller scale, but with an important qualification: small-scale neutralisation is feasible, while full small-scale commercial granulated DAP/MAP production is much less practical.

The core chemistry itself is not especially complex. Both products are made by reacting phosphoric acid with ammonia, and this neutralisation step can be carried out in a relatively small reactor if the feed rates, temperature, and pH are carefully controlled. From a purely chemical standpoint, that means a pilot-scale or limited batch operation is possible.

The difficulty is that commercial DAP and MAP production is not just a reaction step. To make a saleable fertiliser, the process also needs granulation, drying, cooling, screening, recycling of fines, dust control, ammonia handling, and emissions control. Those downstream steps add a great deal of complexity and cost. As a result, while a small plant can make ammonium phosphate material, producing a uniform, hard, low-dust, market-grade granule at small scale is much harder.

MAP is generally more feasible than DAP at smaller scale. It uses a lower ammonia-to-acid ratio, is somewhat easier to control, and is less prone to free ammonia issues. DAP, by contrast, requires tighter control because excess ammonia can lead to volatilisation losses, odour, emissions, and product instability. In practical terms, a small operator is more likely to succeed with MAP or with a simple ammonium phosphate slurry/powder than with high-quality granulated DAP.

A small-scale setup would usually make sense only in certain situations: pilot plants, specialty fertiliser production, custom blends, research facilities, or regional operations that already have access to phosphoric acid and safe ammonia handling systems. It is much less attractive as a standalone small commercial plant because the economics are weak unless there is a niche market or very specific local demand.

The biggest constraints are safety, environmental compliance, and product finishing. Ammonia is toxic and corrosive, phosphoric acid is corrosive, and the neutralisation reaction is strongly exothermic. Even at small scale, the operation needs proper containment, ventilation, gas scrubbing, corrosion-resistant equipment, and trained operators. If granulation is included, the capital cost rises sharply because dryers, coolers, screens, recycle circuits, and dust collection become necessary.

Small Scale Design

Small-scale production of Diammonium Phosphate (DAP) and Monoammonium Phosphate (MAP) is technically feasible when approached at a pilot or small industrial scale, typically in the range of approximately 1,000 to 20,000 tonnes per year. However, it is important to understand the practical realities before undertaking such a project. The core chemical step, neutralisation of phosphoric acid with ammonia, is relatively straightforward and can be scaled down with appropriate equipment. In contrast, downstream processing steps such as granulation, drying, and product finishing introduce significant complexity and cost. In addition, the handling of ammonia and the need for environmental controls represent critical safety and compliance challenges. As a result, most small-scale operations focus on MAP rather than DAP, and often produce slurry, powder, or simple granulated products rather than fully commercial-grade fertiliser.

The process begins with the supply of raw materials, namely phosphoric acid (typically 50–55% P_2O_5) and ammonia, supplied either as a gas or liquid. These two inputs represent the primary cost drivers of the operation. The core reaction takes place in a stirred reactor or ammoniator, where phosphoric acid is neutralised with ammonia. The ratio of ammonia to acid determines whether MAP or DAP is produced. MAP is formed using a lower ammonia ratio, while DAP requires a higher ratio. The reactions produce a slurry or paste of ammonium phosphate, and this neutralisation stage is the simplest part of the overall process, although it remains highly exothermic and must be carefully controlled.

Following the reaction, the slurry is transferred to a conditioning stage where its temperature and consistency are stabilised. Cooling may be required to manage the heat generated during neutralisation, and a mixing or holding tank is typically used to ensure a consistent feed to downstream processes. At this point, the material can be used as a liquid or powder fertiliser, or it can be further processed into granules.

Granulation is optional but essential if a marketable solid fertiliser product is required. In this stage, the slurry is sprayed onto recycled seed particles within a granulator, allowing granules to grow through layering and agglomeration. This step mirrors industrial-scale processes but is more difficult to control at small scale. The granulated material is then dried in a rotary dryer using hot air or a burner system to remove moisture and improve product strength. After drying, the product is screened to achieve the desired particle size distribution. Oversized particles are crushed and recycled, while undersized material is returned to the granulator. Finally, the product is cooled to prevent caking and then stored or bagged for distribution.

The equipment required depends on the level of sophistication of the plant. At the most basic level, a small-scale operation producing slurry or powder MAP requires a reactor (such as a stirred tank or ammoniator), phosphoric acid storage, ammonia storage (often cylinders or a small tank), pumps, piping, and basic ventilation. For operations aiming to produce granular fertiliser, additional equipment is required, including a disc granulator, a small dryer, screens, and conveyors. More advanced small-scale plants, approaching commercial operation, include an ammoniator or pipe reactor, rotary drum granulator, rotary dryer with burner, cooler, cyclones or scrubbers for gas treatment, and dust collection systems. These configurations begin to resemble full industrial flowsheets, albeit at reduced capacity.

The physical setup of the plant is typically organised into functional zones. These include raw material storage areas for acid and ammonia, a reaction section housing the reactor and dosing systems, a granulation area with recycle loops if applicable, drying and cooling sections, environmental control systems such as scrubbers and dust collectors, and storage or bagging facilities. The footprint of such a plant can range from approximately 500 to 2,000 square metres for a pilot-scale facility, up to 2,000 to 10,000 square metres for a small industrial plant.

Capital costs vary significantly depending on the level of processing included. A very small setup that performs only the reaction step without granulation may cost in the range of USD $100,000 to $500,000 (approximately AUD $150,000 to $750,000 or €90,000 to €450,000) and produces slurry or powder products. A more practical small-scale plant with basic granulation typically ranges from USD $500,000 to $2 million (AUD $750,000 to $3 million or €450,000 to €1.8 million). For a more advanced small plant with full granulation, drying, scrubbing, and some automation, costs increase to approximately USD $2 million to $8 million (AUD $3 million to $12 million or €1.8 million to €7.5 million). Costs rise rapidly because the required equipment must be corrosion-resistant, specialised, and capable of handling hazardous materials.

A small-scale MAP plant is most realistic when it is designed around phosphoric acid neutralisation with ammonia, with the option to stop at a slurry/powder product or add a simple granulation line for a more marketable fertiliser. Compared with DAP, MAP is generally easier to control at small scale because it uses a lower ammonia ratio and tends to have fewer free-ammonia handling issues. Industrial references describe MAP/DAP manufacture as a sequence of neutralisation, granulation, drying, cooling, screening, recycle, and gas scrubbing, with either a tank pre-neutraliser or pipe reactor feeding a granulator [168].

A practical small-scale layout can be divided into six zones. First is the raw-materials area, with phosphoric acid storage, ammonia storage, unloading, metering, and bunded containment. Second is the reaction area, containing the neutraliser or ammoniator, dosing lines, agitation, and local ventilation. Third is the granulation area if solids are required, where slurry is deposited onto recycle seed material in a disc or small drum granulator. Fourth is the thermal finishing area, with drying and optional cooling. Fifth is the classification and recycle area, where screens separate on-size product from fines and oversize, with fines recycled and oversize crushed. Sixth is the environmental and dispatch area, including scrubbers, dust collection, product storage, and bagging or bulk loading. This layout follows the same functional logic used in conventional MAP/DAP plants, just at reduced scale [169].

For a minimum viable plant, focused on MAP slurry or powder, the equipment list is relatively short. It includes a phosphoric acid tank, ammonia cylinders or a small pressure vessel, transfer pumps, corrosion-resistant piping, a stirred neutralisation tank or simple ammoniator, a slurry hold tank, ventilation, and a small scrubber for ammonia-bearing off-gas. This setup avoids the most expensive part of the plant, which is granulation and drying, and is the lowest-cost route to product. Industrial sources consistently show that the neutralisation step itself is simple compared with downstream finishing [168].

A more complete small MAP granulation plant adds a disc granulator or small drum granulator, a rotary dryer, optional cooler, vibrating screens, a small crusher for oversize recycle, bucket elevators or conveyors, cyclones, and a tail-gas scrubber. FEECO and FerTech both describe this standard phosphate-fertiliser equipment train: pre-neutralisation, granulation, drying, cooling, screening, and recycle, with gas handling throughout. Rotary dryers and coolers are specifically identified as common finishing equipment for phosphate fertilisers, including MAP and DAP [169].

For a conceptual plant footprint, a pilot-scale unit can often fit within roughly 500–2,000 m^2, while a small industrial setup with granulation, drying, bagging, and environmental controls is more likely to need roughly 2,000–10,000 m^2, depending on storage philosophy and whether curing or additional warehouse space is included. These ranges are illustrative rather than published standards, but they are consistent with the equipment arrangement implied by standard granulation flowsheets and the space needed for materials handling, emission control, and product storage. This should be treated as a planning assumption, not a code requirement [169, 170].

A sensible equipment list for a complete small MAP plant would therefore include: phosphoric acid storage tank and unloading skid; ammonia storage and metering system; neutraliser or pipe reactor; slurry tank; disc or drum granulator; dryer with burner or hot-air system; optional cooler; screens; recycle conveyors/elevators; crusher for oversize; cyclones; ammonia scrubber; dust collector; product bin; bagging system; and basic utilities such as pumps, controls, instrument air, and electrical distribution. Plants using a pipe reactor instead of a tank neutraliser can reduce recycle ratio and moisture load, but they also demand tighter control and more robust materials selection [170].

For budgeting, the clearest public cost references available are not turnkey quotations but industry project-report ranges. IMARC's 2026 MAP plant report [171] explicitly covers plant setup, machinery, raw materials, investment opportunities, cost, and revenue, and its renewable MAP analysis gives a useful operating-cost structure, with raw materials at roughly 60–70% of OpEx and utilities at roughly 10–15% of OpEx. Based on those sources and on the typical equipment scope above, a reasonable conceptual CAPEX envelope for a small MAP project is:

- Reaction-only MAP slurry/powder unit: about USD 0.10–0.50 million, which is roughly AUD 0.15–0.77 million and EUR 0.09–0.46 million.
- Basic small granulation plant: about USD 0.50–2.00 million, which is roughly AUD 0.77–3.07 million and EUR 0.46–1.84 million.
- Advanced small plant with dryer, scrubber, and stronger automation: about USD 2.00–8.00 million, which is roughly AUD 3.07–12.28 million and EUR 1.84–7.36 million [171].

These currency conversions use approximate current cross-rates of 1 USD ≈ 1.535 AUD and 1 USD ≈ 0.920 EUR. They are suitable for concept screening, not investment approval.

A useful CAPEX breakdown by area for a small granular MAP plant is: 15–20% raw-material handling and storage, 15–20% reaction/neutralisation, 20–25% granulation, 15–20% drying and cooling, 10–15% screening/recycle/product handling, and 10–20% utilities, controls, and environmental systems. The exact split depends heavily on whether ammonia is stored as cylinders or bulk, whether a cooler is installed, and how stringent the gas scrubbing and dust collection requirements are. The big cost driver is usually not the reactor but the finishing train plus emissions control [169, 170].

On the operations side, small plants usually run in batch or semi-continuous mode, with the critical control variables being the acid-to-ammonia ratio, slurry temperature, moisture at granulation, and recycle load. The same industrial sources note that neutralisation heat can push slurry toward boiling and that granulation performance depends on the neutralisation level and water balance. In practice, that means MAP is the easier product to make consistently, while DAP becomes more demanding because additional ammoniation and tighter moisture control are needed in the granulator [169, 170].

The biggest safety and compliance requirements are ammonia handling, corrosive acid service, and emissions management. Even a small plant needs compatible materials of construction, local ventilation, ammonia detection, secondary containment, an off-gas scrubber, dust control, and trained operators. The technical literature on ammonium phosphate granulation repeatedly highlights ammonia-bearing fume streams and the need to scrub them before release [169, 170].

The practical takeaway is that the most viable "complete" small-scale MAP project is usually a simple MAP granulation plant, not a DAP plant and not a full world-scale-style complex. If capital is tight, the most economical route is a reaction-only or powder/slurry MAP unit. If a saleable solid is required, the next step up is a disc-granulated MAP plant with modest drying and robust gas scrubbing. The plant concept is commercially plausible, but its economics depend heavily on feedstock cost, especially phosphoric acid and ammonia, and on how much product finishing and environmental control is required [169-171].

Operating costs are dominated by raw materials, particularly phosphoric acid and ammonia, followed by energy for drying and labour. Production costs can vary widely depending on input prices, but a general estimate for DAP production is in the range of USD $600 to $700 per tonne under typical conditions. Small plants are often less efficient than large-scale facilities, resulting in higher unit costs.

Operationally, small plants are usually run in batch or semi-continuous mode, while larger facilities operate continuously. Key control variables include pH (which determines whether MAP or DAP is produced), temperature (due to the exothermic nature of the reaction), ammonia flow rate, and moisture levels in the granulation stage. A small plant may require a workforce of approximately 5 to 20 operators, supported by one or two engineers or technicians.

Safety and compliance are critical considerations. Ammonia is toxic and often stored under pressure, while phosphoric acid is highly corrosive. The process also generates dust and gaseous emissions that must be controlled. Essential systems include gas scrubbers to remove ammonia and other emissions, adequate ventilation, personal protective equipment (PPE), operator training, and spill containment measures.

In practical terms, MAP is generally the preferred product for small-scale production. It is easier to control, involves lower ammonia usage, and presents fewer operational risks. DAP, while widely used industrially, is more difficult to produce at small scale due to tighter process control requirements, higher ammonia handling risks, and more complex granulation behaviour.

Impurities and Environmental Issues (e.g., Phosphogypsum)

Impurities and environmental considerations are central to phosphate fertiliser production, particularly in processes that begin with phosphate rock and employ the wet-process route to phosphoric acid. The impurities present in natural phosphate rock are diverse, including mineral matter such as silica, clays, iron and aluminium oxides, carbonates, and organic matter, as well as trace elements such as fluorine, cadmium, uranium, and other heavy metals. Across processing, these impurities can be partially transferred into the phosphoric acid, partitioned into the by-product phosphogypsum (PG), or released to gas streams, with their fate governed by feedstock quality, process chemistry, and operating conditions [172, 173]. The recognition that rock purity drives both process performance and environmental outcomes is repeatedly emphasized in discussions of WPA production and impurity management, including the need for beneficiation of phosphate rock to reduce undesired constituents prior to digestion [172], as well as the emphasis on understanding how silica, clays, and iron/aluminium oxides influence slurry handling, filtration, and acid consumption [172, 173].

Phosphate rock commonly contains fluorapatite alongside silica, clays, iron and aluminium oxides, carbonates, and organic matter, with fluorine and trace metals such as cadmium and uranium present as challenging impurities. In WPA production by the wet process, fluorine is released during acidulation as fluorine-containing gases, predominantly hydrogen fluoride (HF) and silicon tetrafluoride (SiF_4), which can be partially captured by wet scrubbing and converted into fluosilicic acid (H_2SiF_6), a by-product with its own commercial value. Scrubbing systems are designed to absorb these fluorine-bearing species, thereby mitigating environmental emissions and enabling potential reuse of recovered fluosilicic acid; however, complete elimination of emissions is challenging, and some fluorine-containing effluents persist [174-178]. The fluorine fraction present in the rock is thus a critical environmental and process parameter, influencing both emissions and the composition of WPA and PG, and it is well documented that scrubbing efficiency can markedly reduce airborne fluoride but may still permit some release and downstream handling of H_2SiF_6 [174-178]. In addition, decarbonates and siliceous material in the ore can affect foaming and acid consumption, while iron and aluminium oxides can act as acid consumers, contributing to yield losses and process complexity [172].

Phosphogypsum, the main large-scale environmental footprint from WPA production, forms in stoichiometric opposition to phosphoric acid formation: PG consists primarily of $CaSO_4{\cdot}2H_2O$ but also incorporates residual acids, soluble phosphates, fluorides, and trace radionuclides. The sheer volumes generated, several tonnes of PG per tonne of phosphoric acid produced, pose substantial environmental management challenges. PG stacks are engineered to prevent leachate formation, groundwater contamination, dust emissions, and structural failure; rain infiltration can mobilize contaminants, producing acidic drainage if containment is inadequate, with coastal or high-rainfall settings amplifying the risk [176, 178-180]. In addition to physical containment, the chemical composition of PG, including residual acidity and fluorides, has implications for long-term stewardship and potential revalorisation strategies, such as converting PG constituents into useful products or stabilised waste forms [179, 180]. The environmental burden of PG is closely coupled to the fluorine cycle and the overall fluoride flux from WPA production, which is itself influenced by ore composition and processing practices [175, 176, 178].

Heavy metals, especially cadmium, are salient environmental and regulatory concerns linked to phosphate fertilizer production. Cadmium can accumulate in soils through fertilizer application and enter the food chain, so numerous jurisdictions impose limits on cadmium in fertilisers, shaping feedstock choice and process configurations. Some advanced purification steps aim to reduce heavy metal content, though these add complexity and cost to the process [173, 175, 181]. The overall regulatory landscape thus intertwines with impurity management, since cadmium and other trace elements influence raw material selection, processing strategies, and the permissible quality of final products, as well as the recycling or disposal options for PG and PG-derived materials [173, 175, 181].

From a process perspective, impurities drive operational challenges in WPA production. Silica and clays increase slurry viscosity and complicate filtration, while iron and aluminium oxides can consume phosphoric acid and lower overall yield. Carbonates in the ore can provoke foaming and elevate acid consumption. These issues underscore the value of beneficiation steps aimed at impurity removal before digestion, as clean feedstocks enhance both process efficiency and environmental performance by reducing corrosion, fouling, and reagent consumption, which in turn lowers emissions and waste generation [172, 173]. The role of feedstock quality in determining processing difficulty and environmental footprint is thus well established, highlighting the importance of ore characterization and selective upgrading prior to WPA manufacture [172].

Effluent and wastewater management constitutes another critical pillar of environmental performance in WPA plants. Process water streams may carry dissolved phosphates, fluorides, and acids, demanding treatment prior to discharge or reuse. Closed-loop water systems can minimize effluent volumes but may lead to the accumulation of impurities within the plant, necessitating careful control strategies and periodic purge or treatment steps to prevent buildup of contaminants and foulants [176, 178-180]. In addition, air emissions, dust from handling phosphate rock and fertiliser products, containing phosphate particles, fluorides, and trace metals, must be controlled through engineering controls such as cyclones, bag filters, and scrubbers to meet environmental standards, with emissions profiles intimately tied to feedstock impurity content and process operating conditions [176, 178-180].

Chapter 8

Potash Fertiliser Production

This chapter introduces the production of potash fertilisers, a critical component of modern agriculture and global food security. Potassium is one of the three essential macronutrients required for plant growth, and its role in regulating plant water balance, activating enzymes, and improving crop quality makes it indispensable across a wide range of agricultural systems. As global demand for food continues to rise, the efficient production and application of potassium fertilisers has become increasingly important in sustaining soil fertility and maximising crop yields.

The chapter begins by establishing a clear understanding of what potash fertilisers are, their chemical forms, and their significance in plant physiology. It then explores the primary types of potash products, including muriate of potash (MOP) and sulphate of potash (SOP), highlighting their differences in composition, production pathways, and agricultural applications. Understanding these distinctions is essential, as the choice of fertiliser has direct implications for crop performance, soil health, and economic outcomes.

Building on this foundation, the chapter examines the key production methods used to manufacture potash fertilisers. These include both mining-based approaches, such as conventional and solution mining of potash ores, and chemical or solution-based processes used to produce higher-value products like SOP. The discussion integrates technical, operational, and economic considerations, providing insight into how raw materials are transformed into marketable fertiliser products at both industrial and smaller scales.

In addition, the chapter considers alternative production routes, including processing from naturally occurring sulphate minerals, brine evaporation systems, and chemical conversion processes such as double decomposition. These methods illustrate the diversity of technologies available and the importance of resource availability, climate conditions, and market demand in determining the most appropriate production pathway.

Finally, the chapter links production processes to real-world applications by examining how different potash fertilisers are used across agricultural markets. It highlights how crop requirements, soil conditions, and environmental factors influence fertiliser selection, reinforcing the connection between production technology and end-use performance.

Overall, this chapter provides a comprehensive overview of potash fertiliser production, bridging fundamental chemistry, industrial processes, and agricultural application. It equips the reader with the knowledge needed to understand how potassium fertilisers are produced, why different forms exist, and how they contribute to efficient and sustainable crop production.

Potash Fertiliser

Potash fertiliser refers to a group of fertilisers that supply potassium (K), one of the three primary nutrients required for plant growth alongside nitrogen (N) and phosphorus (P). The term "potash" originates from the historical production of potassium salts from wood ash, but in modern industry it broadly refers to naturally occurring potassium-bearing minerals and the fertilisers derived from them, most commonly in the form of potassium chloride (KCl).

Modern potash fertilisers are produced from mined ores such as sylvinite and are processed into soluble potassium salts that plants can readily absorb. The most widely used form is muriate of potash (MOP), which is potassium chloride (KCl). Other important types include sulphate of potash (SOP), or potassium sulphate (K_2SO_4), which is preferred for chloride-sensitive crops, and potassium nitrate (KNO_3), which supplies both potassium and nitrogen. Fertiliser potassium content is typically expressed as K_2O (potash equivalent), even though the actual compounds do not contain potassium oxide.

Potassium plays a critical role in plant physiology, acting primarily as a regulator and activator rather than a structural component. Unlike nitrogen and phosphorus, it is not incorporated into plant tissues such as proteins or DNA. Instead, it supports a wide range of essential processes. It regulates water balance within the plant by controlling osmotic pressure, helping plants manage water uptake and tolerate drought conditions. It also activates more than 60 enzymes involved in growth and metabolism, enhances photosynthesis and energy transfer, and facilitates the movement of sugars and nutrients throughout the plant. In addition, potassium strengthens plant cell walls, improving resistance to pests and diseases, and contributes significantly to crop quality, including size, colour, taste, and shelf life.

Potash fertilisers are used extensively across almost all agricultural systems to improve crop yield, quality, and resilience. In broadacre cropping systems, such as wheat, maize, soybeans, and rice, potassium improves yield performance and enhances drought tolerance. In horticulture, including crops such as potatoes, tomatoes, citrus, and bananas, potash fertilisers play a key role in improving fruit size, flavour, colour, and storage life. They are also widely used in specialty crops such as coffee, cocoa, sugarcane, and cotton, where they contribute to fibre strength, sugar content, and overall product quality.

Beyond direct crop production, potash fertilisers are important for soil management. They are used to correct potassium deficiencies in soils and can assist in reclaiming saline or degraded soils when used alongside other soil amendments. In turf and landscaping applications, such as lawns, sports fields, and golf courses, potassium improves stress tolerance, wear resistance, and root development, contributing to healthier and more resilient vegetation.

A deficiency in potassium can have significant negative impacts on plant health and productivity. Plants lacking adequate potassium often develop weak stems that are prone to lodging, exhibit reduced tolerance to drought, and show characteristic symptoms such as yellowing or browning along leaf edges, known as leaf scorch. Ultimately, potassium deficiency leads to reduced yields and poorer quality produce, underscoring the essential role of potash fertilisers in modern agriculture.

Potash Ore Types (Sylvinite, Carnallite)

Potash fertiliser production begins with the extraction and processing of naturally occurring potassium-bearing minerals. The two most important commercial potash ore types are sylvinite and carnallite, which differ in composition, processing requirements, and economic implications. Understanding these ore types is critical because they determine the mining method, beneficiation process, and overall efficiency of potassium fertiliser production.

Sylvinite (Most Common Potash Ore)

Sylvinite is the most widely mined and processed potash ore globally. It is a naturally occurring mixture of sylvite (KCl), which is the valuable potassium mineral, and halite (NaCl), which is essentially common salt. The potassium content in sylvinite deposits typically ranges from 20–35% K_2O equivalent, making it a relatively high-grade and economically attractive ore.

Figure 48: Sylvinite (halite-sylvite evaporite rock) (Upper Kama Potash Deposit, Lower Permian; Solikamsk, Russia). James St. John, CC BY 2.0, via Wikimedia Commons.

From a processing perspective, sylvinite is favourable because the two main minerals, sylvite and halite, have different physical and surface properties. This allows them to be separated efficiently using froth flotation or, in some cases, crystallisation methods. In flotation, reagents are used to selectively make sylvite hydrophobic so it attaches to air bubbles and floats, while halite remains in the slurry and sinks.

Sylvinite deposits are typically found in layered evaporite formations, often mined using underground methods such as room-and-pillar mining. In some regions, solution mining may also be used, but conventional mining followed by surface processing is most common.

Sylvinite is typically found where large sedimentary basins experienced repeated cycles of seawater inundation followed by progressive evaporation. In these settings, brines concentrate as water is removed, and minerals precipitate in a predictable sequence dictated by solubility: carbonate and sulphate minerals such as calcite and gypsum/dolomite form first, halite (rock salt) concentrates next, and the most soluble potassium-bearing minerals, including sylvite, crystallize late when brines become highly concentrated. This sequence is well documented in Paleozoic and Mesozoic evaporite settings and underpins the generation of sylvinite ore, the principal source of potash used for fertilizer production [88, 182, 183]. The deposits are typically buried beneath hundreds to more than a thousand metres of overburden, a depth range repeatedly reported for major potash-bearing evaporites in well-studied basins worldwide [184-186].

Geologically, sylvinite forms within restricted or closed evaporite basins where seawater becomes trapped and undergoes protracted evaporation. Evaporite cycles in these basins record a progression from early, less soluble minerals to late-stage potash minerals; the last to crystallize are potassium salts such as sylvite, which precipitate from highly concentrated brines as evaporation proceeds. This conceptual model of evaporite formation and mineral zoning is widely described in discussions of Paradox-type and Michigan/Delaware-type basins, where sylvite commonly appears in the upper portions of halite beds or as intergrown components with halite in polyhalite-rich sequences, reflecting late-stage crystallization of sylvite in the evaporite evolution [186-190]. Deposits of this type are typically located at substantial depths, with many examples in the several-hundred-to-thousand-metre range, consistent with commercially exploited potash districts described in the literature [182, 184, 185].

Sylvinite mining occurs in several world regions that host large, high-quality potash basins. Canada's Saskatchewan Basin is frequently described as the world's largest source of mineable potash, contributing a substantial share of global production; Russia's Ural Basin provides extensive deposits that support major fertilizer production; Belarus and Germany are also key producers, reflecting a long history of potash mining in these regions. The Dead Sea region (Israel–Jordan) is recognized as an important source of potash as well, though much of its production relies on brine extraction from evaporite systems enriched in carnallite and other potassium minerals rather than direct sylvinite mining in a conventional underground mine. This geographic distribution is consistently summarized across major reviews and regional surveys of potash resources and deposits [88, 182, 189, 191-193].

In Canada, the potash beds within the Prairie Evaporite and related Saskatchewan formations are described as the world's largest source of mineable potash, with operations concentrated in flat-lying, relatively undeformed strata that simplify mining geometry and reduce unit costs compared with more complex terrains [182, 191, 194]. Saskatchewan's Saskatchewan Basin is repeatedly highlighted as housing the dominant proportion of known global potash reserves and production, with Canada as a leading producer and exporter of potash fertilizers [182, 191, 194].

Figure 49: Saskatchewan potash mines. Tony Hisgett from Birmingham, UK, CC BY 2.0, via Wikimedia Commons.

In Russia, the Verkhnekamsk deposit within the Permian evaporite sequence of the Ural region is noted as one of the world's largest potash-bearing basins, characterized by layered evaporites with carnallite and sylvite-bearing intervals and extensive mining activity supported by long-standing geologic study and modern geotechnical approaches [190, 195, 196]. The Belarusian Starobin or Starobinskoye deposits and the German Zechstein potash districts are similarly described in historical and contemporary literature as major sources of sylvite- and carnallitite-rich brine and ore bodies, reflecting the global distribution of high-quality potash resources in European and Eurasian [88, 190, 197]. In the Dead Sea region, potash production has historically leveraged brine evaporation in hypersaline systems, though its economic framework often emphasizes brine processing technologies rather than conventional underground mining of sylvinite [182, 184].

The regional distribution of sylvite- and carnallite-bearing evaporites is accompanied by a broader context on potash exploration and resource assessment. Reviews of global potash resources emphasize that approximately 90% of current production comes from the northern hemisphere's major basins, and that large, long-lived potash districts are found in Canada, Russia, and Belarus, with significant activity in Germany and Israel/Jordan, among others. These sources also discuss the evolving mix of mining technologies (underground mining versus in-situ solution mining in deep beds) and the role of brine resources and evaporite sequences in sustaining feeding economies for modern fertilizer industries [88, 182, 189, 194]. Notably, comprehensive overviews of potash exploration in the Americas document extensive sylvinite- and carnallitite-bearing sequences in Saskatchewan, the Paradox and Michigan basins of the United States, and the Carlsbad area of New Mexico, illustrating both the diversity of deposit styles and the shared evaporite-origin framework for sylvite mineralization [185, 186, 191, 197, 198].

The advantages of sylvinite include:

- Relatively simple beneficiation
- High recovery efficiency
- Lower processing costs compared to other ores

Because of these factors, sylvinite is the dominant feedstock for producing muriate of potash (MOP, KCl), the most widely used potassium fertiliser.

Carnallite (More Complex, Lower Grade Ore)

Carnallite is a more complex potash ore composed of hydrated potassium magnesium chloride, with the chemical formula $KMgCl_3 \cdot 6H_2O$. Unlike sylvinite, carnallite contains significant amounts of magnesium, which complicates both mining and processing.

Figure 50: Carnallite. Leon Hupperichs, CC BY-SA 3.0, via Wikimedia Commons.

Carnallite deposits generally have a lower potassium concentration and are more chemically reactive due to their hydrated nature. This makes them less suitable for conventional flotation. Instead, carnallite is commonly processed using solution mining and evaporation techniques. In these operations, water is injected into the deposit to dissolve the soluble salts, and the resulting brine is pumped to the surface. The brine is then processed through evaporation ponds or crystallisation systems to separate potassium salts from magnesium and other impurities.

Carnallite forms in very concentrated evaporite brines and is commonly recovered from evaporite sequences and saline brine systems that develop under extreme evaporation conditions. Its occurrence is typically linked to late-stage ponded brines in which highly soluble minerals crystallize only after more soluble components have already precipitated or been removed; in particular, carnallite constitutes a later, more concentrated stage of evaporite mineral precipitation relative to Sylvite and Halite in classic evaporite sequences [192, 199, 200]. This paragenetic position is reflected in its association with potassium- and magnesium-rich salt layers that form as seawater or hypersaline lake brines reach very high salinities, allowing solutes with high solubility to crystallize; the mineralogical progression in evaporite settings proceeds from early salts toward the most soluble potash-bearing intervals, with carnallite forming toward the terminus of evaporation in many basins [192, 199, 201]. The distribution within evaporite systems is commonly described as "upper" or residual brines where carnallite-bearing horizons occur, rather than within the main sylvinite units, consistent with its late-stage precipitation in the sequence [192, 199, 200] .

Geologically, carnallite forms as evaporation advances beyond the stages that produce halite and sylvite, so that the precipitation sequence in many evaporite basins proceeds (from lowest to highest salinity) roughly as follows: gypsum and anhydrite first, then halite, followed by sylvite, and only at the final, highly concentrated stage do hydrated magnesium- and potassium-rich salts such as carnallite form. In this framework, carnallite crystallizes from brines that remain after substantial halite and sylvite removal or removal of other constituents, and thus is typically hosted in upper evaporite sequences or residual brines rather than the principal sylvinite zones [192, 199, 200]. This interpretation is consistent with detailed stratigraphic work in major potash-bearing basins where the Houston Formation and its upper carnallitite/mixed intervals are described as part of a layered evaporite sequence in which carnallite is a dominant, though sometimes subordinate, horizon at upper levels (often beneath kainite- or sylvite-rich beds) [200, 202-205].

Regional sources and operational contexts for carnallite include both deep evaporite basins and surface brine systems. The Dead Sea region is a prominent historical and contemporary source of carnallite-bearing brines, where solar-evaporation ponds are used to concentrate brines for processing and where carnallitite-related layers have been described within broader evaporite stratigraphy; in addition, sylvinite- and carnallitite-bearing sequences are documented in several near-surface evaporite deposits and associated borehole sequences in other regions [192, 201, 204-206] . In Germany, carnallite occurs in underground evaporite deposits and is accessible via conventional mining or solution mining, illustrating the practical and geological reality that carnallite occurs in both surface brine systems and subsurface evaporite units depending on local tectonic and hydrological regimes [192, 207, 208]. Beyond Europe and the Dead Sea, carnallite-bearing horizons are recognized in Russia and Belarus, where sylvinite associations are common but carnallite is less frequently exploited, indicating the regional variability in economic emphasis related to evaporite mineralogy and brine chemistry [192, 209, 210].

In recent decades, high-profile, modern evaporite basins and their dynamics have been investigated in the Danakil Depression, a region where thick evaporite sequences and potash-bearing horizons are actively studied. The Danakil Depression in Ethiopia and Eritrea hosts multi-hundred-meter-thick evaporite sequences, dominated by halite with subordinate potash minerals such as kainite, carnallite, and bischofite within a stratigraphy that features a series of potash-rich members (e.g., kainite, carnallite, and bischofite layers, followed by sylvite-bearing beds in some intervals). These sequences are associated with intense evaporative settings and complex hydrothermal-tectonic interactions, and they provide a modern analogue for studying the formation and persistence of carnallitic horizons under high evaporation rates and hydrochemical gradients. Recent work on the Danakil depression documents a stacked evaporite succession with a kainite member at the base, a carnallite- and bischofite-rich middle, and a sylvinite-containing upper interval, with Halite and anhydrite interlayering throughout; such compilations illustrate how carnallite sits within the late-evaporite portion of the Houston formation and related units, consistent with its placement in highly concentrated brines that form near the end of the evaporation sequence [199, 200, 204-206, 209-211].

Advances in understanding the Danakil system illuminate the interplay between evaporation, brine chemistry, and tectonics that control carnallite formation and preservation. For example, evaporite sequences in Danakil display multiple potash horizons with interlayered halite, kainite, carnallite, and bischofite, consistent with progressive concentration and dehydration of brines under arid climatic regimes and during episodes of magmatic/tectonic activity. In this setting, carnallite horizons are particularly susceptible to dissolution and alteration under changing pore-fluid conditions, but they also represent stable horizons in which petrographic textures and dehydration pathways can be preserved, especially in relatively recent deposits where burial and diagenesis are limited [199, 200, 205, 212]. The Dallol area specifically exemplifies how evaporitic sequences interact with active hydrothermal systems to produce extreme brines rich in chlorides and sulphates, with carnallite-bearing horizons presenting within a complex succession shaped by dehydration reactions and localized heating that can fluidize or laterally mobilize hydrated salts under magmatic influence; this has implications for carnallite stability and for understanding potential subsurface migration pathways of saline brines in magmatic–evaporitic settings [201, 204, 206, 210, 212].

Beyond Africa, carnallite-related evaporites are documented in other saline lake and saline-brine contexts where rapid evaporation concentrates brines to the point of carnallite crystallization; in some global brine systems, carnallite horizons are described as uppermost evaporite horizons that form in the terminal stages of evaporation and are then exploited through conventional mining or solution mining where feasible [199, 200, 204, 205, 209, 213, 214]. The broad geographic distribution of carnallite-bearing horizons in modern and ancient evaporite basins, tied to arid climates that support high evaporation rates, coupled with tectonically active basins that can subside and trap brines, emphasizes the robust relationship between evaporation-driven brine evolution and carnallite formation, and highlights the ongoing research interest in Danakil and allied regions as modern analogues for carnallitic sedimentology and ore genesis [199, 200, 204, 206, 210-212].

The presence of magnesium chloride introduces several challenges:

- Higher processing complexity
- Increased energy and water requirements
- Corrosion issues in equipment
- More difficult purification steps

However, carnallite can still be economically viable, particularly in regions where large, shallow deposits exist and solar evaporation can be used to reduce processing costs.

Table 7: Key Differences Between Sylvinite and Carnallite.

Feature	**Sylvinite**	**Carnallite**
Main Composition	$KCl + NaCl$	$KMgCl_3 \cdot 6H_2O$
Potassium Content	Higher	Lower
Impurities	Mainly halite ($NaCl$)	Magnesium salts
Processing Method	Flotation / crystallisation	Solution mining + evaporation
Complexity	Relatively simple	More complex
Cost of Processing	Lower	Higher
Typical Product	MOP (KCl)	MOP after additional processing

Mining Methods (Conventional, Solution mining)

Potash ore, primarily sourced from minerals such as sylvinite and carnallite, is extracted using two main mining methods: conventional underground mining and solution mining. The choice between these methods depends on factors such as deposit depth, geological conditions, ore composition, and economic considerations. Each method has distinct operational processes, equipment requirements, and advantages.

Conventional Underground Mining

Conventional mining is the most widely used method for extracting potash, particularly from sylvinite deposits that are relatively thick, laterally extensive, and mechanically stable. This method typically uses a room-and-pillar mining system, where large underground chambers ("rooms") are excavated while leaving behind pillars of ore to support the roof and maintain structural stability.

The process begins with sinking shafts to access the ore body, which may lie several hundred to over 1,000 metres below the surface. Once underground access is established, continuous mining machines are used to cut and collect the ore. These machines grind the ore from the face and load it onto shuttle cars or conveyor systems for transport to the surface.

At the surface, the ore is crushed and processed to separate potassium salts from impurities such as halite. Ventilation systems, ground support, and safety monitoring are critical components of underground operations, ensuring safe working conditions and preventing hazards such as roof collapse or gas accumulation.

The main advantages of conventional mining include high recovery rates, consistent ore quality, and established technology. However, it requires significant capital investment in shafts, underground infrastructure, and ventilation systems. It is also limited to deposits that are geologically stable and accessible.

Solution Mining

Solution mining is used where potash deposits are too deep, too thin, or too unstable for conventional underground mining. It is particularly suited to carnallite-rich deposits or areas where geological conditions make shaft sinking impractical.

In this method, wells are drilled from the surface into the potash-bearing formation. Water or brine is injected into the deposit to dissolve the soluble salts, creating a potassium-rich solution. This brine is then pumped back to the surface, where the dissolved salts are recovered through evaporation and crystallisation processes.

There are two main approaches to solution mining: hot leaching, where heated water increases solubility and extraction efficiency, and cold leaching, which relies on natural temperature conditions. In regions with suitable climates, solar evaporation ponds may be used to concentrate the brine, reducing energy costs.

Solution mining offers several advantages, including lower upfront capital costs (no shafts required), improved safety (minimal underground workforce), and the ability to access otherwise uneconomic deposits. However, it typically has lower recovery efficiency, higher water and energy requirements, and can be slower than conventional mining. Environmental considerations, such as water use and brine disposal, must also be carefully managed.

Dead Sea Potash Plant (Israel and Jordan)

The "potash plant" in the Dead Sea refers to a group of large-scale industrial operations that extract potassium salts from the highly saline waters of the Dead Sea. The two main producers are Dead Sea Works (in Israel) and the Arab Potash Company (in Jordan). These facilities are among the most unique potash operations in the world because they do not rely on underground mining, but instead use solar evaporation of natural brines.

Figure 51: Dead Sea Potash Plant (Israel & Jordan). Heritage Conservation Outside The City Pikiwiki Israel, CC BY 2.5, via Wikimedia Commons.

The Dead Sea is one of the saltiest bodies of water on Earth, rich in dissolved minerals including potassium, magnesium, sodium, and bromine. This natural concentration makes it ideal for potash production using evaporation.

The process begins by pumping Dead Sea brine into vast evaporation ponds located in the southern basin. These ponds cover large areas and are designed to maximise exposure to sunlight and heat. As water evaporates under the hot desert climate, the brine becomes progressively more concentrated, and minerals begin to crystallise in sequence.

Initially, sodium chloride (halite) precipitates and is removed or left behind. As evaporation continues and the brine becomes more concentrated, potassium- and magnesium-rich salts form, particularly carnallite ($KMgCl_3 \cdot 6H_2O$). This intermediate product is then harvested mechanically from the ponds.

The harvested carnallite is transported to processing plants, where it is treated to separate potassium chloride (KCl) from magnesium salts. This typically involves thermal decomposition, dissolution, and recrystallisation steps. The final product is muriate of potash (MOP), which is then dried, sized, and prepared for export.

One of the defining characteristics of Dead Sea potash operations is their reliance on solar energy rather than conventional mining or high-energy processing. The region's high temperatures and low rainfall allow for efficient natural evaporation, significantly reducing energy costs compared to underground mining operations.

These plants operate on a very large scale, producing millions of tonnes of potash annually for global agricultural markets. In addition to potash, they also recover valuable by-products such as bromine and magnesium compounds, making the operations highly integrated chemical production systems.

The Dead Sea method offers several advantages. It avoids the need for underground mining, reduces energy consumption through solar evaporation, and utilises a naturally concentrated mineral resource. However, it also presents environmental and logistical challenges. Large land areas are required for evaporation ponds, and the diversion of water into these ponds has contributed to the declining water level of the Dead Sea, which is a major environmental concern.

Additionally, the process is dependent on climate conditions and can be slower than conventional mining in terms of processing time. The high magnesium content of the brine also requires additional processing steps to produce high-purity potassium chloride.

Processing and Refining Potassium Salts

Processing and refining potassium salts is the stage in potash production where mined or solution-extracted materials are converted into high-purity, marketable fertiliser products, most commonly potassium chloride (KCl). The exact processing route depends on the ore type, such as sylvinite or carnallite, and the mining method used. Regardless of the pathway, the primary objective is to separate potassium-bearing minerals from impurities and produce a consistent, usable fertiliser product.

Once potash ore is brought to the surface through conventional mining, or recovered as brine through solution mining, the first step is ore preparation. For solid ores such as sylvinite, the material is crushed and screened to reduce particle size and liberate potassium minerals from waste materials such as halite (NaCl) and clays. Effective size reduction is essential, as it directly influences the efficiency of subsequent separation processes.

The next stage involves the separation of potassium minerals from impurities. In the case of sylvinite, this is most commonly achieved through flotation. The crushed ore is mixed with water to form a slurry, and chemical reagents such as collectors and frothers are added. Air is introduced into the slurry, creating bubbles that selectively attach to potassium chloride particles, causing them to float to the surface. Meanwhile, sodium chloride and other impurities remain in the slurry and sink. The floating potassium-rich material is then skimmed off as a concentrate.

For solution-mined ores or carnallite-rich brines, separation is typically achieved through evaporation and crystallisation. The brine is progressively concentrated until potassium salts crystallise out of solution. By carefully controlling temperature and concentration, potassium chloride can be selectively crystallised while magnesium salts and other impurities remain dissolved or are removed in later stages.

In the processing of carnallite, additional decomposition and purification steps are required due to the presence of magnesium. The carnallite is heated or dissolved to break down its structure, allowing potassium chloride to be separated from magnesium chloride. This stage is more complex than sylvinite processing and often involves a combination of thermal treatment, leaching, and recrystallisation to achieve the desired product purity.

Following separation, the potassium-rich product typically contains residual moisture that must be removed. Dewatering is carried out using filtration or centrifugation, followed by drying in equipment such as rotary dryers or fluidised bed dryers. This step reduces moisture content, improving the stability, handling, and storage characteristics of the final product.

The dried material is then processed into the required physical form through compaction, granulation, and sizing. Fine particles may be compacted or granulated to produce uniform granules suitable for agricultural application. The product is screened to ensure a consistent size distribution, with oversized and undersized material recycled back into the process. In some cases, anti-caking agents are applied to enhance flowability and prevent clumping during storage.

The final product, typically referred to as muriate of potash (MOP), is stored in bulk facilities or packaged into bags for distribution. It is then transported to agricultural markets worldwide, where it is applied directly to soils or blended with other fertilisers to meet specific crop nutrient requirements.

Throughout processing, several key technical and economic factors must be carefully managed. These include the type and composition of the ore, separation efficiency and recovery rates, energy consumption, particularly in drying and evaporation, impurity removal, and the quality and strength of the final granulated product.

Production of MOP (KCl) and SOP (K_2SO_4)

The production of potassium fertilisers is primarily centred on two key products: muriate of potash (MOP, KCl) and sulphate of potash (SOP, K_2SO_4). While both supply essential potassium to crops, they differ significantly in raw materials, processing methods, and end-use applications. MOP is the most widely produced and used form due to its simplicity and cost-effectiveness, whereas SOP is a higher-value product used for chloride-sensitive crops.

Production of Muriate of Potash (MOP – KCl)

MOP stands for Muriate of Potash, which is the most widely used potassium fertiliser in the world. Chemically, MOP is potassium chloride (KCl), a simple salt that provides potassium in a form that plants can readily absorb.

MOP is a refined product derived from potash ore, such as sylvinite, or from potassium-rich brines. During processing, potassium chloride is separated from other salts like sodium chloride (common salt), purified, and then formed into granules or compacted particles suitable for agricultural use. Although it is commonly referred to as "potash," the nutrient content is typically expressed as K_2O equivalent. Standard MOP products contain around 60–62% K_2O, making them a highly concentrated source of potassium.

The term "muriate" is an older chemical expression used to describe chloride compounds. As such, "muriate of potash" simply refers to potassium chloride. While the term persists in agriculture and industry, potassium chloride is the more accurate modern chemical name.

In agriculture, MOP supplies potassium, an essential nutrient for plant health and productivity. Potassium plays a vital role in regulating water use and improving drought tolerance, activating enzymes required for growth and metabolism, and supporting photosynthesis and energy transfer. It also assists in the movement of nutrients and sugars within the plant, enhances resistance to disease and environmental stress, and contributes to improved crop yield and quality. Because of these benefits, MOP is widely used across a broad range of crops, including cereals, oilseeds, and pasture systems.

MOP is characterised by its chemical formula KCl, a nutrient content of approximately 60–62% K_2O equivalent, and its availability in granular, crystalline, or compacted forms. It is highly soluble in water, allowing for efficient uptake by plants, and is relatively low in cost compared to other potassium fertilisers, which contributes to its widespread use globally.

MOP production begins with potash ores such as sylvinite or with brines obtained from solution mining. The goal is to separate potassium chloride (KCl) from other salts, primarily sodium chloride (NaCl).

For solid ores, the process starts with crushing and grinding, which reduces particle size and liberates potassium minerals. The ore is then processed using froth flotation, the most common separation method. In this step, chemical reagents are added to a slurry of crushed ore, and air is introduced to create bubbles. Potassium chloride particles attach to these bubbles and rise to the surface, forming a concentrate, while sodium chloride and other impurities remain in the slurry.

In solution mining or brine-based operations, the separation is achieved through evaporation and crystallisation. The potassium-rich brine is concentrated until KCl crystallises out, which is then recovered.

After separation, the potassium chloride concentrate is dewatered, dried, and processed into granules or compacted forms. The final product is screened to ensure consistent particle size and may be coated with anti-caking agents. The result is a high-purity MOP product containing approximately 60–62% K_2O equivalent, suitable for direct application or blending into compound fertilisers.

MOP production is relatively straightforward, cost-effective, and highly scalable, which is why it accounts for the majority of global potash fertiliser use.

Production of Sulphate of Potash (SOP – K_2SO_4)

SOP stands for Sulphate of Potash, which is a potassium fertiliser that supplies potassium in the form of potassium sulphate (K_2SO_4). Unlike MOP, SOP is a chloride-free fertiliser, making it particularly important for crops that are sensitive to chloride and where crop quality is a key consideration.

SOP is a refined potassium fertiliser produced either through chemical processes, such as reacting potassium chloride with sulfuric acid, or from naturally occurring sulphate minerals and brines. It provides both potassium (K) and sulphur (S), which are essential nutrients for plant growth. As with other potash fertilisers, its nutrient content is typically expressed as K_2O equivalent. SOP generally contains around 50–52% K_2O and 17–18% sulphur, making it a valuable dual-nutrient fertiliser.

The primary difference between SOP and MOP lies in their chemical composition, particularly the presence or absence of chloride. While MOP (KCl) contains chloride, SOP (K_2SO_4) contains sulphate instead. This distinction is significant because chloride can negatively affect certain crops, reducing yield or quality. As a result, SOP is preferred in situations where chloride sensitivity is a concern.

In agricultural use, SOP supplies potassium, which supports essential plant functions such as regulating water balance, improving drought tolerance, activating enzymes, and enhancing photosynthesis and energy transfer. It also contributes to improved crop yield and overall quality. In addition, the sulphur component of SOP plays a critical role in protein synthesis, enzyme function, and the development of flavour and nutritional quality in crops.

SOP is commonly used for high-value and chloride-sensitive crops, including fruits such as grapes and citrus, vegetables such as tomatoes and potatoes, as well as tobacco, coffee, and tea. It is also widely used in greenhouse production and fertigation systems, where precise control of nutrient supply is required to optimise plant growth and product quality.

In terms of characteristics, SOP has a chemical formula of K_2SO_4 and a nutrient content of approximately 50–52% K_2O equivalent, along with 17–18% sulphur. It contains virtually no chloride, has good solubility making it suitable for various application methods, and is typically more expensive than MOP due to its more complex production processes.

SOP production is more complex and typically involves chemical conversion processes rather than simple physical separation. The most common industrial method is the Mannheim process, although other routes exist.

1. Mannheim Process (Most Common)

In the Mannheim process, potassium chloride (KCl) is reacted with sulphuric acid (H_2SO_4) in a high-temperature furnace. The process occurs in two stages:

- First, KCl reacts with sulfuric acid to form potassium bisulphate ($KHSO_4$) and hydrogen chloride gas (HCl)
- Then, additional KCl reacts with $KHSO_4$ to form potassium sulphate (K_2SO_4)

The reaction is carried out at elevated temperatures (typically 500–600°C) in a Mannheim furnace. The resulting SOP is cooled, crushed, and screened to produce a marketable product. The hydrogen chloride gas produced is recovered as a valuable by-product, often converted into hydrochloric acid.

Overview of the Mannheim Process

The Mannheim process converts potassium chloride into potassium sulphate through a two-stage reaction with sulfuric acid. It also produces hydrogen chloride gas (HCl) as a valuable by-product, which is typically recovered and converted into hydrochloric acid.

The process is continuous or semi-continuous and operates at high temperatures (approximately 500–600°C) in a Mannheim furnace, a specially designed reactor that allows controlled heating, mixing, and gas removal.

Key Chemical Reactions

The process occurs in two main steps:

Stage 1: Formation of Potassium Bisulphate

$$KCl + H_2SO_4 \rightarrow KHSO_4 + HCl\ (gas)$$

In this initial stage, potassium chloride reacts with sulfuric acid to form potassium bisulphate ($KHSO_4$) and hydrogen chloride gas. This reaction occurs at moderate temperatures and is exothermic.

Stage 2: Formation of Potassium Sulphate

$$KCl + KHSO_4 \rightarrow K_2SO_4 + HCl\,(gas)$$

In the second stage, additional potassium chloride reacts with potassium bisulphate to produce potassium sulphate (K_2SO_4) and more hydrogen chloride gas. This stage requires higher temperatures and is endothermic, meaning heat must be supplied.

Step-by-Step Process Description

1. Raw Material Feeding

Potassium chloride (MOP) is fed into the Mannheim furnace along with concentrated sulfuric acid. The materials are introduced in a controlled manner to ensure proper reaction conditions and avoid local overheating or incomplete conversion.

2. Reaction in the Mannheim Furnace

The furnace is a circular, refractory-lined reactor equipped with a rotating mechanical rake or agitator. This rake continuously moves the solid material across the heated surface, ensuring:

- Uniform mixing
- Controlled residence time
- Efficient heat transfer

The process temperature is maintained between 500–600°C, allowing both reaction stages to occur sequentially within the same unit.

3. Hydrogen Chloride Gas Evolution and Recovery

During both reaction stages, hydrogen chloride (HCl) gas is released. This gas is continuously removed from the furnace and directed to a gas absorption system, where it is absorbed in water to produce hydrochloric acid (HCl solution).

This by-product is commercially valuable and improves the overall economics of the process.

4. Discharge of Potassium Sulphate

The solid product, potassium sulphate, is gradually moved to the discharge end of the furnace by the rotating rake. It exits the furnace as a hot, granular or powdery material.

5. Cooling and Size Reduction

The hot SOP product is cooled using air or indirect cooling systems. It may then be crushed, milled, or granulated to achieve the desired particle size for fertiliser use.

6. Screening and Finishing

The product is screened to ensure consistent size distribution. Oversized material is crushed and recycled, while fines may be reprocessed or compacted. Anti-caking agents may be added to improve storage and handling properties.

Equipment Used

- Mannheim furnace (core reactor with rotating rake)
- Burners or external heating system (to maintain high temperatures)
- Gas handling and scrubbing system (for HCl recovery)
- Cooling equipment
- Crushers, mills, and screens
- Conveyors and storage systems

2. Alternative Routes

Other methods for producing SOP include:

- **Langbeinite Processing**: Using naturally occurring minerals such as langbeinite ($K_2SO_4 \cdot 2MgSO_4$), which are processed to separate potassium sulphate
- **Brine Processing**: In some regions, SOP is produced from sulphate-rich brines through evaporation and crystallisation
- **Double decomposition reactions**: For example, reacting potassium chloride with sulphate salts such as gypsum or ammonium sulphate

Langbeinite Processing (Production of SOP from Natural Sulphate Minerals)

Langbeinite processing is an alternative route for producing sulphate of potash (SOP, K_2SO_4) using a naturally occurring mineral rather than relying on chemical conversion processes such as the Mannheim process. Langbeinite, with the chemical formula $K_2SO_4 \cdot 2MgSO_4$, contains both potassium and magnesium sulphates. It is sometimes sold directly as a fertiliser (commonly known as K-Mag), but it can also be further processed to produce higher-purity potassium sulphate suitable for specialised agricultural applications.

Langbeinite is typically found in evaporite deposits and is often associated with other potash minerals such as sylvinite and carnallite. Unlike these chloride-based ores, langbeinite is already a sulphate mineral and is therefore inherently chloride-free. This characteristic makes it particularly valuable as a feedstock for SOP production, especially for crops that are sensitive to chloride. However, because it contains both potassium sulphate and magnesium sulphate ($MgSO_4$), additional processing is required to separate and isolate the potassium component.

The process begins with mining and crushing the langbeinite ore to reduce it to a suitable particle size. This increases the surface area and improves the efficiency of subsequent processing steps. Screening may also be used to ensure a consistent feed size and remove oversized material. The prepared ore is then subjected to leaching, typically using water or a weak solution. During this stage, the langbeinite dissolves into a mixed sulphate solution containing both potassium sulphate and magnesium sulphate. Careful control of temperature and solution chemistry is required to optimise dissolution and prepare the solution for separation.

The separation of potassium sulphate from magnesium sulphate is the key challenge in langbeinite processing. This is achieved through selective crystallisation, taking advantage of the different solubilities of the two salts. By controlling temperature and concentration, potassium sulphate, being less soluble, can be induced to crystallise first. This may be achieved through cooling crystallisation, evaporation crystallisation, or a combination of both in a fractional crystallisation process. Magnesium sulphate remains in solution for longer due to its higher solubility and can be removed separately or recovered as a by-product.

Once potassium sulphate crystals have formed, they are separated from the remaining solution using filtration or centrifugation. The recovered solids are then washed to remove residual impurities, improving the purity of the final product. The material is subsequently dried using equipment such as rotary dryers or fluidised bed dryers to reduce moisture content and improve handling characteristics.

After drying, the product may be milled or granulated to achieve the desired particle size for agricultural use. It is then screened to ensure uniformity, and anti-caking agents may be applied to enhance storage stability and flowability. The final product is a high-quality, chloride-free SOP fertiliser suitable for a wide range of crops.

An additional advantage of langbeinite processing is the generation of magnesium sulphate as a by-product. This material can be marketed as a fertiliser or used in industrial applications, improving the overall economics of the process where suitable markets exist.

Overall, langbeinite processing offers several advantages, including the use of a naturally chloride-free feedstock, reduced risk of chloride contamination in the final product, and the ability to produce both potassium and magnesium fertilisers. It may also involve lower emissions compared to high-temperature chemical processes. However, the process is more complex due to the need to separate magnesium, requires careful control of crystallisation conditions, and depends on the availability of suitable langbeinite deposits. Product purity may also require additional refining steps.

Brine Processing for SOP (Potassium Sulphate) Production

Brine processing is a natural, evaporation-based method used to produce sulphate of potash (SOP, K_2SO_4) from sulphate-rich brines found in salt lakes, inland seas, and underground reservoirs. This approach is most effective in arid regions where high solar evaporation rates make the process economically viable. Unlike high-temperature chemical methods, brine processing relies primarily on solar energy and controlled crystallisation, making it both energy-efficient and well suited to large, open-area operations.

The process begins with the extraction of mineral-rich brines, which are naturally occurring saline waters containing dissolved potassium, magnesium, sodium, chloride, and sulphate ions. These brines are typically sourced from salt lakes, evaporated inland seas, or underground aquifers accessed via wells. In many cases, they originate from ancient seawater that has undergone long-term evaporation, resulting in highly concentrated solutions rich in dissolved salts.

Once extracted, the brine is transferred into a network of large solar evaporation ponds. These shallow ponds allow sunlight and wind to drive the evaporation of water, progressively increasing the concentration of dissolved salts. As the brine becomes more concentrated, different salts begin to crystallise in sequence according to their solubility. The process is carefully managed by moving the brine between ponds at different stages of concentration. Early-stage ponds typically remove less desirable salts such as sodium chloride (NaCl), while later ponds concentrate potassium and sulphate components.

As evaporation continues, salts precipitate in a predictable order. Less soluble minerals such as calcium and carbonate salts crystallise first, followed by sodium chloride (halite), and finally more soluble potassium and magnesium salts. At advanced stages, minerals such as carnallite or mixed sulphate salts may form, depending on the chemistry of the brine. For SOP production, the process is controlled to favour the formation of potassium sulphate-bearing compounds or intermediate salts that can later be converted into potassium sulphate.

In many brine systems, potassium does not initially crystallise as pure K_2SO_4. Instead, intermediate salts—often containing both potassium and magnesium—are produced. These materials are then further processed through leaching, purification, and controlled chemical adjustment of the brine. Secondary crystallisation stages are used to selectively precipitate potassium sulphate while leaving magnesium and other impurities in solution. Careful control of temperature, concentration, and solution chemistry is essential to achieve efficient separation.

Once potassium sulphate crystals, or their precursors, have formed, they are harvested mechanically from the ponds. The material is then washed to remove impurities and separated from residual brine using filtration or centrifugation. Additional refining steps may be applied to improve product purity, depending on the required quality specifications.

The recovered potassium sulphate is then dried to reduce moisture content and improve handling characteristics. Drying may be achieved through solar methods in some operations or through rotary or fluidised bed dryers where higher throughput and consistency are required. After drying, the product is screened to achieve the desired particle size, and it may be granulated or compacted for ease of application. Anti-caking agents are often applied to improve storage stability and flowability. The final product is a high-quality, chloride-free SOP fertiliser suitable for sensitive crops and specialised agricultural applications.

Brine processing also generates valuable by-products, including sodium chloride and various magnesium salts such as magnesium sulphate or magnesium chloride. Efficient operations aim to recover and utilise these materials, improving overall process economics and reducing waste.

This method offers several advantages, including low energy consumption due to reliance on solar evaporation, the ability to produce naturally chloride-free or low-chloride products, and suitability for large-scale, continuous production with minimal chemical inputs. However, it also has limitations. It requires large land areas for evaporation ponds, depends heavily on favourable climatic conditions, and is relatively slow, with evaporation cycles that may take months or even years. Additionally, careful control of brine chemistry is essential to ensure efficient and consistent production.

Double Decomposition Reactions for SOP Production

Double decomposition, also known as double displacement, is a chemical process used to produce sulphate of potash (SOP, K_2SO_4) by reacting potassium chloride (KCl) with a sulphate-containing compound. In this process, ions from two dissolved salts exchange partners in solution, forming new compounds, one of which is potassium sulphate. This method is particularly important in situations where potassium chloride is readily available, but a chloride-free fertiliser is required for agricultural use.

The underlying principle of double decomposition is ion exchange in solution. When two soluble salts are dissolved and mixed, their ions rearrange to form new compounds. The reaction proceeds most effectively when one of the products either precipitates out of solution as a solid or can be removed through crystallisation or other means. In SOP production, the objective is to replace chloride ions (Cl^-) from potassium chloride with sulphate ions (SO_4^{2-}), thereby forming potassium sulphate.

The most common industrial route involves reacting potassium chloride with ammonium sulphate ($(NH_4)_2SO_4$).

One of the most widely used double decomposition reactions involves ammonium sulphate ($(NH_4)_2SO_4$):

$$2KCl + (NH_4)_2SO_4 \rightarrow K_2SO_4 + 2NH_4Cl$$

How the Reaction Works

- Potassium ions (K^+) combine with sulphate ions (SO_4^{2-}) to form potassium sulphate (SOP)
- Ammonium ions (NH_4^+) combine with chloride ions (Cl^-) to form ammonium chloride (NH_4Cl)

This reaction typically occurs in an aqueous solution under controlled temperature conditions.

In this reaction, potassium ions combine with sulphate ions to form potassium sulphate, while ammonium ions combine with chloride ions to form ammonium chloride (NH_4Cl). The reaction typically takes place in an aqueous solution under controlled temperature conditions. Both salts are first dissolved to create a homogeneous solution, allowing the ion exchange to occur efficiently.

Following the reaction, potassium sulphate is separated from the solution through controlled crystallisation. Because potassium sulphate is less soluble than ammonium chloride under certain conditions, it crystallises out first. These crystals are then separated using filtration or centrifugation. The recovered product is subsequently dried, screened, and, if required, granulated to produce a marketable fertiliser. The remaining ammonium chloride solution can be recovered as a by-product and sold or further processed, which can significantly improve the economic viability of the process.

Another possible double decomposition route involves reacting potassium chloride with gypsum (calcium sulphate, $CaSO_4$). In this case, potassium ions combine with sulphate to form potassium sulphate, while calcium combines with chloride to form calcium chloride ($CaCl_2$). Although this reaction is chemically feasible, it is less commonly used in industrial practice. The primary limitation is the low solubility of gypsum, which reduces reaction efficiency and slows the process. As a result, higher temperatures, additives, or additional processing steps may be required to achieve acceptable conversion rates. This route is generally only considered in situations where gypsum is readily available, such as in integrated operations utilising by-products like phosphogypsum.

A typical double decomposition plant includes several key pieces of equipment. Dissolution tanks are used to prepare the initial solutions, while reaction vessels or crystallisers facilitate the ion exchange and product formation. Crystallisation units, either cooling or evaporative, are used to selectively precipitate potassium sulphate. Solid-liquid separation equipment, such as filters or centrifuges, is required to recover the solid product, followed by dryers and granulators to produce the final fertiliser form. Depending on production scale, the process may be operated in either batch or continuous mode.

This method offers several advantages. It enables the production of chloride-free potassium sulphate from widely available potassium chloride, operates at moderate temperatures compared to high-temperature processes like the Mannheim method, and can be integrated with other chemical operations. Additionally, it produces useful by-products such as ammonium chloride or calcium chloride, which can add economic value if suitable markets exist.

However, the process also has limitations. It requires relatively high-purity raw materials to ensure efficient reactions and high-quality output. Careful control of crystallisation and separation conditions is essential, and effective handling or utilisation of by-products is necessary for economic sustainability. While less energy-intensive than some alternatives, the process remains chemically complex and requires precise operational control.

Double decomposition reactions provide a viable chemical pathway for producing potassium sulphate by exchanging ions between potassium chloride and sulphate salts such as ammonium sulphate or gypsum. The ammonium sulphate route is the most widely used due to its favourable reaction and crystallisation characteristics, while the gypsum route is more limited in application. Together, these methods demonstrate how controlled chemical reactions and solubility differences can be used to produce high-value, chloride-free fertilisers from common raw materials.

Small Scale Production Potential

In practice, the small-scale production of muriate of potash (MOP, KCl) is generally not viable unless there is direct access to a suitable resource such as potash ore or brine. This is because MOP production is fundamentally a mining and large-scale mineral processing activity rather than a simple chemical manufacturing process. The production of MOP involves separating potassium chloride from sylvinite ore or potassium-rich brines, which requires significant infrastructure, including underground or solution mining systems, crushing and grinding circuits, flotation units, and, in the case of brines, large evaporation facilities. These operations rely heavily on economies of scale, and the capital investment is dominated by mining, materials handling, and bulk processing rather than chemical transformation.

As a result, MOP production is not suitable for small-scale startup operations. It is only realistically viable where an operator already has access to existing ore or brine deposits, or where there is an opportunity to reprocess waste streams or tailings from larger operations. In most small-scale contexts, MOP is not produced but instead purchased as a feedstock for further processing or blending.

In contrast, sulphate of potash (SOP, K_2SO_4) is significantly more feasible to produce at small scale because several of its production routes are based on chemical or solution processes rather than mining. However, the practicality of small-scale SOP production depends heavily on the specific process selected.

The Mannheim process, which is the dominant industrial method for SOP production, is not suitable for small-scale operations. It requires a high-temperature furnace operating at approximately 500–600°C, along with specialised refractory linings, acid handling systems, and gas scrubbing equipment to manage hydrogen chloride emissions. The process is capital-intensive, operationally complex, and subject to strict safety and environmental controls, making it viable only at large industrial scale.

Langbeinite processing offers a potential alternative, as it involves crushing, leaching, and crystallisation rather than high-temperature reactions. This approach is technically feasible at smaller scale and avoids extreme operating conditions. However, its practicality is entirely dependent on access to langbeinite ore, which is geographically limited. Where such a resource is available, the process can be implemented with moderate complexity.

Brine processing is another option, relying on solar evaporation and crystallisation to produce potassium sulphate. This method is technically simple and has low energy requirements, as it uses natural evaporation rather than thermal input. However, it requires large land areas, suitable climatic conditions (hot, dry environments), and long processing times, often extending over months or years. As such, it is best suited to remote locations with low land costs and long-term operational horizons.

The most practical and widely applicable small-scale method for SOP production is double decomposition. This chemical process involves reacting potassium chloride with a sulphate salt, most commonly ammonium sulphate, to produce potassium sulphate and ammonium chloride. The reaction occurs in solution at relatively low to moderate temperatures and does not require high-pressure systems or specialised furnaces. This makes it well suited to smaller-scale, modular plant configurations.

A typical small-scale setup for double decomposition includes mixing or dissolution tanks, a stirred reactor, a crystallisation unit (either cooling or evaporative), solid-liquid separation equipment such as filters or centrifuges, and optional drying equipment if a solid product is required. The process can produce powder or crystalline SOP, with simple granulation added if needed.

Capital costs for such systems vary depending on scale and level of processing. A very small batch system without granulation may cost in the range of USD 50,000 to 300,000 (approximately AUD 75,000 to 450,000 or EUR 45,000 to 270,000). A more developed small plant with drying capability may range from USD 300,000 to 1.5 million (approximately AUD 450,000 to 2.2 million or EUR 270,000 to 1.3 million). Operating costs are largely driven by raw materials, particularly potassium chloride and ammonium sulphate, as well as energy for drying and labour. The process also generates ammonium chloride as a by-product, which can be sold or utilised, improving overall economics.

From an operational perspective, it is important to recognise that even when SOP production is technically feasible at small scale, raw material costs dominate the economics. Profitability depends on local fertiliser pricing, access to inputs, and the ability to utilise or sell by-products. Granulation and drying significantly increase both capital and operational complexity, so many small producers opt to supply powder or crystalline SOP rather than fully granulated products.

Table 8: Practical Comparison.

Factor	MOP	SOP
Small-scale feasibility	✗ Not viable	✓ Possible
Process type	Mining + physical separation	Chemical / solution
CAPEX scalability	Poor	Moderate
Complexity	Very high	Moderate–high
Best small-scale route	N/A	Double decomposition

Overall, there is a clear distinction between MOP and SOP in terms of small-scale feasibility. MOP production is not practical at small scale due to its dependence on mining and bulk mineral processing, whereas SOP production can be achieved through selected routes. Among these, double decomposition is the most viable and flexible option, while langbeinite and brine processing are feasible only under specific resource or environmental conditions. The Mannheim process remains unsuitable for small-scale implementation.

Market Applications and Crop Requirements

Potassium (K) is a central factor in plant physiology that underpins disease resistance, pest tolerance, and responses to environmental stress. Adequate K nutrition strengthens cell walls and supports robust structural integrity, enabling plants to better endure drought, salinity, and temperature extremes. These functional benefits contribute not only to yield potential but also to crop quality and post-harvest performance across diverse production systems [215-217]. The evidence across horticultural and field crops repeatedly links potassium supply with improvements in fruit size, colour, flavour, sugar content, storage life, and processing quality, especially when chloride-sensitive crops are managed with chloride-free fertilisers such as SOP (sulphate of potash) rather than MOP (muriate of potash) [215, 216, 218, 219].

Potash Fertilisers: Forms, Crop Responses, and Quality Outcomes

MOP and SOP are the two predominant forms of commercial potash fertilisers. MOP (KCl) provides high potassium concentration at relatively low cost and is widely used in broadacre and chloride-tolerant crops, where it can enhance yield potential and nutrient uptake without notable adverse effects on crop quality in tolerant species [215, 217, 220]. In contrast, SOP (K2SO4) is preferred for chloride-sensitive crops and high-value horticultural crops where quality traits such as storage tolerance, flavour, colour, and soluble solids are prioritized, because SOP avoids chloride-related quality reductions and, in sulphur-deficient soils, can supply potassium alongside sulphur to support protein synthesis and crop development [215, 216, 219-221]. The differential use of MOP and SOP according to crop sensitivity and quality objectives is a recurring theme across production systems, with MOP dominating cost-effective, high-volume agriculture and SOP driving quality-focused horticulture and export-oriented crops [215-217, 219, 221].

Crop-specific considerations reveal nuanced choices between fertiliser forms. Cereals (e.g., wheat, barley, maize) and pasture grasses historically tolerate chloride, enabling effective fertilisation with MOP to maximize yield at a low price, provided soil and crop conditions do not impose quality penalties [215, 216]. Conversely, chloride-sensitive crops such as potatoes, grapes, citrus, and many vegetables require chloride-free sources to protect yield and post-harvest quality; in these systems SOP is often preferred despite higher cost due to the premium value of quality attributes and longer marketable shelf life [215, 216, 219]. In high-sulphur-demand crops (oilseeds like canola, legumes, some vegetables), SOP can supply essential sulphur alongside potassium, supporting protein synthesis and overall plant development in sulphur-deficient soils [215, 220].

Greenhouse, fertigation, and precision systems further influence fertiliser choice. In controlled environments, chloride-free fertilisers and highly soluble formulations support precise nutrient management, enabling consistent supply to the root zone and minimizing adverse chloride effects on leaf and fruit quality. SOP and potassium nitrate are commonly used in fertigation to ensure precise, chloride-free delivery, particularly in high-value crops and intensive production settings [217, 218, 220, 222].

Potassium's Role in Post-Harvest Quality and Shelf Life

Across several horticultural crops, adequate potassium nutrition is associated with improved post-harvest qualities and extended shelf life. In tomatoes, potassium supply improves fruit colour, soluble solids, and overall shelf life, while insufficient potassium can compromise post-harvest quality, highlighting the importance of targeted K nutrition during preharvest stages for marketable quality [216, 221]. In leafy and fruiting crops such as strawberries, cucumbers, and citrus, foliar or soil-applied potassium influences carotenoid content, antioxidant capacity, and fruit firmness, contributing to better storage performance and consumer acceptance. While some studies focus on broader preharvest and postharvest factors, the consistent finding is that balanced K nutrition supports post-harvest resilience and market value, particularly when combined with thoughtful irrigation, handling, and storage practices [217, 218, 223, 224].

Potassium in Turf, Landscaping, and Amenity Settings

In turf and landscape applications, potassium enhances stress tolerance (wear, drought, temperature fluctuations), root development, and disease resistance, improving quality and longevity of lawns and sport surfaces. Controlled-release formulations and precision delivery are often employed to sustain a consistent nutrient supply in these contexts, aligning with general 4R nutrient stewardship principles (right source, right placement, right rate, right timing) to optimize performance and minimize environmental impacts [217, 223].

Crop-Specific Market Dynamics and Global Demand

The global potash market is shaped by the demand for broadacre staple crops and high-value horticulture. In broadacre cropping, MOP remains the predominant fertiliser due to its low cost and high K concentration, supporting yield gains in large-scale cereal and legume production on soils with varying fertility [215, 217, 221]. Horticulture, including grape, citrus, tomato, potato, and banana production, relies on potassium to enhance quality attributes and storage performance; SOP is frequently chosen for chloride-sensitive crops or where sulphur supply is needed, despite higher product costs due to quality and export-standard requirements [215, 216, 218, 219]. In greenhouse and fertigation environments, chloride-free fertilisers (SOP or potassium nitrate) are favoured to achieve precise nutrient control and quality outcomes for intensive crops [217, 218, 220]. The rising demand for SOP in high-value agriculture and precision farming reflects shifts toward quality-centric production and export compliance, with potassium nutrition playing a central role in achieving these objectives [215, 216, 218].

Soil and Environmental Considerations in Potash Management

Soil properties (texture, pH, elemental balance) and environmental conditions (rainfall, irrigation) govern potassium mobility, availability, and leaching risk. In saline soils, excessive chloride from MOP can worsen salinity stress, making SOP a more suitable choice in such contexts. Rainfall intensity influences potassium movement and leaching dynamics, necessitating management adjustments to maintain adequate soil K availability. Sulphur-deficient soils provide an additional benefit for SOP by delivering both potassium and sulphur to crops with high sulphur demands, including oilseeds and legumes, further supporting protein synthesis and growth [215, 217, 219, 220].

Practical Implications for Fertiliser Selection and Management:

- For chloride-tolerant, high-volume crops (e.g., wheat, maize, barley), MOP offers a cost-effective potassium source that supports yield without substantial quality penalties when soils and climate permit; routine soil and tissue testing should guide application rates and timing to avoid imbalances with other nutrients [215-217].
- For chloride-sensitive or high-value crops (e.g., grapes, citrus, potatoes, certain vegetables), SOP is frequently preferred to protect storage quality, flavour, and processing attributes; the added sulphur in SOP can benefit sulphur-demanding crops, particularly in sulphur-deficient soils, supporting protein synthesis and overall quality [215, 216, 219, 220].
- In greenhouse, fertigation, and precision farming contexts, highly soluble, chloride-free sources (SOP or potassium nitrate) enable precise nutrient management, improving fruit quality, shelf life, and product consistency in intensive productions [217, 218, 220, 222].
- Across crop types, adequate potassium supports post-harvest performance by contributing to firmness, sugars, soluble solids, and storage resilience; integrated management including irrigation strategy, storage conditions, and handling is essential to realize full benefits [217, 218, 221, 224].
- Soil and environmental conditions must inform potassium management plans, including potential adjustments for leaching in high rainfall regimes and chloride management in saline or chloride-prone soils; a holistic approach aligned with the 4R framework enhances efficiency and sustainability [217, 223, 224].

Chapter 9

Fertiliser Blending and Formulation

This chapter builds on the foundational principles of fertiliser production and nutrient function by shifting focus toward the practical design, formulation, and delivery of multi-nutrient fertilisers, particularly NPK products. Having explored individual fertiliser types and nutrient roles in earlier chapters, the discussion now moves into how these nutrients are combined, engineered, and optimised to meet real-world agricultural demands.

At the core of modern fertiliser practice is the ability to deliver multiple nutrients in a single, efficient product. This chapter begins by examining the two primary manufacturing approaches—bulk blending and chemical granulation—highlighting how each method influences product uniformity, cost, flexibility, and field performance. Understanding these differences is essential for selecting the most appropriate fertiliser type for specific cropping systems and operational constraints.

The chapter then progresses into NPK formulation strategies, where the emphasis shifts from production methods to decision-making. Here, the integration of crop requirements, soil conditions, nutrient sources, and application methods is explored in detail. This section demonstrates how fertiliser design is not simply a matter of mixing nutrients, but a structured process that aligns agronomic science with practical and economic considerations.

To support this, a range of formulation examples is presented, illustrating how nutrient ratios are adjusted for different crops, growth stages, and production systems. These examples bridge theory and application, showing how balanced, nitrogen-rich, phosphorus-rich, and potassium-rich formulations are used strategically to optimise plant growth, yield, and quality.

The chapter also introduces additives and coating technologies, which enhance fertiliser performance by improving handling characteristics, nutrient stability, and release behaviour.

These technologies play an increasingly important role in modern agriculture, particularly in improving nutrient use efficiency and reducing environmental impact.

Equally important is the focus on quality control and uniformity, where the physical and chemical consistency of fertilisers is examined. This section highlights how particle size, granule strength, moisture content, and nutrient distribution directly influence application accuracy and crop response, reinforcing the importance of manufacturing precision.

Finally, the chapter culminates in the concept of custom fertiliser solutions, bringing together all preceding elements into a practical, crop-focused framework. It outlines how fertilisers can be tailored to specific crops, soils, and environments, and provides a step-by-step approach to designing, implementing, and refining these solutions—including considerations for small-scale production and application.

Together, these topics present fertiliser blending and formulation as a multidisciplinary process that combines chemistry, agronomy, engineering, and economics. By the end of this chapter, the reader will have a clear understanding of how fertiliser products are not only produced, but strategically designed and applied to maximise productivity, efficiency, and sustainability in modern agricultural systems.

Bulk Blending vs Chemical Granulation

Bulk blending and chemical granulation are two fundamentally different methods used to produce fertiliser products, particularly compound or multi-nutrient fertilisers such as NPK blends. While both approaches aim to deliver multiple nutrients in a single product, they differ significantly in process, physical characteristics, cost, and performance in the field.

An NPK blend is a fertiliser that contains the three primary plant nutrients, nitrogen (N), phosphorus (P), and potassium (K), combined in specific proportions to support plant growth. These nutrients are essential for different physiological functions within plants, and an NPK fertiliser provides a balanced supply in a single product, making it a convenient and effective solution for a wide range of agricultural and horticultural applications.

Each of the three nutrients plays a distinct role in plant development. Nitrogen promotes leafy growth and is responsible for the green colour of plants through its role in chlorophyll production. Phosphorus supports root development, energy transfer, and flowering, making it critical during early growth stages and reproductive phases. Potassium contributes to overall plant health by regulating water balance, improving resistance to disease, and enhancing tolerance to environmental stress.

NPK fertilisers are typically labelled using three numbers, such as 10-20-10, 15-15-15, or 20-10-5. These numbers represent the percentage of each nutrient in the product. The first number indicates the nitrogen content, the second represents phosphorus (expressed as P_2O_5 equivalent), and the third represents potassium (expressed as K_2O equivalent). For example, a 15-15-15 fertiliser contains 15% nitrogen, 15% phosphorus, and 15% potassium, providing a balanced nutrient ratio.

There are two main types of NPK fertilisers based on how they are produced. Bulk blends are created by physically mixing separate fertiliser granules, such as urea, monoammonium phosphate (MAP), and muriate of potash (MOP). In this type, each granule contains only one nutrient. Bulk blends are cost-effective and flexible but can be prone to segregation if the particles differ in size or density.

In contrast, granulated or compound fertilisers are produced through chemical or slurry-based processes, resulting in granules that contain all three nutrients within each particle. This ensures uniform nutrient distribution and more consistent performance during application, although the production process is more complex and typically more expensive.

NPK blends are widely used because they provide balanced nutrition in a single application, improving crop yield and overall quality. They can be customised to suit specific crops and soil conditions, allowing farmers and growers to optimise nutrient management. Additionally, using a single blended fertiliser reduces the need for multiple applications, improving efficiency and simplifying fertiliser programs.

Bulk blending is the simpler and more cost-effective method. In this process, individual fertiliser materials, such as urea (nitrogen), diammonium phosphate (phosphorus), and potash (potassium, typically MOP or SOP), are physically mixed together in their granular form. Each component remains chemically unchanged, and the final product is simply a mixture of separate granules, each containing a specific nutrient. The blending process typically involves conveying, weighing, and mixing systems designed to achieve a uniform distribution of particles based on size and density.

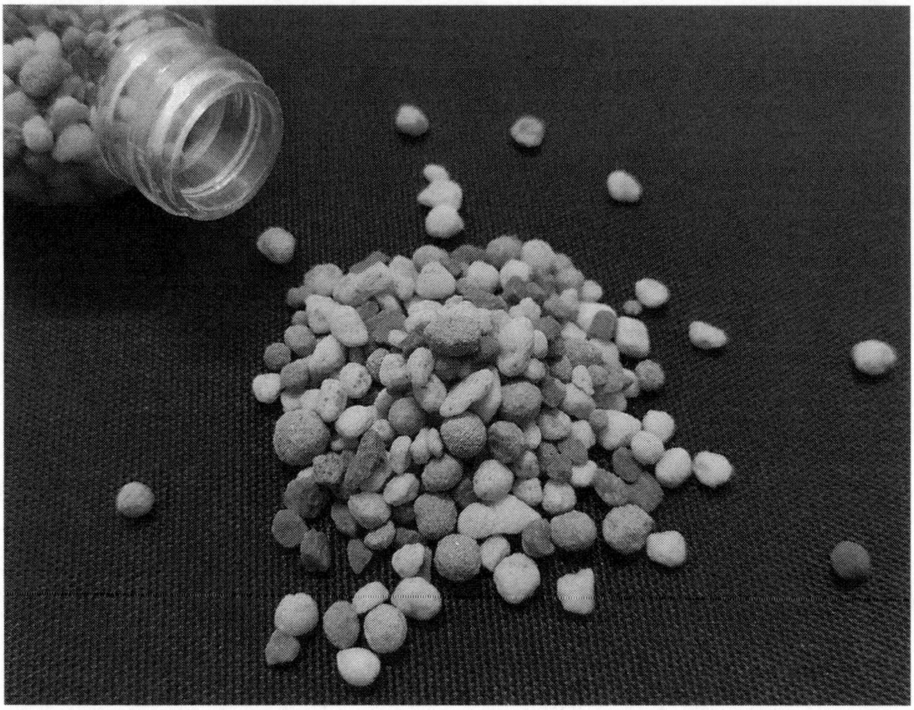

Figure 52: Bulk-blend fertiliser that contains NPK. Wee Hong, CC BY-SA 4.0, via Wikimedia Commons.

The effectiveness of bulk blending depends heavily on the physical compatibility of the materials being mixed. Ideally, all components should have similar particle sizes, densities, and shapes to prevent segregation during handling, transport, or application. If this is not achieved, the mixture can separate, leading to uneven nutrient distribution in the field. This is known as "segregation," and it is one of the main limitations of bulk-blended fertilisers. Despite this, bulk blending remains widely used because it is flexible, relatively low-cost, and allows fertilisers to be customised quickly to meet specific crop or soil requirements.

In contrast, chemical granulation involves combining raw materials in a chemical or slurry-based process to produce uniform granules that contain all required nutrients within each particle. Instead of mixing finished products, the nutrients are reacted or combined during processing, typically in a granulator such as a rotary drum, pan granulator, or fluidised bed system. The result is a homogeneous granule where nitrogen, phosphorus, potassium, and other nutrients are chemically or physically integrated.

The granulation process usually begins with the preparation of a slurry or melt containing the desired nutrient components. This mixture is then fed into a granulator, where it is agglomerated into small particles. As the granules form, they are dried, cooled, and screened to achieve a consistent size range. Oversized particles are crushed and recycled, while undersized particles are returned to the process to improve efficiency. The final product is a uniform, free-flowing fertiliser with consistent nutrient composition in every granule.

One of the key advantages of chemical granulation is uniform nutrient distribution. Because each granule contains the same nutrient ratio, there is no risk of segregation during handling or spreading. This ensures that crops receive a consistent supply of nutrients across the field. Granulated fertilisers also tend to have better physical properties, including improved strength, reduced dust, and enhanced flowability, which makes them easier to handle, store, and apply using modern spreading equipment.

However, chemical granulation is more complex and capital-intensive than bulk blending. It requires specialised equipment, higher energy input (especially for drying and heating), and more sophisticated process control. The production process is less flexible than blending, as changing formulations may require adjustments to the entire system rather than simply altering mix ratios. As a result, granulated fertilisers are typically produced in large-scale industrial facilities.

From an agronomic perspective, granulated fertilisers often provide more predictable nutrient delivery, particularly in precision agriculture systems where uniformity is critical. Bulk blends, on the other hand, are well suited to situations where cost is a primary concern and where application methods can accommodate some variability in nutrient distribution.

NPK Formulation Strategies

NPK formulation strategies refer to the process of designing fertiliser blends or compound products that deliver the appropriate balance of nitrogen (N), phosphorus (P), and potassium (K) to meet specific crop, soil, and environmental requirements. Rather than relying on a single standard fertiliser, formulation involves tailoring nutrient ratios, selecting suitable raw material sources, and considering physical characteristics to optimise plant growth, maximise yield, and improve nutrient use efficiency.

A fundamental aspect of NPK formulation is aligning nutrient supply with the growth stage and physiological needs of the crop. Different crops, and even different stages within the same crop, require varying nutrient ratios. During early growth, higher phosphorus levels are important to support root development. In the vegetative stage, nitrogen becomes more critical for leaf and stem growth. During the reproductive stage, potassium demand increases to support fruiting, improve crop quality, and enhance resistance to stress. For example, a fertiliser such as 12-24-12 may be used at planting, while a formulation like 20-10-10 may be applied later to promote vegetative growth.

Formulation strategies are also tailored to specific crops, as nutrient requirements vary significantly between crop types. Cereals such as wheat and maize typically require higher nitrogen inputs, making nitrogen-rich blends suitable. Root crops like potatoes often benefit from balanced or potassium-rich formulations, while fruits and vegetables require higher potassium levels to improve quality, flavour, and shelf life. Legumes, which can fix atmospheric nitrogen, generally require lower nitrogen inputs but higher phosphorus levels to support root development and nodulation. Customising fertiliser formulations in this way ensures nutrients are supplied efficiently without excess or deficiency.

Soil conditions play a central role in determining the appropriate NPK formulation. Soil testing provides information on existing nutrient levels, allowing fertiliser composition to be adjusted accordingly. For example, nitrogen-deficient soils require increased nitrogen inputs, while soils already rich in phosphorus may require reduced phosphorus content in the fertiliser. In sulphur-deficient soils, the use of sulphate of potash (SOP) or other sulphur-containing fertilisers can provide additional benefits. This soil-based approach improves nutrient efficiency, reduces waste, and minimises environmental impact.

In addition to nutrient ratios, formulation involves selecting appropriate chemical sources for each nutrient. Common nitrogen sources include urea, ammonium nitrate, and ammonium sulphate, while phosphorus is typically supplied through monoammonium phosphate (MAP) or diammonium phosphate (DAP). Potassium is supplied through muriate of potash (MOP, KCl) or sulphate of potash (SOP, K_2SO_4). The choice of source depends on factors such as cost, solubility, compatibility, and crop sensitivity, particularly in the case of chloride-sensitive crops that require SOP instead of MOP.

Physical and chemical compatibility between materials is another critical consideration in NPK formulation. Proper compatibility prevents issues such as caking, moisture absorption, and unwanted chemical reactions during storage and handling. It also ensures that particle size and density are consistent, reducing the risk of segregation in bulk blends and maintaining good flowability for application equipment. In some cases, stabilisers or coatings may be required to improve product stability and handling characteristics.

The intended application method also influences formulation design. Fertilisers used for broadcast spreading require durable, uniform granules to ensure even distribution across the field. In fertigation systems, where fertiliser is delivered through irrigation water, high solubility and purity are essential. Controlled-release systems, on the other hand, use coated or slow-release formulations to provide nutrients gradually over time. Matching the formulation to the application method ensures nutrients are delivered efficiently and effectively.

Modern NPK formulation strategies increasingly incorporate environmental and efficiency considerations. The use of slow-release or stabilised fertilisers can reduce nutrient losses through leaching or volatilisation. Precision blending allows for more accurate nutrient application, minimising overuse and reducing environmental impact. Balanced nutrient ratios also help prevent runoff and soil degradation, which is particularly important in environmentally sensitive regions.

Economic factors are also central to formulation decisions. Cost-effective strategies may involve using MOP instead of SOP where crop tolerance allows, optimising nutrient ratios to avoid unnecessary inputs, and selecting raw materials based on local availability. The overall objective is to achieve the best possible agronomic outcome while maintaining economic viability for producers.

Formulation Examples

NPK formulations vary widely depending on crop type, soil conditions, and the stage of plant growth. These formulations are typically expressed as ratios, such as 10-10-10, which represent the percentage of nitrogen (N), phosphorus (P_2O_5), and potassium (K_2O) in the fertiliser. Different ratios are designed to meet specific agronomic needs, ensuring that plants receive the appropriate balance of nutrients for optimal growth, yield, and quality.

Balanced formulations contain equal or near-equal proportions of the three nutrients and are commonly used for general-purpose fertilisation where soils are moderately fertile or where no specific nutrient deficiency has been identified. Examples include 10-10-10 for general crops, lawns, and vegetables, 15-15-15 for broadacre crops and maintenance fertilisation, and 20-20-20, which is often used as a water-soluble fertiliser in fertigation and hydroponic systems. These formulations provide a consistent and reliable nutrient supply across a wide range of applications.

High-nitrogen formulations are designed to promote rapid vegetative growth, particularly the development of leaves and stems. These blends are commonly used in crops that require strong early growth or high biomass production. For example, 20-10-10 is widely used in cereals, maize, and pasture systems, while 30-10-10 is suitable for turf, lawns, and leafy vegetables. A formulation such as 25-5-5 may be applied during the early vegetative stages of crop development to encourage vigorous growth.

High-phosphorus formulations are primarily used to support root development and early plant establishment. These fertilisers are particularly important during planting and transplanting, as well as in soils that are deficient in phosphorus. Common examples include 12-24-12, which is used as a starter fertiliser, and 10-30-10 for transplanting and early growth stages. A formulation such as 18-46-0, typically based on diammonium phosphate (DAP), provides a very high phosphorus content and is often used for root crops or where rapid root development is required.

High-potassium formulations are used to improve crop quality, enhance disease resistance, and increase tolerance to environmental stress such as drought or temperature extremes. These blends are especially important for fruiting and flowering crops. Examples include 10-10-20 for general fruit and vegetable production, 12-12-36 for high-value horticultural crops, and 15-5-30 for late-stage fruiting, where potassium demand is highest. These formulations are commonly used in systems where crop quality and market value are critical.

Crop-specific formulations are tailored to the nutrient requirements of particular crops. For instance, wheat may use formulations such as 20-10-10 or 18-12-8, which are more nitrogen-focused to support grain production. Maize often requires higher nitrogen inputs, with formulations like 25-10-10. Potatoes typically benefit from higher potassium levels, with blends such as 12-12-18 or 15-10-20, while grapes and other chloride-sensitive crops often use potassium-rich formulations like 10-5-20 or SOP-based blends to maintain quality and avoid chloride-related issues.

Water-soluble or fertigation formulations are designed for use in controlled environments such as greenhouses, hydroponic systems, and precision agriculture. These fertilisers must dissolve completely in water and provide nutrients in a readily available form. Common examples include 20-20-20 for general fertigation, 15-5-30 for flowering and fruiting stages, and 13-40-13 for early growth where high phosphorus is required. These formulations allow precise control over nutrient delivery and are essential in intensive production systems.

Specialty and controlled-release formulations are developed to provide nutrients gradually over time, improving efficiency and reducing nutrient losses. Examples include 16-4-8 slow-release fertiliser for turf and landscaping, 14-14-14 coated fertilisers for long-term crop nutrition, and controlled-release products such as Osmocote-type blends used in horticulture. These formulations are particularly useful in situations where consistent nutrient supply is needed and where minimising leaching or volatilisation is important.

Additives and Coating Technologies

Additives and coating technologies are used in fertiliser production to improve product performance, handling, storage stability, and nutrient efficiency. While the core nutrients in fertilisers such as nitrogen, phosphorus, and potassium provide the agronomic value, additives and coatings enhance how those nutrients are delivered, retained in the soil, and utilised by plants. These technologies are particularly important in modern agriculture, where efficiency, environmental protection, and precision application are key priorities.

Additives are substances incorporated into fertilisers in relatively small amounts to modify their physical or chemical properties. One of the most common types is anti-caking agents, which prevent fertiliser granules from sticking together during storage and transport. Fertilisers are often hygroscopic, meaning they absorb moisture from the air, which can lead to clumping. Anti-caking additives, such as clays, silicates, or surface coatings, help maintain free-flowing granules and ensure consistent application through spreading equipment.

Another important category is dust control agents, which reduce the formation of fine particles during handling. Dust can lead to product loss, health risks for workers, and uneven nutrient distribution. Oils, waxes, or polymer-based additives are often applied to bind fine particles to granules, improving safety and product quality. Similarly, flow conditioners are used to enhance the movement of fertiliser through hoppers and spreaders, ensuring uniform delivery in the field.

Chemical additives are also used to improve nutrient stability and efficiency. For example, urease inhibitors are added to urea-based fertilisers to slow the conversion of urea into ammonia, reducing nitrogen losses through volatilisation. Nitrification inhibitors slow the conversion of ammonium to nitrate, reducing leaching and denitrification losses. These additives help retain nitrogen in forms that are more available to plants over a longer period, improving fertiliser efficiency and reducing environmental impact.

In addition to additives, coating technologies play a critical role in controlling how nutrients are released. Coatings are applied to fertiliser granules to create a physical barrier between the nutrient and the surrounding environment. One of the most widely used approaches is controlled-release fertilisers (CRFs), where nutrients are released gradually over time. These coatings are typically made from polymers, resins, or sulphur, and they regulate nutrient release based on factors such as temperature, moisture, and microbial activity.

Polymer-coated fertilisers are among the most advanced controlled-release systems. The coating acts as a semi-permeable membrane that allows water to enter the granule, dissolve the nutrients inside, and then slowly release them back into the soil. This controlled diffusion process ensures a steady supply of nutrients that matches plant demand, reducing losses and improving uptake efficiency. Sulphur-coated fertilisers, on the other hand, use a layer of sulphur around the granule, often combined with a sealant such as wax. These coatings are less precise than polymer coatings but are generally more cost-effective and can also provide sulphur as a secondary nutrient.

Coating technologies also include stabilised fertilisers, where the coating or additive works to protect nutrients from rapid chemical transformation in the soil. These products do not necessarily control release physically but instead slow down processes such as volatilisation, nitrification, or leaching. This approach is particularly important for nitrogen fertilisers, where losses can be significant if not managed properly.

Another application of coatings is in improving the compatibility and uniformity of blended fertilisers. Coatings can be used to standardise particle size, reduce segregation in bulk blends, and improve the durability of granules during handling. This ensures that nutrients are distributed evenly during application, which is critical for achieving consistent crop performance.

From an environmental perspective, additives and coating technologies play a key role in reducing nutrient losses to the atmosphere and waterways. By improving nutrient retention in the soil and synchronising release with plant uptake, these technologies help minimise issues such as nitrate leaching, ammonia volatilisation, and eutrophication of water bodies. As a result, they are increasingly important in sustainable agriculture and in meeting regulatory requirements related to nutrient management.

However, these technologies also introduce additional cost and complexity. Coated and enhanced-efficiency fertilisers are typically more expensive than conventional products, and their performance can vary depending on environmental conditions such as temperature and soil moisture. Therefore, their use must be carefully matched to crop requirements, management practices, and economic considerations.

Quality Control and Uniformity

Quality control and uniformity in fertiliser production refer to the processes and standards used to ensure that fertiliser products consistently meet specified nutrient content, physical properties, and performance requirements. These factors are critical because fertilisers must deliver precise nutrient amounts evenly across a field. Poor quality or lack of uniformity can lead to uneven crop growth, reduced yields, inefficient nutrient use, and potential environmental impacts.

At the core of quality control is the verification of nutrient composition. Fertilisers are labelled with guaranteed analysis values (e.g. NPK ratios), and manufacturers must ensure that each batch meets these specifications. This involves regular sampling and laboratory testing to confirm the concentration of nitrogen, phosphorus, potassium, and any secondary or micronutrients. Consistency is essential, as variations in nutrient content can result in over- or under-application in the field.

Uniformity is particularly important in relation to granule composition and distribution. In high-quality fertilisers—especially chemically granulated or compound products—each granule contains the same proportion of nutrients. This ensures that when fertiliser is spread across a field, every part of the soil receives a balanced nutrient supply. In contrast, bulk-blended fertilisers may contain separate granules for each nutrient, making them more susceptible to segregation if particle size, density, or shape differs between components. Maintaining uniform particle characteristics helps minimise this risk.

Another key aspect of quality control is particle size distribution. Fertiliser granules must fall within a specified size range to ensure even spreading and predictable application rates. If particles are too fine, they may create dust and drift during application; if too large, they may not spread evenly or may segregate from smaller particles. Manufacturers use screening processes to classify granules and recycle out-of-spec material to maintain consistency.

Granule strength and durability are also critical quality parameters. Fertilisers must withstand handling, transport, and storage without breaking down into fines. Weak granules can degrade, leading to dust formation, caking, and uneven nutrient distribution. Crushing strength tests and abrasion resistance testing are commonly used to assess these properties.

Moisture content is another important factor. Excess moisture can lead to caking, microbial activity, or chemical degradation, while very low moisture levels may increase dustiness.

Quality control processes ensure that moisture levels are maintained within optimal ranges to preserve product stability and handling characteristics.

Manufacturers also monitor flowability and handling properties. Fertilisers must move freely through storage systems, conveyors, and application equipment. Additives such as anti-caking agents and coatings are often used to improve flow characteristics. Testing may include angle of repose measurements and flow tests to ensure consistent performance.

In addition to physical properties, chemical stability and compatibility are assessed to prevent unwanted reactions during storage or blending. Some fertiliser components can react with each other, leading to nutrient loss or product degradation. Quality control procedures ensure that compatible materials are used and that storage conditions are appropriate.

Quality assurance systems typically include process control during manufacturing, such as monitoring temperature, reaction conditions, and granulation parameters. Automated systems and sensors are often used to maintain consistent production conditions. Final product inspection and certification ensure that only materials meeting specifications are released to market.

From an agronomic perspective, uniformity directly affects fertiliser efficiency and crop performance. Even distribution of nutrients leads to consistent plant growth, while variability can create areas of deficiency or excess. This is especially important in modern precision agriculture, where accurate nutrient placement is essential.

Custom Fertiliser Solutions for Crops

Custom fertiliser solutions for crops refer to the practice of designing and supplying fertilisers that are specifically tailored to the nutrient requirements of a particular crop, soil type, and growing environment. Instead of applying standard, one-size-fits-all fertilisers, custom solutions aim to deliver the right nutrients, in the right proportions, at the right time to maximise crop performance, improve efficiency, and reduce waste.

A key foundation of custom fertiliser solutions is soil and plant analysis. Soil testing identifies existing nutrient levels, pH, salinity, and other characteristics that influence nutrient availability. Plant tissue analysis can further indicate how well crops are absorbing nutrients during growth. By combining this data, agronomists can determine which nutrients are deficient, sufficient, or excessive, and design fertiliser programs that address specific gaps without over-application.

Crop type and growth stage are also central to customisation. Different crops have unique nutrient demands, and these requirements change throughout the growing cycle. For example, cereals such as wheat and maize require higher nitrogen inputs during vegetative growth, while fruiting crops such as tomatoes or grapes require increased potassium during flowering and fruit development. Custom fertiliser solutions adjust nutrient ratios over time, often through staged applications, to match these changing demands and optimise yield and quality.

Another important aspect is the selection of appropriate nutrient sources. Custom fertilisers may incorporate different forms of nitrogen, phosphorus, and potassium depending on crop sensitivity, soil chemistry, and application method. For instance, chloride-sensitive crops may require sulphate of potash (SOP) instead of muriate of potash (MOP), while ammonium-based or nitrate-based nitrogen sources may be selected based on soil conditions and environmental factors. Secondary nutrients such as sulphur, calcium, and magnesium, as well as micronutrients like zinc, boron, and iron, can also be included where required.

Custom fertiliser solutions are often designed to suit specific application methods. In broadacre farming, fertilisers may be formulated for uniform spreading and durability. In contrast, greenhouse and hydroponic systems require highly soluble fertilisers for fertigation, allowing precise control of nutrient delivery through irrigation systems. Controlled-release or coated fertilisers may be used in situations where a steady nutrient supply is needed over an extended period, such as in horticulture or landscaping.

Precision agriculture technologies play an increasingly important role in custom fertilisation. Tools such as GPS-guided equipment, variable rate application systems, and remote sensing enable fertilisers to be applied at different rates across a field based on soil variability. This allows farmers to target specific areas that require more or less nutrients, improving efficiency and reducing environmental impact.

Environmental and sustainability considerations are also integral to custom fertiliser strategies. By applying nutrients more accurately and only where needed, custom solutions reduce the risk of nutrient losses through leaching, runoff, or volatilisation. This helps protect water quality and reduces greenhouse gas emissions associated with fertiliser use. In addition, improving nutrient use efficiency can lower overall fertiliser consumption and reduce input costs.

From an economic perspective, custom fertiliser solutions aim to maximise return on investment. Although they may involve higher upfront costs for testing, formulation, or specialised products, they often result in improved yields, better crop quality, and more efficient use of inputs. This can lead to greater profitability, particularly in high-value cropping systems.

Creating a Custom Fertiliser Solution

Creating a custom fertiliser solution is a structured process that starts with understanding the crop and ends with monitoring results and adjusting the program over time. The goal is to supply the right nutrients, in the right form, at the right rate, and at the right time for a specific crop and growing environment.

The first step is to clearly define the crop, production system, and performance target. This means identifying what crop is being grown, whether it is broadacre, horticulture, greenhouse, hydroponic, or turf, and what the production objective is. For example, the aim may be maximum grain yield, improved fruit quality, faster vegetative growth, better storage life, or correcting a visible nutrient deficiency. A fertiliser program for wheat will differ greatly from one for tomatoes, grapes, or potatoes.

The second step is to assess the site conditions. This includes collecting soil samples and having them tested for pH, electrical conductivity, organic matter, and nutrient levels such as nitrogen, phosphorus, potassium, sulphur, calcium, magnesium, and trace elements where relevant. If the crop is already established, plant tissue analysis can also be useful to show what nutrients are actually being taken up. This step provides the baseline information needed to avoid guessing.

The third step is to identify the nutrient demand of the crop across its growth stages. Different crops require different nutrient ratios, and those ratios change over time. Early growth often needs more phosphorus to support root development. Vegetative growth usually requires more nitrogen. Flowering and fruiting stages often need more potassium, and some crops also need extra calcium, magnesium, or sulphur. At this stage, the aim is to map out the crop's nutrient demand from planting through to harvest.

The fourth step is to compare crop demand with soil supply. Once the soil and tissue data are available, the next task is to determine what the soil can already provide and what must be added through fertiliser. If soil phosphorus is already high, the fertiliser may need less phosphorus. If potassium is low, the blend may need to be potassium-rich. If sulphur is deficient, that should also be included. This gap analysis is what turns a general fertiliser plan into a custom one.

The fifth step is to choose the nutrient sources. Nitrogen may come from urea, ammonium sulphate, ammonium nitrate, calcium ammonium nitrate, or nitrate-based liquid sources. Phosphorus may come from MAP, DAP, phosphoric acid, or superphosphate. Potassium may come from MOP or SOP, depending on crop chloride sensitivity. Secondary nutrients and micronutrients may also be added if needed. The selection depends on crop sensitivity, soil chemistry, solubility requirements, compatibility, cost, and method of application.

The sixth step is to determine the formulation ratio. This is where the NPK ratio and any additional nutrients are calculated. For example, one crop may need a nitrogen-focused blend, such as 20-10-10, while another may need a potassium-focused blend, such as 12-12-24. In fertigation systems, the solution may be built around nutrient concentration in the irrigation water rather than a simple bag ratio. The formulation should reflect the nutrient gaps identified earlier and the crop stage being targeted.

The seventh step is to match the fertiliser to the application method. A custom fertiliser for broadcast spreading must have suitable granule size, strength, and flowability. A custom fertiliser for fertigation must be highly soluble and chemically compatible in solution. A starter fertiliser placed near the seed may need lower salt risk. A controlled-release program may be more appropriate where fewer applications are possible. The application method affects both the product design and the nutrient source selection.

The eighth step is to calculate the application rate. Once the nutrient ratio is set, the total amount to apply must be determined based on the crop target, soil test results, and area to be treated. This step converts nutrient requirements into kilograms per hectare, tonnes per field, grams per plant, or concentration in irrigation water. Care is needed to ensure that the application supplies enough nutrients without causing toxicity, salt stress, or unnecessary losses.

The ninth step is to check compatibility and physical quality. If the fertiliser is a bulk blend, the materials should have similar particle size and density to reduce segregation. If it is a liquid or soluble blend, the materials must remain stable in solution and not precipitate. If the product contains additives, coatings, or stabilisers, they should suit the storage and application conditions. This step is essential for product consistency and effective field performance.

The tenth step is to implement the fertiliser program in stages rather than treating it as a single fixed input. In many cases, it is better to split the program into several applications across the season. For example, a starter blend may be used at planting, a nitrogen-rich blend during vegetative growth, and a potassium-rich blend during fruiting or grain filling. This improves nutrient use efficiency and better matches crop uptake.

The eleventh step is to monitor crop response. After application, the crop should be observed for growth, colour, vigour, leaf symptoms, yield components, and quality indicators. Follow-up soil or tissue testing can confirm whether the nutrient program is working as intended. This is especially important in high-value crops, greenhouse systems, and precision agriculture programs.

The final step is to review and refine the formulation. A custom fertiliser solution should not be static. Results from one crop cycle should be used to improve the next one. If a nutrient was still deficient, the formulation may need adjustment. If nutrient levels were excessive, rates or sources may need to be reduced. Over time, this creates a more accurate, efficient, and economically sound fertiliser strategy.

In simple terms, the process is: define the crop and goal, test the soil and plant, identify nutrient demand, calculate the nutrient gap, choose the right raw materials, create the formulation, match it to the application method, set the rate, check compatibility, apply in stages, monitor results, and refine the program. This step-by-step approach turns fertiliser use from a generic input into a targeted crop management tool.

Establishing Small Scale Production

A practical small-scale approach to custom fertiliser solutions begins with focusing on data-driven decisions rather than complex infrastructure. At this scale, there is no need for an in-house laboratory; instead, reliable baseline information can be obtained by sending soil samples to a local laboratory for analysis of nitrogen (N), phosphorus (P), potassium (K), pH, electrical conductivity (EC), and sulphur. For higher-value crops, optional plant tissue testing can provide additional insight into nutrient uptake. Alongside this, basic records should be kept on crop type, growth stage, and yield or quality targets. In practice, a single soil test per field or production block is often sufficient. Indicative costs for this stage are relatively low, typically around AUD $50–$150 (USD $30–$100 / €25–€90) per soil test, and AUD $80–$200 (USD $50–$130 / €45–€120) for tissue analysis.

Once baseline data is available, the next step is to define the nutrient gap, which is where true customisation occurs. This involves comparing what the soil can supply with what the crop requires at a given growth stage. For example, if soil potassium is low, a potassium-rich formulation is required; if phosphorus is already high, it can be reduced in the blend. A fruiting crop such as tomatoes may require increased potassium during later stages. The outcome of this step is a target nutrient ratio, such as 15-5-25 instead of a generic 15-15-15. This stage involves minimal direct cost, aside from agronomic advice if required, which may range from AUD $100–$500 (USD $65–$330 / €60–€300) depending on complexity.

A critical enabler of small-scale feasibility is the use of off-the-shelf fertiliser materials rather than manufacturing from scratch. Common inputs include urea, ammonium sulphate, calcium nitrate (nitrogen sources), MAP or DAP (phosphorus), and MOP or SOP (potassium), along with secondary nutrients such as gypsum or magnesium sulphate and trace elements. These materials are widely available and relatively affordable. Indicative raw material costs vary with market conditions, but typical ranges might be AUD $400–$900 per tonne (USD $260–$600 / €240–€550) for bulk fertiliser inputs, with specialty products like SOP or micronutrients costing more.

The next stage is selecting a production method, with three practical options for small-scale operations. The most common is dry bulk blending, where granular fertilisers are physically mixed. A basic setup includes accurate scales, a small mixer (such as a cement mixer or ribbon blender), and storage bags or bins. The process involves weighing components, mixing thoroughly, and bagging or applying the product. This is the lowest-cost option, with initial setup costs typically ranging from AUD $1,000–$10,000 (USD $650–$6,500 / €600–€6,000) depending on equipment quality and scale.

For higher-value systems such as greenhouses or hydroponics, liquid or soluble fertiliser mixing is often preferred. This requires a mixing tank, agitation (pump or stirrer), a clean water source, and filtration. Soluble fertilisers such as MAP, potassium nitrate, or SOP are dissolved, adjusted to the required concentration, and applied via irrigation systems. This approach offers high precision but requires careful management of chemical compatibility. A small-scale liquid system may cost approximately AUD $2,000–$20,000 (USD $1,300–$13,000 / €1,200–€12,000) depending on tank size, automation, and dosing equipment.

A third and highly practical approach is program-based customisation, where instead of producing a single blended product, a staged fertiliser schedule is developed. For example, a high-phosphorus fertiliser may be used at planting, followed by nitrogen during vegetative growth and potassium during fruiting. This approach requires minimal equipment and is widely used in small farming operations, with costs largely limited to fertiliser inputs and basic handling.

Regardless of the production method, the fertiliser must be matched to the application method. Broadcast applications require durable, uniform granules, while fertigation requires fully soluble materials. Banding near the seed requires low salt index products to avoid plant damage. At small scale, simplicity is important—trying to optimise for too many application methods at once can introduce unnecessary complexity and cost.

Application rates must then be calculated carefully. This involves converting nutrient requirements into practical units such as kilograms per hectare, grams per plant, or concentration levels (ppm or EC) for fertigation systems. This step is critical, as incorrect rates can lead to nutrient deficiencies, toxicity, or wasted inputs. There is no major capital cost here, but errors can be economically significant.

Compatibility and product quality must also be managed. In dry blends, differences in particle size and density can cause segregation, while moisture can lead to caking. In liquid systems, incompatible salts may precipitate. Addressing these issues may involve simple controls such as selecting similar particle sizes, maintaining dry storage conditions, or using clean water. Minor additional costs for handling and storage improvements may range from AUD $500–$5,000 (USD $330–$3,300 / €300–€3,000).

One of the major advantages of small-scale systems is the ability to apply fertiliser in stages. Instead of a single large application, nutrients can be applied more frequently in smaller amounts and adjusted based on crop response. This improves nutrient use efficiency and reduces risk. Monitoring is typically done through visual crop assessment, yield tracking, and occasional follow-up testing, with ongoing costs similar to initial soil testing.

A simple example illustrates how this works in practice. For a tomato crop in the fruiting stage with low soil potassium and adequate phosphorus, the fertiliser strategy would reduce phosphorus inputs and increase potassium, typically using SOP. A blend might include urea for nitrogen, a reduced amount of MAP for phosphorus, and SOP for potassium and sulphur. This could be applied through weekly fertigation, allowing precise adjustment based on plant response.

It is equally important to understand what is not feasible at small scale. Manufacturing base fertilisers such as MOP or DAP, operating high-temperature processes like the Mannheim process, or undertaking chemical synthesis is not practical due to high capital costs, safety requirements, and regulatory complexity. Instead, small-scale operations should remain focused on blending, dissolving, and program-based nutrient management.

Chapter 10

Fertiliser Application Technologies

This chapter builds on the principles of nutrient management by focusing on how fertilisers are effectively delivered to crops in real-world agricultural systems. While previous discussions have explored fertiliser types, formulations, and nutrient behaviour in soil, the practical success of any fertiliser program ultimately depends on how, when, and where those nutrients are applied. Application methods are not merely operational choices, they directly influence nutrient availability, uptake efficiency, crop performance, and environmental outcomes.

Modern agriculture requires a shift from uniform, convenience-based application toward more deliberate and strategic approaches that align with crop demand, soil characteristics, and production systems. Whether nutrients are spread across the soil surface, placed precisely near the root zone, or delivered through irrigation systems, each method carries distinct advantages, limitations, and suitability depending on the context. Understanding these differences is essential for improving fertiliser efficiency and reducing losses.

This chapter examines the major fertiliser application methods, including broadcasting, banding, and fertigation, along with the equipment, technologies, and decision-making processes that support them. It also considers how application strategies interact with factors such as nutrient mobility, soil conditions, crop type, and environmental risk. By linking theory with practical implementation, the chapter provides a comprehensive framework for selecting and applying fertilisers in a way that maximises productivity while supporting sustainable agricultural practices.

Broadcasting, Banding, Fertigation

Broadcasting, banding, and fertigation are three common fertiliser application methods used to deliver nutrients to crops. Each method differs in how nutrients are placed in the soil or delivered to plants, and each has specific advantages, limitations, and ideal use cases depending on the crop, soil conditions, and production system.

Broadcasting

Broadcasting is the simplest and most widely used method of fertiliser application. In this approach, fertiliser is spread evenly across the entire soil surface, either by hand or using mechanical spreaders such as spinner spreaders or boom spreaders. The fertiliser may be left on the surface or incorporated into the soil through cultivation or irrigation.

Figure 53: Amazone mounted fertiliser spreader ZA-M 1001 in operation, equipped with a safety set, with spreading widths up to 52 m. Amazone GmbH & Co. KG, CC BY-SA 3.0, via Wikimedia Commons.

This method is commonly used in broadacre cropping systems, pastures, and large-scale farming operations where speed and coverage are important. It ensures that nutrients are distributed across the whole field, making them accessible to plant roots as they grow and spread.

However, broadcasting is relatively inefficient compared to more targeted methods. Because nutrients are spread over the entire area, some may not be immediately accessible to plant roots, particularly in the early stages of growth. This can lead to nutrient losses through leaching, runoff, or volatilisation, especially for nitrogen-based fertilisers. Despite this, broadcasting remains popular due to its simplicity, low cost, and suitability for large areas.

Broadcasting equipment is designed to spread fertiliser evenly across the soil surface over a wide area. The goal is uniform coverage, consistent application rate, and efficient operation over large areas. The type of equipment used depends on scale, terrain, fertiliser type (granular or liquid), and level of precision required.

1. Spinner (Centrifugal) Spreaders

Spinner spreaders are the most widely used broadcasting equipment in agriculture. They operate using one or two rotating discs fitted with fins or vanes that throw fertiliser outward in a wide pattern.

Key components include:

- Hopper → holds fertiliser
- Metering gate → controls flow rate
- Spinning discs → distribute fertiliser
- Drive system → powered by PTO, hydraulics, or electric motors

These spreaders can achieve spreading widths from 10 to over 50 metres, making them ideal for large-scale farming. Twin-disc models provide more even distribution than single-disc units.

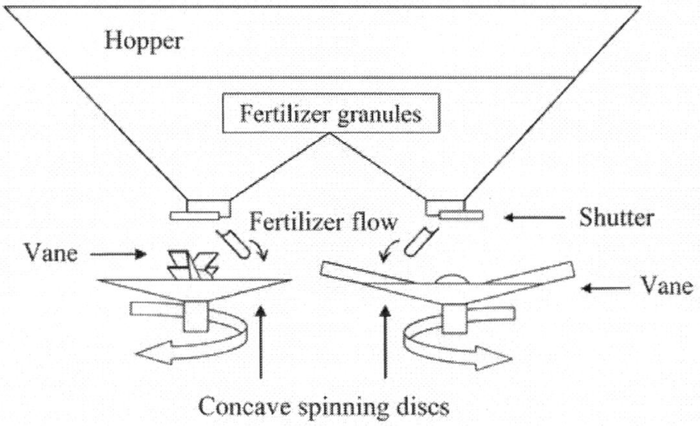

Figure 54: Centrifugal spreader principle using two spinning discs. Adapted from Villette, et al. (2008) [225].

Advantages:

- High work rate and efficiency
- Simple design and relatively low cost
- Suitable for most granular fertilisers

Limitations:

- Distribution affected by wind and granule quality
- Requires calibration for accuracy

2. Drop Spreaders

Drop spreaders release fertiliser directly below the hopper through a series of openings, creating a narrow, controlled application band.

They are commonly used in:

- Turf management
- Lawns and landscaping
- Small-scale agriculture

Advantages:

- Very accurate placement
- Minimal drift or loss
- Ideal for precise applications

Limitations:

- Narrow working width
- Slower operation compared to spinner spreaders

3. Air (Pneumatic) Spreaders

Air spreaders use pressurised air to transport fertiliser through hoses and distribute it evenly across a boom.

Key components:

- Hopper or tank
- Air fan or blower
- Distribution manifold
- Boom with outlets

These systems provide:

- Highly uniform distribution
- Reduced sensitivity to wind
- Precise application across wide booms

They are often used in:

- Precision agriculture
- High-value cropping systems

Limitations:

- Higher cost and complexity
- More maintenance required

4. Liquid Broadcast Sprayers

Liquid fertilisers can be broadcast using sprayers similar to those used for pesticides.

Components include:

- Tank for liquid fertiliser
- Pump and pressure system
- Spray boom with nozzles
- Flow control system

This method is used for:

- Nitrogen solutions (e.g. UAN)
- Foliar feeding
- Rapid nutrient application

Advantages:

- Even coverage
- Easy rate control
- Compatible with precision systems

Limitations:

- Requires soluble fertilisers
- Risk of leaf burn if misapplied

5. Hand-Held and Small-Scale Spreaders

For small-scale use, simple manual or push spreaders are common.

Types include:

- Hand-crank spreaders
- Push broadcast spreaders
- Small hopper spreaders

These are used in:

- Gardens
- Small farms
- Landscaping

Advantages:

- Low cost
- Easy to use
- Portable

Limitations:

- Limited capacity and coverage
- Less precise than larger systems

Banding

Banding is a more targeted fertiliser application method where nutrients are placed in concentrated strips (bands) in the soil, typically close to the seed or root zone. This can be done at planting (starter fertiliser) or during crop growth using specialised equipment.

The key advantage of banding is improved nutrient use efficiency. By placing fertiliser close to where roots will develop, plants can access nutrients more quickly and effectively, particularly during early growth stages. This is especially important for nutrients like phosphorus, which do not move easily through the soil.

Banding also reduces nutrient losses compared to broadcasting, as fertiliser is not exposed across the entire soil surface. This can improve fertiliser efficiency and reduce environmental impact. However, banding requires more precise equipment and careful placement. If fertiliser is placed too close to the seed, it can cause salt injury or "fertiliser burn," which can damage seedlings.

Banding requires specialised equipment designed to place fertiliser precisely in the soil near the seed or root zone, rather than spreading it across the surface. The equipment used varies depending on the cropping system, scale, and timing (at planting vs during growth), but all banding systems share the same goal: accurate placement and controlled delivery.

1. Seed Drills and Planters with Fertiliser Attachments

The most common banding equipment is a seed drill or planter fitted with fertiliser delivery systems. These machines allow fertiliser to be applied at the same time as seeding, often referred to as *starter fertiliser banding*.

They typically include:

- Fertiliser hoppers or bins for storing granular or liquid fertiliser
- Metering systems to control application rate
- Delivery tubes that carry fertiliser to the soil
- Openers or coulters that create a trench for placement

Fertiliser can be placed:

Directly below the seed (*in-furrow*)

Slightly to the side and below (*side-banding*, often called "2x2 placement")

This equipment is widely used in cereals, maize, and row crops because it ensures nutrients, especially phosphorus, are immediately available to emerging roots.

Figure 55: Direct drilling with seed drill. Merilijoost, CC BY-SA 4.0, via Wikimedia Commons.

2. Fertiliser Coulters and Knife Applicators

Coulters and knife applicators are soil-penetrating tools used to place fertiliser in bands either during planting or as a separate operation.

- Disc coulters cut through soil and residue, creating a narrow slot for fertiliser placement
- Knife applicators penetrate deeper and are often used for injecting fertiliser below the surface

These systems are commonly used for:

- Side-dressing crops during growth
- Deep banding nutrients such as nitrogen or potassium
- Applying fertiliser in minimum tillage or no-till systems

They are particularly useful where surface application would lead to losses (e.g. nitrogen volatilisation).

3. Side-Dress Applicators

Side-dress applicators are used after the crop has emerged, placing fertiliser alongside growing plants.

Typical features include:

- Row-aligned applicator units
- Adjustable spacing to match crop rows
- Tanks or bins for liquid or granular fertiliser
- Injection knives or coulters

These machines are widely used in crops like maize and vegetables to:

- Apply nitrogen during peak demand
- Supplement nutrients mid-season
- Avoid early-season losses

<u>4. Air Seeders with Banding Systems</u>

Air seeders are advanced systems that use pneumatic (air-driven) delivery to distribute seed and fertiliser simultaneously.

Key components:

- Large air carts holding seed and fertiliser
- Distribution heads that divide product into multiple lines
- Dual-shoot systems allowing separate placement of seed and fertiliser

These systems can:

- Place fertiliser below or beside the seed
- Apply multiple products at different depths
- Cover large areas efficiently

They are widely used in large-scale grain farming due to their precision and capacity.

Figure 56: Bourgault 7950 Air Seeder with a Bourgault 3320 QDA Paralink Hoe Drill. Rfagnou, CC BY-SA 4.0, via Wikimedia Commons.

5. Liquid Banding Systems

Liquid banding systems deliver fertiliser in solution form directly into the soil band.

These systems include:

- Storage tanks
- Pumps and flow controllers
- Injection tubes or coulters

They are commonly used for:

- Starter fertilisers
- Precision nutrient placement
- High-value crops and controlled systems

Liquid systems allow very accurate dosing but require:

- Fully soluble fertilisers
- Clean water

- Maintenance to prevent clogging

Fertigation

Fertigation is the process of applying fertiliser through an irrigation system, delivering nutrients directly to the plant root zone in dissolved form. This method is commonly used in drip irrigation, sprinkler systems, greenhouses, and hydroponic operations.

Fertigation offers a high level of precision and control. Nutrients can be applied in small, frequent doses that match the crop's growth stage and nutrient demand. This improves nutrient uptake efficiency and reduces losses, as fertiliser is delivered directly to the active root zone with water.

This method is particularly effective for high-value crops such as fruits, vegetables, and greenhouse plants, where consistent nutrient supply is critical for quality and yield. It also allows for rapid adjustments to nutrient programs based on plant response.

However, fertigation requires more complex infrastructure, including pumps, injection systems, and filtration. It also requires high-quality, fully soluble fertilisers and careful management to avoid issues such as clogging, precipitation, or nutrient imbalance. Initial setup costs are higher than broadcasting or banding, but the efficiency and precision often justify the investment in intensive production systems.

Fertigation equipment is designed to deliver dissolved fertilisers through an irrigation system, allowing nutrients to be applied directly to the plant root zone with water. Unlike broadcasting or banding, fertigation relies on controlled injection, mixing, and distribution of nutrient solutions, making it one of the most precise fertiliser application methods. The equipment used can range from simple manual systems to fully automated, computer-controlled setups depending on scale and crop value.

<u>1. Fertiliser Tanks and Stock Solution Systems</u>

The starting point of any fertigation system is the fertiliser tank, where nutrients are dissolved to create a stock solution.

These systems typically include:

- Storage tanks (plastic, fiberglass, or stainless steel)
- Agitation systems (mechanical stirrers or recirculation pumps)
- Measuring and dosing markings

In more advanced systems, multiple tanks are used (e.g. A/B tank systems) to keep incompatible fertilisers separate (such as calcium and phosphates).

✓ Function: Prepare and store nutrient solutions

⚠ Key requirement: Complete solubility and correct mixing

2. Injection Systems (The Core of Fertigation)

Injection systems are the most critical component—they introduce fertiliser into the irrigation water.

Common types:

A. Venturi Injectors

- Use pressure difference in the pipeline to draw fertiliser into the water
- Simple, low-cost, no external power

✓ Cheap and reliable

⚠ Less precise control

B. Dosing Pumps (Electric or Hydraulic)

- Inject fertiliser at a controlled rate
- Can be automated and highly accurate

✓ High precision
✓ Suitable for commercial systems
⚠ Higher cost

C. Proportional Injectors

- Maintain a fixed ratio of fertiliser to water flow

✓ Consistent nutrient delivery
✓ Widely used in horticulture

3. Irrigation Delivery Systems

Fertigation relies on irrigation systems to transport nutrients to plants.

Main types:

- Drip irrigation → delivers nutrients directly to root zone (most efficient)
- Sprinkler systems → wider coverage, less precise

- Micro-sprayers → used in orchards and horticulture
- Hydroponic systems → nutrients delivered in solution without soil

✓ Function: Uniform distribution of water + nutrients

⚠ Requires clean water and clog-free operation

4. Filtration Systems

Filtration is essential to prevent clogging of emitters and pipes, especially in drip systems.

Types include:

- Screen filters
- Disc filters
- Sand/media filters

✓ Protects system integrity

✓ Ensures uniform nutrient delivery

⚠ Poor filtration leads to blocked emitters and uneven fertilisation

5. Control Systems and Automation

Modern fertigation systems often include control and monitoring equipment.

These may include:

- Timers and controllers → schedule irrigation and fertiliser application
- EC (electrical conductivity) meters → measure nutrient concentration
- pH sensors → ensure nutrient availability
- Automated dosing systems → adjust fertiliser levels in real time

✓ Enables precision agriculture

✓ Reduces labour and error

6. Backflow Prevention and Safety Devices

These components ensure that fertiliser does not contaminate the water supply.

Common devices:

- Check valves
- Backflow preventers
- Pressure regulators

✓ Essential for safety and compliance

✓ Protects water sources

<u>7. Pumps and Pressure Systems</u>

Pumps are used to:

- Move water through the system
- Maintain pressure for uniform delivery
- Support injection systems

Types:

- Centrifugal pumps
- Booster pumps

✓ Ensures consistent flow and distribution

Application by Fertiliser Type

Broadcasting, banding, and fertigation are not simply different ways of applying fertiliser; they are closely linked to the chemical and physical properties of the fertilisers being used, as well as to the needs of the crop and the production system. The suitability of each method depends on factors such as nutrient mobility in the soil, fertiliser solubility, potential for volatilisation or leaching, and crop sensitivity. As a result, the most effective fertiliser strategy is often based on matching the right product to the right application method.

Broadcasting is generally best suited to granular fertilisers that are stable, relatively low in cost, and applied in large quantities. Common examples include nitrogen sources such as urea and ammonium sulphate, as well as liquid nitrogen solutions like UAN when applied through sprayers. Phosphorus sources such as single superphosphate (SSP) and triple superphosphate (TSP) are also used in broadcast systems, particularly where incorporation into the soil follows application. For potassium, both muriate of potash (MOP, KCl) and sulphate of potash (SOP, K_2SO_4) are commonly broadcast. These fertilisers work well in broadcasting because they can be spread efficiently across large areas, and nutrients become available as roots expand through the soil. Typical examples include broadcasting urea or ammonium sulphate onto pastures before rainfall, applying pre-plant NPK blends such as 15-15-15 in broadacre cereals, or broadcasting potash as a maintenance fertiliser across entire fields. However, broadcasting is less suitable for expensive or highly soluble fertilisers, for micronutrients that are needed only in small quantities, or for phosphorus in cold soils where placement near the root zone is more important.

Banding is best suited to nutrients that are relatively immobile in soil or that need to be placed close to the root zone for efficient uptake, especially in early crop development. Phosphorus is the most important example, and fertilisers such as monoammonium phosphate (MAP) and diammonium phosphate (DAP) are commonly banded at planting. Nitrogen sources such as urea, ammonium nitrate, and UAN can also be used in banded or injected form, particularly where more targeted placement is required or where loss risks are high. Potassium fertilisers such as MOP or SOP may also be banded in low-potassium soils, and certain micronutrients, including zinc, boron, and iron, can be applied in small bands when targeted delivery is needed. These fertilisers are well suited to banding because phosphorus does not move readily through the soil, so placing it near emerging roots improves access. Banding can also reduce nutrient losses, especially for nitrogen, and support stronger early plant establishment. Common examples include banding MAP or DAP with maize at planting as a starter fertiliser, placing phosphorus with wheat seed, side-dressing vegetables with nitrogen during crop growth, or deep banding nutrients in no-till systems. Care is needed, however, because fertiliser placed too close to the seed can cause salt injury or fertiliser burn. Banding also requires more precise equipment and is generally slower than broadcasting.

Fertigation is most suitable for fertilisers that are fully soluble, high in purity, and stable in solution. Common nitrogen sources used in fertigation include calcium nitrate, potassium nitrate, and fully dissolved urea. Phosphorus can be supplied through phosphoric acid or soluble-grade MAP, while potassium is often provided through potassium nitrate or soluble-grade SOP. Secondary nutrients such as magnesium sulphate (Epsom salts) and calcium nitrate are also widely used, and micronutrients are typically supplied in chelated forms such as Fe-EDTA or Zn-EDTA. These fertilisers are ideal for fertigation because they dissolve completely in water, can be delivered directly to the active root zone, and are suitable for small, frequent applications that match plant demand. Typical applications include greenhouse tomatoes receiving potassium nitrate and calcium nitrate through drip irrigation, hydroponic systems using complete nutrient solutions, orchards supplied via drip fertigation with soluble fertilisers, and high-value crops receiving staged nutrient programs through irrigation. Fertigation is not well suited to poorly soluble fertilisers such as rock phosphate, and it can also be problematic where incompatible materials are mixed, such as calcium with phosphates, which may precipitate and clog the system. While MOP can technically be used in some fertigation programs, it is less desirable for chloride-sensitive crops.

In practical terms, broadcasting is most appropriate for bulk granular fertilisers such as urea or MOP where wide coverage and speed are the main goals. Banding is most effective for phosphorus, targeted nitrogen, and certain micronutrients where placement efficiency is critical. Fertigation is best reserved for high-purity, fully soluble fertilisers where precision and control are essential.

Real-world farming systems often combine these methods to improve overall nutrient efficiency. A broadacre wheat farm may broadcast urea across the paddock while banding MAP at planting to improve early phosphorus availability. Vegetable production systems may use banding for starter fertiliser and fertigation for ongoing nutrient supply throughout the growing cycle. In greenhouse tomato production, fertigation alone may be used, with calcium nitrate and potassium nitrate supplying nutrients in a highly controlled manner. These examples show that the most effective fertiliser program is often not based on a single application method, but on using the strengths of each method in combination.

Precision Agriculture and Variable Rate Technology (VRT)

Precision agriculture is a farming paradigm that leverages data, technology, and spatial analysis to manage within-field variability rather than treating a paddock as uniform. This perspective recognizes that soil conditions, nutrient availability, moisture, topography, and historical management create significant heterogeneity, which can be exploited to optimize inputs and outputs. The central aim is to apply inputs such as fertiliser, water, and chemicals more precisely and site-specifically, thereby improving efficiency, reducing waste, and maximizing crop performance [226-228]. Across diverse agricultural systems, the data-driven nature of precision agriculture emerges from the routine collection and integration of spatial information including soil sampling mapped into management zones, yield data from harvest, remote sensing indices (e.g., NDVI), and real-time sensor measurements of soil moisture and crop condition, which together reveal patterns of high productivity, nutrient deficiency, or water stress within fields [229-233]. When these patterns are layered, management can transition from blanket applications to targeted interventions, enabling continuous improvement as decisions are refined by observed outcomes [234-236].

Variable Rate Technology (VRT) lies at the core of operationalizing precision agriculture, providing the means for machines to adjust input application rates automatically as they traverse a field. Rather than delivering a single uniform rate, VRT enables spatially varying inputs such as fertiliser, seed, and agrochemicals in real time, guided by pre-defined prescription maps or live sensor data. This approach aligns input delivery with the specific needs of each field portion, potentially raising yields while reducing input costs and environmental losses [234-236]. In practice, VRT systems are commonly categorized as map-based or sensor-based. Map-based VRT uses pre-constructed prescription maps, derived from soil tests, yield maps, remote sensing, or integrated datasets, to assign rates to each management zone; the machinery then uses GPS positioning to apply the prescribed rates as it moves across the paddock [230, 231, 234, 235, 237]. Sensor-based VRT relies on real-time measurements from mounted sensors that assess crop or soil status (e.g., biomass, chlorophyll indices, or soil moisture) and adjust input rates instantly, which can be particularly beneficial for dynamic nitrogen management where crop demand can fluctuate rapidly [230, 231, 234, 235, 237].

The equipment ecosystem supporting VRT encompasses GPS-guided tractors or harvesters, variable-rate controllers, and compatible application tools such as spreaders, sprayers, and planters. These components are integrated by on-board computing systems that ingest prescription maps or sensor streams and regulate actuators to modulate flow rates or application speeds accordingly. For example, a VRT-enabled fertiliser spreader can vary metering openings or disc speeds to different rates across zones, while sprayers adjust nozzle output and planters modify seeding density to conform with the management map [230, 231, 234, 238]. The investment in such technology, data collection, and analytical capabilities is non-trivial and involves a learning curve associated with data interpretation and system integration; nonetheless, adoption patterns show a growing uptake of precision agriculture in large-scale and high-value cropping systems where the benefits tend to outweigh costs [235, 239-241].

Quantitative evidence across regions supports the economic and environmental advantages of precision agriculture and VRT when correctly implemented. Precision management can reduce input costs and improve nutrient use efficiency, with positive implications for yields and crop quality in heterogeneous fields. Environmentally, VRT and precision application reduce nutrient losses through leaching, runoff, and volatilisation, contributing to more sustainable farming systems by lowering overall fertiliser and chemical use. These conclusions are supported by broad reviews and empirical studies that trace how site-specific management improves input efficiency and reduces environmental impact, while acknowledging that these benefits are contingent on data quality, robust management zones, and the appropriate selection between map-based and sensor-based VRT strategies [226, 227, 235, 236, 242].

Despite clear advantages, several challenges influence adoption and effectiveness. High upfront costs for hardware, data collection, and analytical capabilities, as well as required technical know-how and data management skills, can impede rapid uptake, particularly for smallholders or in regions with limited extension support. Studies of adoption patterns indicate that while GPS guidance and yield mapping have achieved relatively high penetration, the adoption of VRT remains lower, and variability exists across farm size, crop type, and regional institutional frameworks. Barriers include capital intensity, perceived complexity, and insufficient institutional support; conversely, adopter groups tend to be larger, younger, and more commercially oriented, with training and ongoing technical support cited as critical enablers for sustained use of VRT and site-specific management [228, 235, 239-241].

The broader scholarly discourse emphasizes that precision agriculture is not a single technology but a toolkit for SSCM (site-specific crop management). Central to SSCM is the delineation of field management zones, which are informed by an integration of soil, yield, and remote sensing data, sometimes enhanced by crop models and decision-support systems. Prescription maps generated from this multi-source data are essential for guiding VRT, whether map-based or sensor-based, and for enabling adaptive input applications that reflect spatial and temporal variability. Reviews and meta-analyses highlight that successful adoption hinges on the ability to translate data into actionable prescriptions, the reliability and accessibility of software tools, and the alignment of technology with farmer needs, training, and policy incentives that encourage investment in digital agriculture [233, 235, 242-244].

In essence, precision agriculture and VRT transform fertiliser and input application from a uniform, input-driven process into a data-driven, site-specific management framework. By capturing spatial variability and applying inputs according to actual field conditions and crop needs, these technologies offer the potential to optimize productivity while reducing waste and environmental impact. Realizing these benefits requires high-quality spatial data, appropriate management zone delineation, robust prescription maps, and well-calibrated VRT systems; it also demands sustained investment in hardware, software, and human capital to overcome adoption barriers and to ensure that the data-informed decisions translate into measurable agronomic and economic gains [226, 229, 234-236, 241, 242].

A practical way to understand precision agriculture and Variable Rate Technology (VRT) is to consider a real-world broadacre wheat paddock in southern Australia, where soil variability is common due to changes in soil type, elevation, and historical management. Rather than treating the entire paddock as uniform, the farmer begins by collecting spatial data over several seasons. This includes grid or zone-based soil sampling (showing variations in nitrogen, phosphorus, potassium, and pH), yield maps from previous harvests, and satellite imagery such as NDVI that highlights differences in crop vigour across the field. When these datasets are layered together, clear patterns emerge—for example, sandy ridges with low nutrient-holding capacity, heavier clay areas with higher fertility, and lower-lying zones prone to waterlogging.

Using this information, the paddock is divided into management zones. A high-performing zone may consistently produce strong yields and show adequate nutrient levels, while a poorer zone may exhibit low nitrogen and lower biomass. Instead of applying a uniform fertiliser rate across the entire paddock, the farmer develops a prescription map. This map might specify, for example, 120 kg/ha of nitrogen in low-performing zones, 80 kg/ha in moderate zones, and only 50 kg/ha in high-performing areas where residual fertility is already sufficient.

During application, a tractor equipped with GPS guidance and a VRT-enabled fertiliser spreader follows the prescription map. As the machine moves across the paddock, the system automatically adjusts the fertiliser rate in real time. When the spreader enters a low-nutrient sandy zone, the metering system opens to increase the application rate. As it transitions into a more fertile clay zone, the system reduces the rate accordingly. From the operator's perspective, this process is seamless—the adjustments are controlled by onboard software using spatial positioning data.

In a more advanced scenario, the farmer may also use sensor-based VRT for in-season nitrogen application. For example, optical sensors mounted on the boom measure crop greenness (a proxy for nitrogen status) as the machine travels through the field. If a section of the crop shows poor vigour or low chlorophyll levels, the system increases nitrogen application in that area. Conversely, where the crop is already healthy and nitrogen-sufficient, the system reduces or skips application. This allows the fertiliser program to respond dynamically to actual crop conditions rather than relying solely on pre-season assumptions.

The outcomes of this approach become evident at harvest. Yield maps often show improved consistency across the paddock, with previously underperforming zones lifting in productivity due to targeted inputs. At the same time, input costs are reduced because fertiliser is not wasted on areas that do not require it. Environmentally, there is less risk of excess nitrogen leaching from already fertile zones or being lost through volatilisation, as applications are better aligned with crop demand.

This example demonstrates how precision agriculture and VRT translate from theory into practice. The process follows a clear cycle: collect data → identify variability → define management zones → create prescription maps → apply inputs variably → monitor results → refine the approach. Rather than a single technology, it is a system of integrated tools and decisions that allows farmers to manage each part of a paddock according to its specific characteristics. In doing so, precision agriculture shifts fertiliser use from a uniform, input-driven model to a targeted, data-driven strategy that improves both economic and agronomic outcomes.

Timing and Efficiency of Nutrient Uptake

Timing and efficiency of nutrient uptake refer to how well plants are able to absorb and utilise nutrients at the right stage of growth, and how effectively applied fertiliser is converted into plant biomass, yield, or quality. In practical terms, it is not just about how much fertiliser is applied, but when it is applied, in what form, and whether the plant can access it when needed. Poor timing can lead to nutrient losses or reduced crop performance, while well-timed applications significantly improve fertiliser use efficiency (FUE) and economic returns.

Plant nutrient demand is not constant throughout the growing cycle. Instead, crops follow a dynamic uptake pattern, where different nutrients are required in varying amounts at specific growth stages. During early establishment, phosphorus is critical for root development and energy transfer, enabling seedlings to establish quickly. As the crop enters vegetative growth, nitrogen demand increases to support leaf area development, chlorophyll production, and photosynthesis. Later, during flowering and fruiting or grain filling, potassium becomes more important for water regulation, enzyme activity, and product quality. If nutrients are applied too early, they may not be available when the plant needs them; if applied too late, yield potential may already be limited.

Efficiency of nutrient uptake depends heavily on synchronising fertiliser supply with crop demand. This concept is often described as supplying nutrients at the "right time", one of the core principles of nutrient stewardship. For example, applying large amounts of nitrogen well before peak crop demand increases the risk of losses through leaching (especially in sandy soils), volatilisation (in the case of urea), or denitrification in waterlogged conditions. In contrast, splitting nitrogen applications, such as applying some at planting and the remainder during active growth, can significantly improve uptake efficiency because nutrients are supplied when the plant can actively use them.

Synchronising fertiliser supply with crop demand, commonly described as applying nutrients at the "right time", is a practical, step-by-step process rather than a theoretical concept. It involves understanding when a crop requires specific nutrients, aligning fertiliser applications with those growth stages, and adjusting decisions based on soil conditions and weather. When done correctly, this approach ensures that nutrients are available when plants can actively use them, improving both efficiency and crop performance.

A key starting point is understanding crop growth stages and how nutrient demand changes over time. Most crops follow predictable patterns. During establishment, phosphorus is critical for root development and early growth. As the crop enters the vegetative stage, nitrogen demand increases to support leaf and biomass production. Later, during flowering, grain fill, or fruiting, potassium becomes more important for yield formation and quality. Rather than relying on complex modelling, farmers can use simple crop guides, such as wheat tillering stages or maize V6 growth, to identify peak nutrient demand periods and plan applications accordingly.

One of the most common mistakes in fertiliser management is "front-loading," where most or all fertiliser is applied at planting. This approach is inefficient because nutrients, particularly nitrogen, may remain in the soil long before the plant needs them. During this time, nitrogen can be lost through leaching (especially in sandy soils), volatilisation (in the case of surface-applied urea), or denitrification in waterlogged conditions. A more effective approach is to apply only the nutrients required for early growth and retain the remainder for later stages when plant demand increases.

Splitting fertiliser applications is the most practical way to achieve this synchronisation. Instead of applying a single large dose, fertiliser is divided into smaller amounts applied at key growth stages. For example, in wheat production, a small amount of nitrogen and phosphorus may be applied at planting, followed by additional nitrogen during tillering and a larger application at stem elongation. In maize, a starter fertiliser such as MAP or DAP is applied at planting, with the main nitrogen application at the V6 stage and a possible top-up before tasselling. This approach ensures that nutrients are available when roots are active, reduces the time nutrients remain unused in the soil, and significantly improves uptake efficiency.

The choice of fertiliser type should also align with timing. Different nutrient forms behave differently in the soil. Urea, for instance, should be applied close to rainfall or incorporated into the soil to minimise volatilisation losses. Nitrate-based fertilisers are immediately available but prone to leaching, so they are best applied closer to peak crop demand. Phosphorus fertilisers such as MAP or DAP are relatively immobile and should be applied early and placed near the root zone, often through banding. Potassium sources such as MOP or SOP are typically applied before or during reproductive stages when demand is highest. As a general rule, fast-release fertilisers should be applied closer to when the crop needs them, while immobile nutrients should be placed early and strategically.

Application method also plays a role in effective timing. Broadcasting is suitable for pre-plant or large-area applications, while banding is ideal at planting because it places nutrients close to emerging roots. Fertigation offers the highest level of control, allowing nutrients to be applied in small, frequent doses throughout the growing cycle. Even at a small scale, timing can be improved by simply splitting applications manually and applying fertiliser before irrigation or rainfall events to enhance nutrient availability.

Soil and weather conditions must also be considered when deciding when to apply fertiliser. Rainfall can help incorporate urea and reduce volatilisation losses, while dry soil conditions may limit nutrient uptake and make application less effective. Waterlogged soils increase the risk of nitrogen losses through denitrification, and cold soils can restrict phosphorus availability, making banding more important. The key principle is to apply fertiliser when the plant can actively use it, rather than applying it based solely on convenience or schedule.

Monitoring crop response is a simple but powerful way to refine timing decisions. Visual indicators such as leaf colour, growth rate, and crop uniformity can provide valuable insights into nutrient status. Pale or yellowing leaves may indicate nitrogen deficiency, while uneven growth can signal variability in nutrient availability. More advanced tools such as NDVI imagery or plant tissue testing can provide additional precision, particularly in high-value crops. Observations from the crop can then be used to adjust subsequent fertiliser applications.

A practical example can be seen in tomato production. Instead of applying all fertiliser at planting, a staged approach is used. A small starter fertiliser is applied during the first one to two weeks, followed by increased nitrogen during vegetative growth. As the crop enters flowering, nitrogen is reduced and potassium is increased to support fruit development. During fruiting, a potassium-focused program is maintained. This staged approach results in better nutrient uptake, reduced waste, and improved fruit quality.

In essence, synchronising nutrient supply with crop demand comes down to a simple but powerful principle: apply nutrients when the plant needs them, not before and not after. This means avoiding single large applications, splitting fertiliser across growth stages, matching fertiliser types to timing, and adjusting based on environmental conditions and crop response. When implemented effectively, this approach increases nutrient use efficiency, reduces losses, and improves both yield and crop quality. It transforms fertiliser use from a fixed, routine input into a responsive and strategically managed component of crop production.

The form of the nutrient also influences uptake timing and efficiency. Nitrogen, for instance, can be applied as ammonium, nitrate, or urea, each behaving differently in soil. Nitrate is immediately available but prone to leaching, while ammonium is more stable but must be converted before uptake in some cases. Phosphorus is relatively immobile in soil, so its efficiency depends on placement close to the root zone, particularly during early growth. Potassium is more mobile than phosphorus but still benefits from timely application aligned with peak demand periods. Choosing the right nutrient form ensures that availability matches plant uptake patterns.

Soil conditions play a major role in nutrient availability and uptake efficiency. Factors such as soil moisture, temperature, pH, and microbial activity influence how nutrients are released, transformed, and absorbed by plants. For example, dry soils limit nutrient movement and root activity, reducing uptake even if fertiliser is present. Cold soils can restrict phosphorus availability, which is why starter fertilisers are often banded near the seed. Waterlogged soils can lead to nitrogen losses through denitrification. Effective nutrient timing must therefore consider not only crop stage but also prevailing environmental conditions.

Application method is another critical factor influencing timing and efficiency. Broadcasting may delay nutrient access because roots must grow into the fertilised zone, whereas banding places nutrients closer to emerging roots, improving early uptake. Fertigation allows nutrients to be delivered in small, frequent doses directly to the root zone, providing the highest level of control and efficiency. In precision agriculture systems, Variable Rate Technology (VRT) can further refine timing and placement by adjusting applications based on spatial variability and crop condition.

From an efficiency perspective, the goal is to maximise the proportion of applied nutrients that are actually taken up by the crop. This is often expressed as nutrient use efficiency (NUE), which can be significantly improved through proper timing. For example, nitrogen use efficiency may increase from less than 50% under poorly timed applications to over 70% when applications are split and aligned with crop demand. Higher efficiency not only improves yield but also reduces environmental impacts such as nutrient runoff, groundwater contamination, and greenhouse gas emissions.

In practice, improving timing and efficiency involves strategies such as split applications, stage-specific fertiliser programs, use of controlled-release fertilisers, and real-time monitoring of crop condition. For instance, a farmer may apply a small amount of starter fertiliser at planting, follow with nitrogen during vegetative growth, and then apply potassium during reproductive stages. Monitoring tools such as plant tissue testing or remote sensing can help fine-tune these decisions during the season.

Ultimately, timing and efficiency of nutrient uptake are about aligning three key elements: crop demand, nutrient availability, and environmental conditions. When these factors are synchronised, plants can absorb nutrients more effectively, leading to improved growth, higher yields, better quality, and reduced waste. This alignment transforms fertiliser from a simple input into a precisely managed resource within the broader crop production system.

Minimising Losses (Leaching, Volatilisation)

Minimising nutrient losses, particularly through leaching and volatilisation, is a central objective of efficient fertiliser management. These loss pathways reduce the proportion of applied nutrients that are actually taken up by crops, lowering fertiliser use efficiency (FUE), increasing costs, and contributing to environmental issues such as groundwater contamination and greenhouse gas emissions. Effective management focuses on understanding how and why these losses occur, and then applying practical strategies to reduce them.

Leaching occurs when water moves through the soil profile and carries soluble nutrients, especially nitrate (NO_3^-), below the root zone. This is most common in sandy or well-drained soils, or during periods of heavy rainfall or excessive irrigation. Because nitrate is highly mobile in soil, it can be lost quickly if it is present in large amounts when the crop is not actively taking it up. Minimising leaching therefore depends on reducing the time that soluble nutrients remain unused in the soil. This can be achieved by splitting nitrogen applications so that smaller amounts are applied closer to peak crop demand, rather than applying a large amount early. Matching fertiliser timing to weather forecasts is also critical; for example, avoiding application just before heavy rainfall reduces the risk of nutrients being washed beyond the root zone. In irrigated systems, careful water management is equally important as over-irrigation can drive nutrients downward, while controlled irrigation helps retain them within the root zone.

Another important strategy for reducing leaching is selecting appropriate fertiliser forms. Ammonium-based nitrogen (NH_4^+) is less prone to leaching than nitrate because it binds to soil particles, although it will eventually convert to nitrate through nitrification. The use of nitrification inhibitors can slow this conversion, keeping nitrogen in the less mobile ammonium form for longer. Similarly, controlled-release fertilisers can provide a gradual supply of nutrients, reducing the risk of large concentrations being lost at once. In some systems, placing fertiliser deeper in the soil (e.g. through banding) can also reduce surface losses and improve retention within the root zone.

Volatilisation refers to the loss of nitrogen to the atmosphere as ammonia gas (NH_3), most commonly from surface-applied urea or ammonium-based fertilisers. This process is influenced by factors such as temperature, soil pH, moisture, and wind. Volatilisation losses are typically highest when urea is applied to the soil surface without incorporation, particularly under warm, dry, and windy conditions or on alkaline soils. To minimise volatilisation, one of the most effective strategies is to incorporate fertiliser into the soil through cultivation or rainfall/irrigation shortly after application. Even 5–10 mm of rainfall can significantly reduce losses by moving urea into the soil where it is less exposed to the atmosphere.

Timing is again critical for reducing volatilisation. Applying urea immediately before rainfall or irrigation helps ensure rapid incorporation. Alternatively, using fertiliser products treated with urease inhibitors can slow the conversion of urea to ammonia, reducing the rate of volatilisation and allowing more time for incorporation. In situations where incorporation is not possible, switching to less volatile nitrogen sources—such as ammonium nitrate or liquid nitrogen solutions applied under suitable conditions—can also reduce losses.

Application method plays a key role in minimising both leaching and volatilisation. Broadcasting fertiliser onto the soil surface without incorporation increases the risk of both types of loss, whereas banding or injection places nutrients below the surface, reducing exposure to runoff, evaporation, and gas loss. Fertigation offers even greater control by delivering nutrients in small, frequent doses directly to the root zone, minimising the accumulation of excess nutrients in the soil at any one time. These targeted methods improve nutrient retention and availability while reducing environmental risk.

Soil and environmental conditions must always be considered. For example, in sandy soils prone to leaching, more frequent and smaller applications are preferred. In alkaline soils, volatilisation risk is higher, so incorporation or inhibitor use becomes more important. In high rainfall regions, careful scheduling of applications is essential to avoid periods of heavy precipitation. Understanding these site-specific factors allows fertiliser strategies to be tailored to minimise losses under local conditions.

In practice, minimising nutrient losses comes down to applying the principles of right source, right rate, right time, and right place. This includes selecting appropriate fertiliser types, timing applications to coincide with crop demand and favourable weather, placing nutrients where roots can access them, and avoiding conditions that promote loss. When these principles are applied together, nutrient retention improves, fertiliser efficiency increases, and both economic and environmental outcomes are enhanced.

Chapter 11

Environmental and Sustainability Considerations

This chapter examines the environmental dimensions of fertiliser use within modern agricultural systems, highlighting the complex relationship between productivity, resource efficiency, and ecological impact. As global food production intensifies to meet growing demand, fertiliser inputs play a critical role in maintaining crop yields; however, their use is increasingly scrutinised due to associated greenhouse gas emissions, water pollution, waste generation, and resource depletion. Understanding these impacts is essential for developing agricultural systems that are both productive and environmentally responsible .

The chapter begins by exploring the role of fertilisers in contributing to greenhouse gas emissions, particularly nitrous oxide (N_2O) and carbon dioxide (CO_2), and the mechanisms through which these emissions occur in both production and field application. It then examines the movement of nutrients through agricultural landscapes and the resulting impacts on water systems, including eutrophication and harmful algal blooms, which have significant ecological, economic, and public health consequences.

Attention is also given to waste management across the fertiliser lifecycle, from industrial by-products generated during production to on-farm losses and packaging waste. These issues are considered alongside emerging opportunities to improve resource efficiency through recycling, recovery, and the integration of alternative nutrient sources. In this context, the chapter introduces circular economy principles as a framework for transforming waste streams into valuable inputs and reducing reliance on finite or energy-intensive resources.

The discussion further extends to innovations in fertiliser production, particularly the development of green ammonia and low-carbon fertilisers, which offer pathways to significantly reduce the carbon footprint of agriculture. These technologies, when combined with improved nutrient management practices, have the potential to align agricultural productivity with climate change mitigation objectives.

Overall, this chapter positions sustainable fertiliser management as a critical component of modern agriculture. By integrating environmental considerations into nutrient management strategies, producers can enhance efficiency, reduce environmental harm, and contribute to the long-term resilience of agricultural systems.

Greenhouse Gas Emissions (e.g., N_2O, CO_2)

From an environmental and sustainability perspective, greenhouse gas (GHG) emissions associated with fertiliser use are a critical consideration in modern agriculture. The two most significant gases linked to fertiliser management are nitrous oxide (N_2O) and carbon dioxide (CO_2), both of which contribute to climate change, albeit through different pathways and processes. Understanding how these emissions are generated—and how they can be reduced—is essential for developing more sustainable nutrient management systems.

Nitrous oxide (N_2O) is the most environmentally significant greenhouse gas associated with fertiliser use. Although it is emitted in smaller quantities than CO_2, it has a global warming potential approximately 265–300 times greater than CO_2 over a 100-year period, making it a major contributor to agricultural emissions. N_2O is primarily produced in soils through microbial processes, specifically nitrification and denitrification. During nitrification, soil microbes convert ammonium (NH_4^+) into nitrate (NO_3^-), releasing small amounts of N_2O as a by-product. During denitrification, typically under waterlogged or low-oxygen conditions, nitrate is converted into nitrogen gases (N_2 and N_2O), with N_2O being released into the atmosphere. These processes are strongly influenced by soil moisture, temperature, oxygen availability, and the amount of nitrogen present in the soil.

Fertiliser practices can significantly influence N_2O emissions. Applying large amounts of nitrogen, particularly in forms that are rapidly converted to nitrate, increases the pool of nitrogen available for microbial transformation and loss. Conditions such as heavy rainfall, over-irrigation, or poorly drained soils further exacerbate emissions by promoting denitrification. From a sustainability perspective, this highlights the importance of optimising nitrogen use efficiency, ensuring that nitrogen is taken up by crops rather than lost to the atmosphere. Strategies such as splitting applications, avoiding over-application, using nitrification inhibitors, and improving soil drainage can all help reduce N_2O emissions.

Carbon dioxide (CO_2) emissions associated with fertilisers arise from both direct and indirect sources. Indirectly, CO_2 is emitted during the manufacture and transport of fertilisers, particularly nitrogen fertilisers produced via the energy-intensive Haber–Bosch process, which relies heavily on fossil fuels. This makes nitrogen fertilisers one of the most carbon-intensive inputs in agriculture. Direct CO_2 emissions can also occur in the field through processes such as soil respiration and the breakdown of organic matter, which may be influenced by fertiliser application and soil disturbance.

In addition to manufacturing emissions, certain fertiliser reactions in the soil can contribute to CO_2 release. For example, the application of urea leads to the formation of carbon dioxide during its hydrolysis in soil. While this is generally a smaller component compared to industrial emissions, it still forms part of the overall carbon footprint of fertiliser use. From a sustainability standpoint, reducing reliance on high-emission fertiliser inputs, improving application efficiency, and integrating organic or alternative nutrient sources can help lower total CO_2 emissions.

The environmental implications of these greenhouse gases extend beyond climate change alone. Elevated N_2O emissions represent a loss of valuable nitrogen from the farming system, reducing fertiliser efficiency and increasing input costs. Similarly, high CO_2 emissions reflect energy-intensive production systems that rely on non-renewable resources. Together, these factors underscore the need for integrated nutrient management approaches that balance productivity with environmental responsibility.

Sustainable fertiliser management therefore focuses on reducing emissions through improved practices aligned with the principles of "right source, right rate, right time, and right place." This includes using enhanced-efficiency fertilisers, adopting precision agriculture technologies to avoid over-application, improving soil health to support nutrient retention, and aligning fertiliser use with crop demand and environmental conditions. In doing so, agriculture can maintain productivity while reducing its contribution to greenhouse gas emissions and supporting broader climate change mitigation goals.

In essence, greenhouse gas emissions from fertiliser use highlight the interconnected nature of agricultural productivity and environmental sustainability. By recognising how fertiliser practices influence N_2O and CO_2 emissions, and by implementing targeted management strategies, producers can play a key role in reducing agriculture's environmental footprint while maintaining efficient and profitable production systems.

Water Pollution and Eutrophication

Water pollution and eutrophication from fertiliser use are well-established risks when nutrients are lost from agricultural systems and enter surrounding water bodies. A consensus across multiple sources emphasizes that nitrogen (N) and phosphorus (P) are the primary nutrients driving these processes, moving through soil, groundwater, surface water, and aquatic ecosystems in ways that disrupt ecological balance and degrade water quality [245-247]. When crops do not efficiently utilise applied fertilisers, nutrients reach rivers, lakes, and coastal zones via runoff, leaching, or erosion, leading to ecosystem alterations, harmful algal blooms (HABs), hypoxia, and heightened water treatment costs [245, 247-249]. This synthesis draws on evidence from nutrient cycling, transport pathways, eutrophication dynamics, HABs biology and health risks, and management strategies designed to keep fertilisers within the crop root zone and protect water resources.

Nutrient dynamics underpin the link between fertiliser use and water quality impacts. Excess fertiliser N and P can leave fields through surface runoff, sediment- and soil-bound particulates, or subsurface leaching, depending on soil properties, rainfall, irrigation practices, and land slope. Surface runoff is particularly impactful on sloped, compacted, or erodible soils and during heavy rainfall, with phosphorus often transported bound to eroding soil particles, while leaching tends to mobilise soluble nitrate (NO_3^-) deeper into the soil profile and groundwater, especially in sandy soils or under intensive irrigation [245, 246, 250]. The differential fate of P and N in runoff versus leaching has consequences for downstream water quality, as P is frequently associated with sediment transport while NO_3^- more readily migrates through drainage to aquifers and rivers [245, 246, 249]. The consequences include elevated nutrient levels in surface and groundwater, which fuel eutrophication and HABs in receiving waters [245, 247].

Eutrophication is the enrichment of water bodies with nutrients, leading to excessive growth of algae and aquatic plants. While nutrients are essential, overabundance, especially of N and P, drives rapid algal blooms that reduce water clarity, limit light to submerged vegetation, and alter habitat quality. The decomposition of algal blooms consumes dissolved oxygen, creating hypoxic or anoxic conditions that harm fish and invertebrates and can generate "dead zones" with reduced biodiversity [245-247]. Algal blooms may also produce toxins, giving rise to HABs with implications for drinking water safety, livestock, wildlife, and human health, and they can pose broader economic costs through fisheries disruption, tourism decline, and increased water treatment expenditures [248, 251, 252]. The relationship between dual nutrient loading and HABs is increasingly recognised: control of HABs often requires addressing both N and P inputs rather than focusing on P alone, because dual enrichment can sustain blooms even when one nutrient is relatively scarce [247]. There are nuanced disagreements in the literature about whether N, P, or the N:P balance is the primary limiting factor in all systems, underscoring the context-dependent nature of management responses [246-248].

HABs and eutrophication threaten drinking water safety through toxin production (cyanotoxins) and taste/odour compounds, as well as through the broader ecological disruption of aquatic ecosystems. HABs can lead to fish kills, reduced habitat quality, and elevated water treatment costs to remove toxins and tastes/odours, with downstream effects on fisheries and tourism [248, 251, 252]. The health risks extend beyond the environment to humans and animals via contaminated drinking water and recreational exposures, and HAB events often incur substantial economic damages for water utilities, fisheries, and coastal communities [248, 251, 252]. Importantly, these issues are interlinked with climate change and land-use changes, which can amplify nutrient loading, water temperatures, and HAB frequency and intensity, compounding management challenges [248, 252].

From a sustainability perspective, eutrophication signals environmental and efficiency failures in nutrient management: nutrients intended for crop production are lost to the environment, representing inefficiency and external costs. Improving nutrient use efficiency (NUE) and keeping fertilisers within the root zone are central to preventing nutrient losses and safeguarding water quality, while maintaining crop yields [245, 247-249]. This aligns with conceptual shifts in nutrient management paradigms that advocate dual-nutrient control (N and P) for HAB prevention in many freshwater and downstream coastal systems, recognizing that single-nutrient strategies may fail in the presence of complex lake and river dynamics [247].

Management strategies to minimise pollution and eutrophication Effective mitigation requires an integrated suite of practices that address timing, rate, placement, and delivery of fertilisers, as well as landscape design and irrigation optimisation. Key strategies supported by the literature include:

- Rate and timing alignment: Applying fertilisers to match crop demand and avoid surplus nutrients in the soil reduces the potential for nutrient losses through runoff and leaching [245, 249].
- Buffer zones and riparian strips: Vegetated zones between farmland and waterways trap sediment and associated nutrients before they reach water bodies, thereby reducing nutrient export and protecting water quality [245, 249].
- Conservation tillage and soil erosion control: Reduced tillage and conservation agriculture minimise soil erosion and associated phosphorus loss, improving soil structure and increasing nutrient retention in the root zone [245, 249].
- Irrigation management: Optimising irrigation to avoid overwatering reduces nitrate leaching, while precise irrigation scheduling minimises water flux that can mobilise nutrients [247, 249].
- Cover crops: Planting cover crops during fallow periods captures residual nutrients and mitigates nutrient losses through leaching and runoff [247, 249].
- Application methods: Placement strategies such as banding or incorporating fertilisers below the surface reduce exposure to surface runoff; broadcasting increases runoff risk when followed by rainfall [249].
- Fertigation and precision agriculture: Controlled, site-specific nutrient delivery using fertigation or variable-rate application reduces over-fertilisation and helps keep nutrients within the root zone, particularly in intensive systems where inputs are higher [246, 248, 249].
- Landscape and ecological measures: Buffer strips, wetlands, and restoration of natural features help filter nutrients, while targeted land-use planning can limit nutrient export from high-risk areas [247-249, 253, 254].
- Decision-support and monitoring: Early warning systems, regular water quality monitoring, and predictive modelling (including machine learning approaches) support proactive management of HAB risks and eutrophication dynamics [248, 251, 253, 255].

Waste Management and By-products

From an environmental and sustainability perspective, waste management and by-products in fertiliser production and use are critical considerations, as they influence resource efficiency, environmental impact, and the overall sustainability of agricultural systems. Fertiliser supply chains generate a range of wastes, from mining residues and industrial by-products to packaging and unused or misapplied fertilisers in the field. Effective management of these materials is essential to minimise pollution, reduce resource depletion, and support circular economy principles.

In fertiliser production, particularly for phosphate and potash, significant quantities of industrial by-products are generated. For example, the production of phosphoric acid (used in many phosphate fertilisers) produces phosphogypsum as a by-product. This material contains calcium sulphate along with trace impurities, including heavy metals and naturally occurring radioactive elements. While phosphogypsum can be reused in some applications, such as soil amendment, construction materials, or land rehabilitation, large volumes are often stored in stacks, posing long-term environmental risks if not properly managed. Similarly, potash and other mineral processing operations generate tailings and brines that must be carefully handled to prevent soil and water contamination.

Nitrogen fertiliser production also has associated by-products and waste streams. The Haber–Bosch process, while primarily producing ammonia, is energy-intensive and generates emissions that must be managed through process optimisation and emissions control technologies. In addition, manufacturing facilities produce wastewater and chemical residues that require treatment before discharge. From a sustainability standpoint, improving process efficiency and capturing or reusing by-products (e.g. heat recovery, carbon capture) can significantly reduce environmental impact.

Beyond production, on-farm waste management is equally important. Fertiliser losses, whether through spills, improper storage, or over-application, represent both economic waste and environmental risk. Excess fertiliser that is not taken up by crops can contribute to issues such as leaching, runoff, and greenhouse gas emissions, effectively becoming a "waste stream" within the agricultural system. Proper storage (e.g. dry, covered conditions for granular fertilisers; sealed tanks for liquids), accurate calibration of equipment, and adherence to application plans are essential to minimise these losses.

Packaging waste is another consideration, particularly in smaller-scale or high-value agricultural systems where fertilisers are supplied in bags, containers, or drums. Improper disposal of packaging can lead to environmental contamination. Sustainable practices include recycling programs, bulk handling systems that reduce packaging requirements, and the use of reusable or biodegradable materials where feasible. In many regions, industry stewardship programs support the collection and recycling of fertiliser containers to reduce environmental impact.

A key sustainability trend is the increasing use of recycled and alternative nutrient sources, which transform waste into valuable fertiliser inputs. Organic wastes such as manure, compost, biosolids, and food processing residues can be processed and applied as fertilisers or soil amendments, returning nutrients to the soil and reducing reliance on synthetic fertilisers. Industrial by-products, such as ammonium sulphate from chemical processes or recovered phosphorus from wastewater treatment, are also being integrated into fertiliser supply chains. These approaches support a circular economy, where nutrients are reused rather than discarded.

However, the use of recycled materials requires careful management to ensure safety and quality. Potential contaminants, such as heavy metals, pathogens, or excess salts, must be monitored and controlled to prevent adverse effects on soil health, crops, and human health. This highlights the importance of regulatory frameworks, quality assurance systems, and proper treatment processes in managing fertiliser by-products.

Ultimately, sustainable waste management in fertiliser systems involves reducing waste generation, safely managing unavoidable by-products, and maximising resource recovery.

This includes improving production efficiency, preventing on-farm losses, recycling packaging, and integrating alternative nutrient sources. By viewing waste not simply as a disposal problem but as a potential resource, agricultural systems can reduce environmental impact, improve nutrient efficiency, and move toward more sustainable and resilient production models.

Green Ammonia and Low-Carbon Fertilisers

The transition to green ammonia and low-carbon fertilisers represents a paradigmatic shift in how essential agricultural inputs are produced and consumed. A broad evidence base ties fertiliser production to substantial greenhouse gas emissions, largely because conventional nitrogen fertilisers are anchored in fossil-fuel–dependent ammonia production via the Haber–Bosch process. Green ammonia, produced with hydrogen from renewable energy and water electrolysis, offers a pathway to decarbonise both upstream fertiliser production and downstream use when paired with efficient agronomic practices. Blue ammonia, hydrogen derived from natural gas with carbon capture and storage (CCS), is frequently discussed as a near-term bridging step, while green ammonia is positioned as the long-term sustainable option. Low-carbon fertilisers extend beyond green or blue ammonia to include stabilised fertilisers and enhanced-efficiency formulations that reduce emissions during storage, handling, and after application, particularly through inhibitors and controlled-release technologies. The synthesis below integrates multiple strands of the literature to show how green ammonia and low-carbon fertilisers could reshape environmental outcomes in agriculture, the technology and policy levers involved, and the practical challenges and opportunities for adoption.

Nitrogen fertilisers are a major source of agricultural emissions, driven both by emissions from production and by losses during use. Analyses of global fertiliser systems consistently identify fertiliser production and field N losses as substantial contributors to climate and environmental footprints, with N_2O and NH_3 being principal byproducts of traditional nitrogen management [72, 256, 257]. The production of ammonia for Haber–Bosch is energy-intensive and emits CO_2, linking fertiliser emissions to fossil-energy systems [72, 257]. Green ammonia replaces fossil-fuel–based hydrogen with hydrogen produced through electrolysis powered by renewables, thereby decoupling ammonia synthesis from fossil fuels and offering near-term and long-term decarbonisation potential for the fertiliser supply chain [72, 258]. Blue ammonia sits between these poles, using natural gas-derived hydrogen with CCS to reduce emissions, but still leaving some fossil-energy dependence and CO_2 capture considerations [72, 259].

The Haber–Bosch production route remains energy-intensive and CO_2 emitting when hydrogen is sourced from natural gas, a reality highlighted across multiple studies. Green ammonia eliminates the direct CO_2 emissions associated with hydrogen production by using electrolysis powered by renewable electricity, thereby achieving substantial reductions in the embedded carbon of fertiliser production when powered by low-carbon grids [72, 258, 259].

Blue ammonia reduces, but does not eliminate, emissions via CCS; its value resides in bridging investment, renewable energy scale-up, and infrastructure development needed to support green hydrogen ramp-up when renewables and grids are not yet fully capable of supplying electrolytic hydrogen at scale [72, 259]. The overall climate benefit of green ammonia is greatest when the renewable energy supply is reliable and the ammonia synthesis process is integrated with efficient nitrogen use in crops, enabling a system-level decline in emissions across upstream production and downstream application [72, 259].

Low-carbon fertilisers encompass products produced from green or blue ammonia, as well as enhanced-efficiency fertilisers that limit losses and emissions during production, storage, transport, and field application. Stabilised nitrogen fertilisers, often formed by incorporating urease and/or nitrification inhibitors, couple emission reductions from volatilisation and nitrous oxide formation with potential improvements in nitrogen use efficiency (NUE) and crop yield, though responses are context-dependent [45, 257, 260, 261]. Strategic use of inhibitors such as NBPT (a urease inhibitor) and DMPP or DCD (nitrification inhibitors) has repeatedly been shown to lower NH_3 volatilisation and N_2O emissions under certain soils and climate conditions, while sometimes displaying limited or variable effects on yield or long-term N retention depending on timing, rate, and soil type [256-258, 261-263]. Meta-analytic and region-specific assessments indicate that nitrification and urease inhibitors can significantly reduce emission factors for N2O and NH3 in some contexts, though the magnitude of benefit is sensitive to management and environmental factors and may entail trade-offs such as NH_3 emissions in some cases [256, 258, 264, 265]. Stabilised fertilisers and controlled-release formulations can also reduce N losses by better matching N supply with crop demand, with some studies reporting substantial reductions in N_2O emissions and improvements in NUE, particularly under deep soil conditions or sprinkler irrigation where management can be precisely tuned [259, 261, 266, 267].

Adoption of green ammonia-based fertilisers and low-carbon fertilisers does not inherently change nutrient application rules in the field; rather, it shifts the carbon footprint of inputs and enables decarbonised fertiliser supply chains. When combined with precision agriculture, split applications, and improved fertiliser choice (e.g., nitrification and/or urease inhibitors, controlled-release products), green ammonia-based fertilisers can meaningfully reduce the environmental footprint of crop production while preserving or enhancing yields in many contexts [45, 261, 265, 268, 269]. The ongoing transition also relies on policy instruments, carbon pricing, and investment in renewable energy and electrolyser capacity to close the cost gap between green and conventional ammonia [72, 258, 259].

Cost differentials currently favour conventional ammonia production in many markets, though green hydrogen and green ammonia can become cost-competitive through scale, learning effects, and carbon pricing mechanisms that internalise climate externalities [72, 258, 259]. Policy support for renewable energy, electrolyser deployment, and CO_2 pricing can catalyse the uptake of green ammonia and incentivise low-carbon fertiliser production and use. In addition, the emission-reduction potential of low-carbon fertilisers, from reduced fertiliser production emissions to lower field emissions through inhibitors and improved NUE, suggests substantial benefits under scenarios where decarbonisation policies accompany incentives for adoption [72, 256, 258, 265].

Region-specific factors influence the effectiveness of inhibitors and stabilisers and the economics of green ammonia adoption. Brazilian studies show region-specific emission factors for N_2O and NH_3 that highlight the potential reductions achievable with nitrification and urease inhibitors, though with context-dependent variability and regional inventories that can significantly re-adjust national emission budgets [264]. In temperate grasslands and pasture systems, inhibitor-based stabilised fertilisers have demonstrated potential to mitigate N_2O emissions in field trials, though responses can vary with climate, soil type and management practices [257, 258, 269]. Mediterranean and sprinkler-irrigated systems show that nitrification inhibitors can reduce N_2O emissions and that management practices such as no-till or deficit irrigation interact with fertiliser form to influence outcomes [265]. Across these studies, the central message is that efficacy and economics are highly context-specific, and adoption should be guided by site-specific NUE and emission reduction targets, supported by decision tools and policy incentives [45, 261, 265, 269].

Key barriers to wide-scale adoption include the substantial capital required for renewable-energy infrastructure and electrolysers, the need for safe hydrogen and ammonia handling/es storage, and the relative premium of green ammonia against conventional products in the near term. However, these barriers are counterbalanced by the potential for economies of scale, policy support, and the growing demand for climate-smart inputs. Blue ammonia can serve as a transitional stepping stone, enabling gradual infrastructure build-out while reducing emissions relative to conventional ammonia, and thereby enabling the later transition to green ammonia as renewables, grids, and electrolyser costs improve [72, 258, 259]. In addition, the use of low-carbon fertilisers, whether via green or blue ammonia or stabilised formulations, offers direct co-benefits such as improving NUE, enabling better alignment with precision-application strategies, and reducing downstream environmental impacts, including N_2O and NH_3 emissions [45, 257, 258, 261, 263].

Green ammonia and low-carbon fertilisers offer a viable and increasingly well-supported pathway toward sustainable intensification in agriculture. The central argument is that decarbonising fertiliser production (upstream) and optimising nitrogen use (downstream) can cohesively reduce climate and environmental impacts without compromising agricultural productivity. Blue ammonia represents a pragmatic interim strategy to lower emissions while renewable energy and green hydrogen capacities scale, whereas green ammonia constitutes the long-term objective for fully decarbonised fertiliser systems. The integration of stabilised fertilisers with nitrification and urease inhibitors, controlled-release formulations, and improved nutrient management practices enhances the potential environmental benefits and supports climate targets. However, realizing this transition at scale requires continued investment, supportive policy frameworks, and careful attention to regional agronomic conditions to avoid unintended trade-offs. The evidence consistently demonstrates meaningful emission reductions from decarbonised ammonia production pathways and from well-designed low-carbon fertiliser products, albeit with site-specific performance variations and economic considerations that must be addressed through targeted implementation strategies and policy incentives [72, 256-259, 261-263, 266, 269].

To make the concept of green ammonia and low-carbon fertilisers more tangible, it is useful to examine how these ideas translate into real-world farming and industry scenarios. The key shift is not simply theoretical decarbonisation, but a practical transformation in how fertilisers are produced, selected, and applied within agricultural systems. In most cases, the crops, machinery, and overall farming systems remain unchanged; what evolves is the carbon intensity of inputs and the efficiency with which nutrients are managed.

In a broadacre grain farming context, such as wheat production in Australia, growers traditionally rely on urea derived from conventional ammonia, which carries a high embedded carbon footprint due to fossil fuel–based production. A low-carbon transition involves switching to urea produced from green ammonia, generated using renewable energy. At the same time, nitrogen applications are split across key growth stages—such as planting, tillering, and stem elongation—rather than applied all at once. The addition of a urease inhibitor (e.g. NBPT) further reduces ammonia volatilisation. The practical result is a fertiliser program that remains agronomically familiar but delivers lower upstream emissions, reduced field losses (NH_3 and N_2O), and improved nitrogen use efficiency (NUE). Importantly, the farmer does not change what they grow—only the efficiency and environmental impact of how nutrients are supplied.

In intensive vegetable production systems, particularly those using fertigation, growers already operate with a high degree of control over nutrient delivery. By sourcing fertilisers derived from green ammonia and applying them in small, frequent doses aligned with crop growth stages, nutrient delivery becomes highly precise. This reduces excess nitrogen in the soil, lowering N_2O emissions and minimising nutrient losses. The outcome is both environmental and economic: less fertiliser is required overall, and nutrient uptake is maximised. This example highlights how combining green ammonia with precision application systems amplifies sustainability benefits.

For maize production in high-rainfall regions, where nitrogen losses are a persistent challenge, stabilised fertilisers provide a practical pathway to emissions reduction. A grower may use urea treated with nitrification inhibitors such as DMPP or DCD, combined with split nitrogen applications at planting and key growth stages (e.g. V6). Where green ammonia is not yet fully available, fertiliser sourced from blue ammonia can serve as a lower-carbon alternative. These strategies reduce nitrate leaching and N_2O emissions while maintaining nitrogen availability in the root zone, often resulting in stable or improved yields. This demonstrates that even without full adoption of green ammonia, significant environmental gains can be achieved through improved fertiliser formulation and timing.

In high-control environments such as greenhouse tomato production, the integration of green ammonia-based fertilisers with advanced nutrient management systems offers even greater efficiency. Hydroponic systems using complete nutrient solutions—such as calcium nitrate and potassium nitrate—can operate in closed loops where drainage water is captured and reused. Real-time monitoring of electrical conductivity (EC), pH, and nutrient levels allows precise adjustment of nutrient supply. The result is near-zero nutrient loss, extremely high nutrient use efficiency, and a substantial reduction in emissions per unit of production. These systems demonstrate the full potential of combining low-carbon inputs with advanced technology.

At the industrial level, fertiliser manufacturing facilities are also undergoing transition. A typical pathway begins with the adoption of blue ammonia, where carbon capture and storage (CCS) reduces emissions from natural gas–based hydrogen production. Over time, facilities invest in electrolysers powered by renewable energy to produce green hydrogen, which is then used to synthesise green ammonia. This stepwise transition enables gradual emissions reduction while maintaining production reliability and supporting the development of infrastructure for a fully decarbonised future. Blue ammonia thus acts as a practical bridge toward long-term sustainability.

Mixed farming systems provide another example of how low-carbon fertilisers can be integrated into broader sustainability strategies. Farms that combine cropping and livestock can recycle nutrients through manure, reducing reliance on synthetic fertilisers. When supplemented with green ammonia-based fertilisers and supported by precision application techniques, these systems lower overall fertiliser demand, reduce emissions from both production and use, and improve soil health. This highlights the importance of combining low-carbon inputs with circular nutrient management approaches.

Policy and economic drivers also play a significant role in adoption. In regions where carbon pricing is introduced for fertiliser production emissions, conventional fertilisers become more expensive, making green ammonia-based alternatives more competitive. This encourages farmers to adopt low-carbon fertilisers and enhanced-efficiency products, while also improving nutrient management practices. The result is a market-driven transition toward reduced agricultural emissions, demonstrating how policy frameworks can accelerate change.

Across these scenarios, a consistent pattern emerges: what changes is the source and efficiency of fertiliser inputs, while the fundamental agricultural systems remain largely the same. Broadacre farming adopts lower-carbon fertilisers and inhibitors; horticulture integrates precision fertigation; industry shifts to renewable energy; and mixed systems incorporate organic nutrient recycling. The chemistry of ammonia and the nutrient requirements of crops remain unchanged, but the environmental footprint is significantly reduced.

In practice, the transition to green ammonia and low-carbon fertilisers is not about reinventing agriculture—it is about refining it. This involves decarbonising fertiliser production through green and blue ammonia, improving nutrient efficiency through better timing, placement, and stabilisation, and integrating circular approaches to nutrient use. When these elements are combined, agriculture can maintain productivity while significantly reducing its environmental impact, making low-carbon fertilisers a practical and achievable pathway toward sustainable intensification.

Circular Economy Approaches

Circular economy approaches in fertiliser systems aim to move away from the traditional "linear" model of extract → produce → use → dispose, toward a system where nutrients are continuously recovered, reused, and recycled. In agriculture, this is particularly important because nutrients such as nitrogen (N), phosphorus (P), and potassium (K) are finite or energy-intensive to produce, and significant losses occur at multiple points in the supply chain. A circular approach seeks to retain these nutrients within productive systems for as long as possible, reducing waste, environmental impact, and reliance on virgin resources.

A central concept in circular nutrient management is nutrient recovery from waste streams. Organic wastes, such as livestock manure, crop residues, food waste, and municipal biosolids, contain valuable nutrients that can be returned to the soil as fertilisers or soil amendments. Instead of being treated as waste requiring disposal, these materials are processed (e.g. composting, anaerobic digestion, drying, or pelletising) to stabilise nutrients, reduce pathogens, and improve handling. This not only reduces the need for synthetic fertilisers but also improves soil structure, organic matter content, and long-term soil health.

In addition to organic recycling, circular approaches increasingly focus on recovering nutrients from industrial and urban systems. For example, phosphorus can be recovered from wastewater treatment plants in forms such as struvite (magnesium ammonium phosphate), which can be used as a slow-release fertiliser. Similarly, nitrogen can be captured from waste streams and converted into usable fertiliser products. These technologies help close the loop between urban consumption and agricultural production, reducing nutrient discharge into waterways and mitigating issues such as eutrophication.

Another important aspect of circular systems is improving resource efficiency within fertiliser use itself. This includes reducing losses through better application methods, precision agriculture, and enhanced-efficiency fertilisers. By ensuring that a greater proportion of applied nutrients is taken up by crops, less is lost to the environment, and fewer inputs are required overall. In this sense, efficiency is a key enabler of circularity, minimising the need for new inputs by maximising the value of those already in use.

The integration of crop–livestock systems is a practical example of circular nutrient flows. In such systems, manure from livestock is returned to crop fields, providing nutrients that support plant growth, while crops or crop residues are used as animal feed. This creates a closed-loop system at the farm level, reducing external input requirements and improving overall system resilience. Similarly, cover crops can capture residual nutrients in the soil and recycle them for subsequent crops, preventing losses and enhancing nutrient cycling.

From a sustainability standpoint, circular economy approaches offer multiple benefits. They reduce dependence on non-renewable resources such as phosphate rock and fossil fuels used in fertiliser production. They lower greenhouse gas emissions by decreasing both production-related emissions and nutrient losses. They also reduce environmental pollution by diverting nutrients away from waterways and landfills, while improving soil health and long-term agricultural productivity.

However, implementing circular systems also presents challenges. Variability in the composition of organic and recycled materials can make nutrient management less predictable compared to synthetic fertilisers. There are also logistical and economic considerations, such as collection, transport, processing costs, and the need for infrastructure and regulatory frameworks to ensure safety and quality. Contaminants (e.g. heavy metals, microplastics, pathogens) must be carefully managed to avoid unintended environmental or health risks.

Despite these challenges, circular economy approaches are increasingly recognised as essential for sustainable agriculture. Advances in technology, policy support, and growing awareness of resource limitations are driving innovation in nutrient recovery and reuse. When combined with efficient fertiliser practices and low-carbon production methods, circular systems contribute to a more resilient, resource-efficient, and environmentally responsible agricultural sector.

A small-scale producer can actively participate in a circular economy by treating waste streams not as disposal problems, but as valuable nutrient resources that can be recovered, processed, and reused within the farm system. In practice, this means shifting from a linear model of input use and waste disposal to a closed-loop approach where nutrients are continuously cycled. The following example illustrates how a mixed smallholder operation can integrate multiple fertiliser production options into a sustainable, circular system.

Consider a small-scale market garden of approximately 2–5 hectares, producing vegetables such as leafy greens, tomatoes, and root crops, alongside a small number of chickens and goats. Rather than relying entirely on purchased fertilisers, the farmer adopts a circular nutrient management approach that combines on-farm waste recycling, locally sourced organic inputs, and targeted use of low-carbon fertilisers. This integrated system allows the farm to reduce external inputs while maintaining productivity.

The foundation of this system is composting. The farmer collects crop residues such as leaves, stems, and unsold produce, along with animal manure from chickens and goats, and kitchen scraps sourced from the household or local cafés. These materials are composted to produce a stable organic fertiliser rich in nutrients and organic matter. This compost provides essential macronutrients (N, P, K), improves soil structure and water retention, and reduces reliance on synthetic fertilisers. In this way, waste is transformed into a reusable soil input, forming the core of the circular system.

To complement solid compost, the farmer also produces liquid fertilisers from on-farm inputs. Compost tea, an aerated extract of compost, and diluted, stabilised manure-based liquids are applied through watering systems or simple fertigation setups. These liquid fertilisers supply readily available nutrients, support soil microbial activity, and enable low-cost, frequent nutrient applications. By cycling nutrients in liquid form, the system enhances nutrient availability and efficiency.

For further optimisation, the farmer may implement a small anaerobic digester. In this system, animal manure and organic waste are processed to produce biogas, which can be used for cooking or heating, and digestate, a nutrient-rich fertiliser. This approach not only reduces reliance on external energy sources but also produces a more stable, plant-available fertiliser. As a result, waste streams are converted into both energy and nutrients, closing multiple loops within the farm system.

While organic inputs provide a strong nutritional base, some crops, particularly high-value or nutrient-demanding crops such as tomatoes, require more precise nutrient management. To address this, the farmer supplements with small amounts of low-carbon fertilisers, such as nitrogen products derived from green ammonia, along with soluble fertilisers used in fertigation. These inputs are applied strategically at key growth stages to ensure optimal crop performance. In this context, external inputs are not eliminated but minimised and used efficiently.

Biological nutrient cycling further strengthens the system. The farmer incorporates crop rotation and cover cropping, including legumes such as beans or clover, which fix atmospheric nitrogen, and cover crops that capture residual nutrients in the soil. These practices reduce the need for synthetic nitrogen fertilisers, prevent nutrient losses through leaching, and improve overall soil health. Nutrients are thus captured and reused through natural biological processes.

Efficient water and nutrient management is achieved through a simple drip irrigation or micro-fertigation system. This system allows the farmer to apply liquid fertilisers directly to the root zone while minimising losses. Where possible, runoff water is captured and reused, further enhancing efficiency. In this way, both water and nutrients are retained within the system rather than being lost to the environment.

The circular approach can also extend beyond the farm boundary through local resource integration. The farmer partners with nearby cafés and landscaping services to collect food waste and green waste, which are then incorporated into the composting system. This reduces landfill waste while strengthening local nutrient cycles, effectively linking the farm to a broader community-based circular economy.

Overall, the system operates as a continuous loop: waste is converted into compost or digestate, returned to the soil, taken up by crops, and ultimately becomes food, with residual materials re-entering the cycle. Targeted low-carbon fertiliser inputs are used only where necessary to support crop performance. The practical benefits of this approach include reduced fertiliser costs, improved soil health and productivity, lower greenhouse gas emissions, reduced nutrient losses through leaching and runoff, and greater farm self-sufficiency.

A small-scale producer does not require advanced or expensive technology to participate in a circular economy. By combining on-farm recycling practices such as composting and manure use, biological processes like legumes and cover crops, efficient application methods such as drip irrigation, and targeted use of low-carbon fertilisers, the farm becomes a closed-loop nutrient system. This approach reduces waste, improves efficiency, and supports both environmental sustainability and long-term economic resilience.

Chapter 12

Industry Trends, Regulation, and Future Innovations

This final chapter brings together the key themes explored throughout the book, positioning fertilisers within a rapidly evolving global agricultural system shaped by regulation, market forces, technological innovation, and sustainability imperatives. While earlier chapters have examined the science, production, and application of fertilisers, this chapter shifts focus toward the broader industry context, how fertilisers are governed, traded, optimised, and reimagined in response to emerging challenges and opportunities.

As agriculture moves deeper into the 21st century, fertiliser use is no longer viewed solely as a technical input decision, but as part of a complex, interconnected system involving environmental stewardship, economic resilience, and global food security. Increasing regulatory scrutiny, supply chain volatility, and climate-related pressures are reshaping how fertilisers are produced and used, requiring producers, manufacturers, and policymakers to adopt more integrated and adaptive approaches.

At the same time, advances in digital agriculture, biotechnology, and enhanced-efficiency fertilisers are transforming nutrient management practices. These innovations are enabling more precise, data-driven, and biologically informed systems that improve efficiency while reducing environmental impact. Alongside these developments, the concept of sustainable intensification is emerging as a central framework for balancing productivity with long-term resource protection.

This chapter therefore explores the intersection of regulation, market dynamics, and innovation, providing both a systems-level perspective and practical examples across different scales of operation. It highlights how global fertiliser systems are transitioning from traditional, input-intensive models toward more efficient, circular, and sustainable approaches.

Ultimately, this chapter serves as both a synthesis and a forward-looking discussion, equipping readers with an understanding of how fertiliser use is evolving, and how these changes will shape the future of agriculture, nutrient management, and global food systems.

Global Fertiliser Regulations and Standards

Global fertiliser regulations and standards form the governance framework that ensures fertilisers are safe, effective, environmentally responsible, and accurately marketed. While requirements vary by country, most regulatory systems share common objectives: protecting human health, maintaining soil and water quality, ensuring product quality, and supporting fair trade. These frameworks operate across the entire fertiliser lifecycle, from production and importation to labelling, sale, and on-farm use.

At a global level, fertiliser regulations are designed to address several critical risks and responsibilities associated with the production and use of nutrient inputs. Fertilisers contain concentrated nutrients essential for crop growth, but they may also include trace contaminants such as heavy metals, for example cadmium in some phosphate fertilisers. Without appropriate regulatory oversight, the use and distribution of fertilisers could lead to significant negative outcomes across environmental, economic, and health domains.

Unregulated fertiliser use has the potential to cause environmental harm, including eutrophication of water bodies and long-term soil degradation. It may also pose risks to human and animal health through contamination of food systems or water supplies. In addition, the absence of standards can result in fraud or misrepresentation of nutrient content, where products do not meet claimed specifications, leading to reduced effectiveness and financial loss for users. Inefficient agricultural practices may also arise, as incorrect or misleading information can result in poor application decisions and increased nutrient losses.

Regulatory frameworks are therefore essential to ensure that fertilisers meet defined quality, safety, and performance standards. They also ensure that accurate and transparent information is provided to users, enabling fertilisers to be applied correctly and efficiently. In doing so, regulations support both agricultural productivity and environmental protection.

Key Components of Fertiliser Regulation

Fertiliser regulation is built around several key components that collectively ensure products are safe, effective, and responsibly used. One of the primary elements is product registration and approval. In most countries, fertilisers must be registered before they can be sold on the market. This process typically requires manufacturers to declare the nutrient composition, such as nitrogen (N), phosphorus (P), and potassium (K) content, identify the raw materials and manufacturing processes used, and demonstrate that the product meets safety requirements, including limits on contaminants like heavy metals. For example, within the European Union, fertilisers must comply with the Fertilising Products Regulation (EU) 2019/1009, which establishes strict criteria for product categories and allowable materials.

Another critical component is the establishment of nutrient content and quality standards. Regulations define minimum nutrient levels and allowable tolerances to ensure that fertilisers perform as claimed. These standards prevent both under- and over-concentration of nutrients, ensure consistency between production batches, and support reliable agronomic outcomes. For instance, a fertiliser labelled as 20-10-10 (NPK) must fall within specified tolerance ranges for each nutrient to ensure accuracy and effectiveness in application.

Contaminant limits are also a fundamental aspect of fertiliser regulation. To protect soil health, crop safety, and human wellbeing, regulatory authorities set thresholds for potentially harmful substances such as heavy metals (including cadmium, lead, and arsenic), pathogens in organic fertilisers or biosolids, and chemical residues. Countries such as Australia and Canada enforce strict contaminant limits, particularly for recycled or organic fertiliser products, to ensure they are safe for agricultural use.

Clear labelling and product information requirements are universally enforced to promote transparency and informed decision-making. Fertiliser labels typically include nutrient analysis (NPK values), recommended application rates and instructions, safety warnings, and details of the manufacturer or importer. This information enables farmers and users to apply fertilisers correctly, aligning nutrient supply with crop requirements and reducing the risk of misuse.

Regulations also address the safe storage, transport, and handling of fertilisers. Certain products, particularly nitrogen-based fertilisers like ammonium nitrate, can pose significant risks if not managed properly. As a result, regulatory frameworks specify requirements for safe storage conditions, including temperature control and segregation from incompatible materials, as well as transport classifications in line with hazardous goods legislation. Emergency response procedures are also mandated, often aligned with international safety frameworks developed by the United Nations for the transport of dangerous goods.

Finally, modern fertiliser regulations increasingly extend beyond product control to include environmental use requirements. These regulations govern how fertilisers are applied in agricultural systems, with measures such as limits on application rates, seasonal restrictions to avoid application before heavy rainfall, and the establishment of buffer zones near waterways. In regions particularly vulnerable to nutrient pollution, such as nitrate-sensitive areas in Europe, strict controls are imposed on nitrogen application to reduce the risk of groundwater contamination and protect environmental quality.

International Standards and Harmonisation

Although fertiliser regulations are primarily developed and enforced at the national level, there are significant international efforts aimed at harmonising standards to support global trade, improve consistency, and promote sustainable agricultural practices. These efforts help reduce regulatory fragmentation between countries, making it easier for fertiliser products to move across borders while ensuring that environmental and safety standards are maintained [13, 45, 270].

The Food and Agriculture Organization (FAO) plays a central role in guiding sustainable fertiliser use worldwide. It provides frameworks and best-practice guidance on nutrient management and soil health, with a strong emphasis on improving efficiency and reducing environmental impacts. One of its most widely adopted approaches is the 4R Nutrient Stewardship framework, which promotes applying the right nutrient source, at the right rate, at the right time, and in the right place to optimise crop uptake and minimise losses [13, 45, 270, 271].

The International Organization for Standardization (ISO) contributes by developing technical standards that support consistency and quality across fertiliser production and testing. These standards include methods for sampling and laboratory analysis, quality assurance systems to ensure product reliability, and environmental management frameworks such as ISO 14001, which guide organisations in reducing their environmental footprint and improving sustainability performance [13, 45, 270, 272, 273].

In addition, the Organisation for Economic Co-operation and Development supports the harmonisation of regulatory approaches and testing methodologies among member countries. Its work focuses particularly on environmental safety, product evaluation, and data consistency, helping to align regulatory practices and improve the comparability of results across jurisdictions. Together, these international organisations contribute to a more coordinated and sustainable global fertiliser system [13, 45, 270, 274].

Regional Regulatory Approaches

Regional approaches to fertiliser regulation vary significantly, reflecting differences in governance structures, environmental priorities, and agricultural systems. In Europe, the European Union adopts a highly centralised regulatory framework that applies across member states. This model places strong emphasis on environmental protection, the promotion of circular economy principles, such as the use of recycled fertilisers, and the enforcement of strict contaminant limits. Products that meet these standards are eligible for "CE marking," which allows them to be freely traded within the EU, supporting both regulatory consistency and market integration.

In Australia, fertiliser regulation is primarily managed at the state and territory level, although it is supported by national guidelines and industry standards. These frameworks focus on maintaining fertiliser quality, ensuring accurate labelling, and protecting the environment through measures such as runoff control and nutrient management. There is also a strong emphasis on industry stewardship programs and sustainable practices, particularly in environmentally sensitive areas such as the Great Barrier Reef catchment, where nutrient runoff can have significant ecological impacts.

In United States and Canada, fertiliser regulation is typically administered at the state or provincial level, with mandatory requirements for product registration and labelling to ensure transparency and product integrity. Environmental controls in these regions are largely focused on reducing nutrient runoff and protecting water quality. In Canada, this system is complemented by a strong federal framework under the Fertilizers Act, which sets national standards for safety, efficacy, and environmental protection, providing an additional layer of regulatory oversight.

Emerging Trends in Fertiliser Regulation

Global fertiliser regulations are undergoing a profound shift as environmental pressures, technological advances, and sustainability imperatives drive policymakers to reframe fertiliser policy from a product-centric approach to a holistic nutrient lifecycle governance framework. Across jurisdictions, regulators are increasingly prioritising the full environmental footprint of fertiliser production, distribution, and use, rather than solely focusing on product composition and safety. This trend is documented in recent sectoral analyses and policy reviews, which emphasise a transition from regulatory silos to integrated systems that encompass emissions, nutrient losses, and broader climate and water-quality objectives [275-277]. The synthesis of these sources indicates that government action is converging on several coherent pillars: stricter environmental controls on nitrogen emissions and nutrient losses; support for low-carbon and alternative nutrient sources; expanded regulation of recycled and bio-based fertilisers; and the introduction of digital traceability to enhance transparency and governance across the nutrient lifecycle [275, 278, 279].

One significant trend is the tightening of environmental controls on fertiliser use, particularly regarding nitrogen emissions, nutrient runoff, and associated environmental externalities. Several regulatory trajectories converge on setting tighter application rates, emission limits, and runoff controls designed to reduce greenhouse gas releases and water quality degradation. These measures align with broader climate strategies and water protection programmes and are repeatedly identified as central to contemporary fertiliser governance in diverse regulatory contexts [275, 277, 280]. The literature shows that nitrogen management is a focal point for reducing environmental impacts of fertiliser use, with policymakers linking emissions performance to regulatory compliance and, in some cases, to performance-based incentives or standards that drive reduced nutrient losses along the value chain [275, 277]. At the same time, there is growing policy momentum behind low-carbon fertilisers, including products derived from green ammonia, as a path to decarbonise agricultural supply chains and align fertiliser use with decarbonisation targets in the food system and energy transitions [281-283]. While much of the attention remains at the macro-policy level, regulatory analyses also highlight the need for robust safety and environmental standards for new low-carbon inputs to ensure environmental and consumer protection while enabling market uptake [281-283].

In parallel with stricter emission controls, regulatory frameworks are expanding to embrace recycled and bio-based fertilisers as part of circular economy strategies. Biosolids, composts, and recovered phosphorus products feature prominently in contemporary regulatory discussions as alternative nutrient sources that can substitute conventional fertilisers, subject to safety and quality regimes. The regulation of these products is framed around ensuring safety, quality, and environmental protection while enabling broader adoption and market acceptance. This orientation reflects a broader shift toward nutrient and material recycling in the agriculture sector as a means to reduce waste, enhance resource efficiency, and lower lifecycle emissions [275, 277, 284]. The literature notes that while opportunities are significant, regulatory design must balance innovation with risk management, requiring clear standards, permitting regimes, and consistent enforcement across regions [275, 277, 284].

A third strand concerns the digital governance of fertilisers through traceability and reporting systems. The adoption of digital traceability is increasingly framed as a governance instrument to improve transparency, accountability, and environmental performance across the fertiliser lifecycle. Regulators and industry actors are leveraging digital identifiers, barcoding, RFID, and blockchain-based record-keeping to track fertiliser composition, movement, and application, enabling better compliance monitoring, fraud prevention, and lifecycle assessment. The literature emphasises that digital traceability is not merely a technical upgrade but a governance innovation that can support regulatory inspections, compliance verification, and evidence-based policy adjustment. These developments are linked to broader governance objectives, including public sector transparency, sustainability reporting, and better data-driven policy design [275, 277, 279, 285]. Some analyses also discuss the role of digital governance in enabling cross-border trade compliance and the harmonisation of standards, while noting challenges around data ownership, interoperability, and privacy that regulators must address [275, 277, 285].

A further development is the integration of fertiliser regulation with wider climate and sustainability policies. Fertiliser decisions are increasingly connected to national emissions trajectories, water-quality strategies, sustainable land management, and forestalling nutrient-related pollution. This regulatory integration reflects a shift from product-by-product oversight toward nutrient lifecycle management that encompasses production, distribution, and application along with environmental consequences. The synthesis of sources indicates a trend toward aligning fertiliser policy with national decarbonisation plans, water-management targets, and soil stewardship initiatives, reinforcing the role of fertilisers within broader environmental and climate governance frameworks [275, 277, 286, 287]. In some regions, this integrated approach also interacts with sectoral decarbonisation strategies for energy-intensive industries, given the substantial energy and emission footprint of fertiliser production and distribution chains, including ammonia-based fertilisers [284, 286].

In sum, global fertiliser regulation is moving toward an integrated nutrient lifecycle framework anchored in environmental protection, circularity, digital governance, and climate policy alignment. Stricter nitrogen emission controls and nutrient-loss regulations, the deployment of low-carbon and recycled nutrient alternatives, the expansion of digital traceability for transparency and compliance, and the integration of fertiliser policy with national climate and sustainability agendas collectively reframe fertiliser governance from a product-centric regime to a systems-oriented governance architecture. These shifts reflect an evolving understanding that sustainable agriculture requires managing the full environmental footprint of fertiliser production and use, from cradle to field and beyond, with robust data-driven oversight and cross-cutting policy coherence [275, 277, 279, 282-285].

Practical Implications for Industry and Producers

For fertiliser manufacturers and agricultural producers, regulatory compliance has direct and practical implications for daily operations and long-term business viability. Ensuring that fertiliser products meet established quality and safety standards is a fundamental requirement, encompassing accurate nutrient composition, adherence to contaminant limits, and consistency in production. This is supported by maintaining clear and accurate documentation and labelling, which provides transparency to users and regulators while enabling informed application decisions.

In addition to product compliance, producers are expected to adopt best management practices that minimise environmental impact. This includes aligning fertiliser application with crop requirements, reducing nutrient losses, and implementing measures that protect soil and water resources. Staying informed about evolving regulations, environmental expectations, and sustainability requirements is also essential, as regulatory frameworks continue to change in response to new scientific knowledge and policy priorities.

Failure to comply with fertiliser regulations can result in significant consequences, including financial penalties, product recalls, reputational damage, and environmental harm. Conversely, compliance supports market access, builds trust with stakeholders, and contributes to more sustainable and resilient agricultural systems over the long term.

Overall, global fertiliser regulations and standards provide a critical foundation for balancing agricultural productivity with environmental protection and human safety. Although systems vary across regions, they share common principles of quality assurance, risk management, transparency, and sustainability. As agriculture moves toward lower emissions and more circular nutrient systems, regulatory frameworks are becoming more integrated, focusing not just on what fertilisers are, but how they are produced, distributed, and used within the broader environmental system.

Implementation Examples

To understand how global fertiliser regulations and standards are applied in practice, it is useful to examine real-world scenarios across different scales of operation. While regulatory frameworks are often described at a policy or systems level, their true impact is reflected in how producers, both small and large, interpret, implement, and comply with these requirements in day-to-day activities. The following case studies provide practical examples of how fertiliser regulations operate across the full lifecycle, from production and quality control to labelling, distribution, and environmental management.

Agricultural Fertiliser Production: Principles, Processes, and Practice

The first case study explores a small-scale fertiliser producer in Australia, demonstrating how regulatory principles can be applied in a localised, resource-efficient context. It highlights how even low-volume producers must meet requirements for product safety, traceability, labelling, and environmental protection, while also integrating circular economy practices.

The second case study examines a large-scale fertiliser manufacturer operating within the European Union. This example illustrates the complexity of compliance at an industrial level, including product registration, strict quality assurance systems, international standards alignment, emissions management, and digital traceability across global supply chains.

Together, these case studies provide a comparative perspective on how the same core regulatory principles (quality, safety, transparency, and environmental responsibility) are implemented at different scales. They demonstrate that while the level of complexity may vary, the underlying objective remains consistent: to ensure that fertilisers are produced and used in a way that supports agricultural productivity while protecting human health and the environment.

Case Study 1: Small-Scale Fertiliser Producer (Organic Compost and Liquid Fertiliser – Australia)

A small-scale fertiliser producer operates a 3-hectare market garden in regional Australia, producing compost-based fertilisers and liquid nutrient solutions derived from farm waste and local organic inputs. The business sources raw materials such as animal manure, crop residues, and food waste from nearby cafés. While the operation is relatively small, it must still comply with state-based fertiliser regulations, environmental protection laws, and industry guidelines to ensure product safety and market credibility.

To meet regulatory requirements, the producer begins by testing compost batches through a certified laboratory to determine nutrient composition (N, P, K values) and ensure contaminant levels, such as heavy metals and pathogens, are within allowable limits. This supports compliance with product quality and safety standards. Even though the fertiliser is not mass-produced, the producer documents raw material sources, composting processes, and treatment methods (e.g. temperature control for pathogen reduction), aligning with expectations for traceability and safe production.

Labelling is also implemented in a simplified but compliant manner. Each product includes nutrient analysis, application guidelines, and safety information. This ensures transparency and helps customers apply the fertiliser correctly, reducing the risk of overuse or environmental harm. The producer adopts best management practices such as aligning application recommendations with crop needs and avoiding sales or use during high rainfall periods, thereby supporting environmental use regulations and minimising runoff.

Storage and handling practices are adapted to meet safety expectations, including covered storage areas to prevent nutrient leaching and contamination of nearby waterways. Although hazardous materials are minimal, the producer still follows safe handling procedures consistent with broader regulatory expectations. Additionally, the business participates in a local sustainability program that promotes nutrient recycling and circular economy practices, aligning with emerging regulatory trends.

Through these measures, the small-scale producer demonstrates that even low-volume operations can effectively implement global fertiliser regulation principles. Compliance not only ensures environmental protection and product safety but also builds trust with customers and enables participation in local and regional markets.

<u>Case Study 2: Large-Scale Fertiliser Manufacturer (Granular NPK Production – European Union)</u>

A multinational fertiliser company operates a large-scale production facility within the European Union, manufacturing granular NPK fertilisers for distribution across multiple countries. Due to the scale and complexity of operations, the company must comply with comprehensive regulatory frameworks, including the Fertilising Products Regulation (EU) 2019/1009, environmental directives, and international standards.

The company begins by registering each fertiliser product, providing detailed documentation on nutrient composition, raw material sourcing, and manufacturing processes. Advanced quality control systems are implemented to ensure that each batch meets strict nutrient tolerances and contaminant limits, particularly for substances such as cadmium in phosphate inputs. Continuous monitoring and laboratory testing ensure consistency and compliance across production volumes.

To meet labelling and transparency requirements, all products carry detailed labels including nutrient composition, safety instructions, traceability codes, and CE marking, enabling free trade within EU member states. Digital traceability systems are integrated into the supply chain, allowing regulators and customers to track products from production through to distribution. This supports both compliance and emerging trends in digital governance.

Environmental regulations are a major focus of the operation. The company invests in emissions reduction technologies, including carbon capture systems and the use of lower-carbon inputs such as green ammonia where feasible. Nutrient management guidelines are embedded into product design, with enhanced-efficiency fertilisers (e.g. controlled-release or inhibitor-treated products) developed to reduce nitrogen losses and improve uptake efficiency in the field.

Storage and transport are managed under strict hazardous goods regulations, particularly for nitrogen-based fertilisers such as ammonium nitrate. Facilities are designed with temperature control, segregation systems, and emergency response protocols aligned with international safety standards developed by the United Nations. Regular audits and compliance checks ensure adherence to these requirements.

Beyond production, the company actively supports sustainable fertiliser use by providing agronomic guidance aligned with frameworks promoted by the Food and Agriculture Organization, such as the 4R Nutrient Stewardship approach. This ensures that fertilisers are not only compliant as products but are also used responsibly in agricultural systems.

This case demonstrates how large-scale producers implement fertiliser regulations across the entire lifecycle—from raw material sourcing and production to distribution and end-use. Compliance at this level is complex but essential, enabling international market access, reducing environmental impact, and supporting global sustainability objectives.

<u>Key Insight Across Both Cases</u>

Both small and large-scale producers operate within the same fundamental regulatory principles, quality assurance, safety, transparency, and environmental responsibility, but apply them at different scales and levels of complexity. While small producers focus on practical, localised compliance and traceability, large manufacturers integrate advanced systems, technologies, and international standards. Together, these examples illustrate how global fertiliser regulations function as a flexible yet comprehensive framework that supports safe, effective, and sustainable nutrient management across the entire industry.

Supply Chain and Market Dynamics

Supply chain and market dynamics in the fertiliser industry describe how fertilisers are produced, transported, traded, priced, and ultimately delivered to end users, as well as the economic and external factors that influence these processes. Because fertilisers are essential inputs for global food production, their supply chains are highly interconnected, energy-intensive, and sensitive to geopolitical, environmental, and economic conditions.

At the upstream level, the fertiliser supply chain begins with the extraction and production of raw materials. Key inputs include natural gas for nitrogen fertilisers, phosphate rock for phosphorus-based products, and potash ore for potassium fertilisers. These resources are geographically concentrated, meaning that a relatively small number of countries dominate global supply. For example, natural gas availability directly influences ammonia production costs, while phosphate and potash reserves are controlled by specific regions, creating dependencies in global markets. This concentration introduces supply risk, particularly during geopolitical disruptions or trade restrictions.

The manufacturing stage converts raw materials into usable fertiliser products such as urea, ammonium nitrate, and NPK blends. This stage is highly energy-intensive, especially for nitrogen fertilisers produced via the Haber–Bosch process. As a result, fertiliser production costs are closely linked to energy prices, particularly natural gas. When energy prices rise, fertiliser production becomes more expensive, often leading to reduced output or higher market prices. This creates a strong connection between global energy markets and fertiliser affordability.

Figure 57: Flexible intermediate bulk containers (FIBC, big bags). Material: 600 kg Lithactyl N Pro fertilizer by TIMAC Agro, Roullier group. Burgundy, France. Cjp24, CC BY-SA 3.0, via Wikimedia Commons.

Once produced, fertilisers move through complex distribution networks involving bulk transport (ships, rail, and trucks), storage facilities, wholesalers, and retailers. Efficient logistics are critical because fertilisers are typically required in large volumes and must be delivered in time for planting seasons. Seasonal demand creates logistical pressures, with peak periods requiring coordinated supply chain management to ensure availability. Disruptions such as port congestion, transportation delays, or infrastructure limitations can significantly affect supply reliability and pricing.

Figure 58: Fertiliser train, Castlerock. Albert Bridge / Fertiliser train, Castlerock, CC BY-SA 2.0, via Wikimedia Commons.

Market dynamics in the fertiliser sector are shaped by the interaction of supply and demand, as well as external factors such as weather patterns, crop prices, and government policies. Demand for fertilisers is closely tied to agricultural cycles and commodity prices. When crop prices are high, farmers are more likely to invest in fertilisers to maximise yields. Conversely, when crop prices are low, fertiliser use may decline as farmers reduce input costs, affecting overall demand.

Global trade plays a major role in balancing supply and demand. Many countries rely heavily on fertiliser imports due to limited domestic production capacity. International trade flows are therefore essential to maintaining supply, but they are also vulnerable to tariffs, export bans, and political instability. Events such as conflicts, sanctions, or trade policy changes can disrupt supply chains and lead to price volatility in global fertiliser markets.

Another important aspect of market dynamics is price volatility. Fertiliser prices can fluctuate significantly due to changes in raw material costs, energy prices, supply disruptions, and demand shifts. For example, spikes in natural gas prices can rapidly increase nitrogen fertiliser costs, while supply constraints in potash-producing regions can drive up potassium fertiliser prices. This volatility can create uncertainty for farmers and impact agricultural planning and profitability.

In recent years, sustainability and environmental considerations have also begun to influence supply chain and market dynamics. There is growing demand for low-carbon fertilisers, such as those produced using green ammonia, as well as increased interest in recycled and bio-based nutrient sources. These emerging products are reshaping markets by introducing new supply chains, technologies, and regulatory requirements. At the same time, policies aimed at reducing emissions and nutrient losses are influencing both production practices and market demand.

Overall, fertiliser supply chains and market dynamics reflect a complex and interconnected system influenced by resource availability, energy markets, logistics, global trade, and environmental policy. Understanding these dynamics is essential for producers, policymakers, and farmers, as they directly affect fertiliser availability, pricing, and the sustainability of agricultural production systems.

Small Scale Production Approaches

For small-scale fertiliser producers, participating in supply chain and market dynamics does not require operating at the same scale or complexity as large industrial manufacturers. Instead, it involves adapting the same principles (sourcing, production, distribution, pricing, and responsiveness to market forces) at a local and manageable level. Small producers typically operate within short, localised supply chains, reducing reliance on global inputs while still being influenced by broader market conditions.

At the upstream level, small-scale producers often avoid dependence on globally traded raw materials such as natural gas, phosphate rock, or potash ore. Instead, they source locally available inputs, including animal manure, crop residues, compostable organic waste, and sometimes regionally available mineral amendments (e.g. lime, rock phosphate). This reduces exposure to global supply risks and price volatility. However, small producers are still indirectly affected by global markets, for example, if synthetic fertiliser prices rise, demand for locally produced organic alternatives often increases.

In terms of production, small-scale operations rely on low-energy, biologically based processes rather than energy-intensive industrial methods like the Haber–Bosch process. Composting, vermiculture, anaerobic digestion, and simple blending are common approaches. These methods reduce production costs and environmental impact while aligning with growing demand for sustainable and circular fertiliser products. Although production volumes are smaller, value is often added through customisation, niche products, or organic certification, allowing small producers to compete in specialised markets.

Distribution for small-scale producers is typically direct and decentralised. Rather than relying on bulk shipping, rail networks, or large storage facilities, products are distributed through local channels such as farm gate sales, farmers' markets, cooperatives, or direct supply agreements with nearby growers. Some producers also use small-scale logistics, such as utility vehicles or local delivery services, to supply customers within a defined geographic area. This short supply chain reduces transport costs and improves responsiveness to customer needs, particularly during peak seasonal demand.

Market dynamics for small producers are closely tied to local agricultural conditions and customer relationships. Demand is influenced by seasonal planting cycles, crop types, and local farming practices. Small producers often have a closer connection with end users, allowing them to adjust products and recommendations based on specific crop requirements. For example, they may develop customised compost blends or liquid fertilisers tailored to vegetable growers, vineyards, or horticultural producers.

Price setting in small-scale operations is influenced by both local competition and broader market trends. While global fertiliser prices (e.g. urea or NPK) may fluctuate due to energy costs or supply disruptions, small producers can position their products based on value rather than volume—emphasising benefits such as soil health, organic certification, or sustainability. In periods of high global fertiliser prices, small-scale producers often become more competitive, as farmers seek cost-effective or alternative nutrient sources.

Small producers must also remain aware of external factors such as weather, policy changes, and sustainability trends. For example, heavy rainfall can increase demand for soil-improving amendments, while government incentives for sustainable agriculture may support the adoption of organic or recycled fertilisers. Increasing demand for low-carbon and circular products creates new market opportunities for small producers, particularly those already using waste-derived inputs.

Ultimately, while small-scale fertiliser producers operate on a much smaller footprint, they still participate in the same fundamental supply chain and market dynamics as large producers. The key difference lies in scale, localisation, and flexibility. By leveraging local resources, maintaining short supply chains, and responding quickly to customer needs and environmental conditions, small producers can operate efficiently within the broader fertiliser market while reducing exposure to global volatility and contributing to more sustainable agricultural systems.

Practical Example: Small-Scale Fertiliser Producer Operating Within Local Supply Chain and Market Dynamics (United States)

A small-scale fertiliser producer operates a mixed vegetable and composting business on approximately 10 acres in rural United States. The business produces compost, worm castings, and liquid fertilisers, supplying nearby vegetable farms, small-scale growers, and a local vineyard. Although the operation is relatively small, it actively participates in supply chain and market dynamics by applying the same core principles as larger fertiliser companies—adapted to a local and manageable scale.

At the sourcing (upstream) stage, the producer avoids reliance on globally traded inputs such as imported potash or synthetic nitrogen fertilisers. Instead, they form partnerships with a nearby dairy farm to collect manure, source green waste from a local landscaping business, and gather food scraps from restaurants and farmers' markets in a nearby town. When global fertiliser prices increase, driven by rising natural gas costs, local farmers begin seeking more affordable and sustainable alternatives. As a result, demand for the producer's compost and organic fertilisers increases, illustrating how global market pressures influence even local supply chains.

In production, the business relies on low-energy, biologically based methods such as windrow composting, vermiculture (worm farming), and small-scale liquid fertiliser brewing. Rather than producing standardised products, the producer develops customised blends tailored to local needs, for example, a high-organic-matter compost for vegetable growers and a potassium-enhanced mix using locally sourced wood ash for vineyard clients. This approach allows the business to compete through product quality, specialisation, and performance rather than volume.

Distribution is managed through short, local supply chains. Products are sold directly from the farm, delivered in bulk using a small truck within a regional radius, and stocked in a nearby agricultural supply store. During peak planting seasons in spring and early summer, demand increases significantly, and the producer adjusts production schedules and delivery routes accordingly. Because distribution is localised, the business can respond quickly to customer needs without relying on complex national logistics networks.

Market dynamics are closely linked to local agricultural conditions. For instance, after a season of heavy rainfall, several farms in the area experience nutrient leaching and poor soil structure. In response, the producer promotes a soil-rebuilding compost blend, leading to increased sales. Close relationships with customers enable the producer to provide tailored advice and adjust product offerings based on real-time farming challenges.

Pricing strategies are influenced by both local competition and broader market trends. When the cost of synthetic fertilisers rises, the producer increases prices modestly due to higher demand but remains competitive by emphasising long-term soil health benefits and sustainability. Rather than competing directly with bulk fertiliser suppliers, the business positions its products as premium, environmentally friendly alternatives, with value based on performance, local sourcing, and regenerative agriculture benefits.

External factors such as state and federal sustainability programs, as well as growing interest in organic and regenerative farming, further support the business. The producer may access grants or incentives to expand composting infrastructure and markets their products as part of a circular economy model, appealing to environmentally conscious growers and consumers.

This example demonstrates how a small-scale fertiliser producer in the United States can actively participate in supply chain and market dynamics without operating at an industrial scale. By leveraging local resources, maintaining short supply chains, and responding quickly to both local and global market signals, the business remains competitive, resilient, and aligned with broader trends in sustainability and resource efficiency.

Digital Agriculture Integration

Digital agriculture integration refers to the use of digital technologies, data systems, and automation to improve how fertilisers are planned, applied, monitored, and managed within farming systems. It represents a shift from traditional, uniform farming practices toward data-driven, site-specific decision-making, where inputs such as fertilisers are applied precisely according to crop needs, soil conditions, and environmental factors.

Digital agriculture integrates multiple technologies to collect and analyse data. These include GPS-enabled machinery, remote sensing (such as satellites and drones), soil sensors, weather stations, and farm management software. Together, these tools provide detailed, real-time information about field variability, such as differences in soil fertility, moisture levels, crop health, and growth stages. This allows farmers and fertiliser providers to move away from "one-size-fits-all" applications and instead tailor nutrient inputs to specific areas within a field.

One of the most important applications of digital agriculture is precision fertiliser application, often enabled through Variable Rate Technology (VRT). Using digital maps and data analytics, fertiliser spreaders or sprayers can automatically adjust application rates as they move across a field. For example, areas with lower soil nutrient levels can receive higher fertiliser rates, while more fertile zones receive less. This improves nutrient use efficiency, reduces waste, and minimises environmental impacts such as runoff and greenhouse gas emissions.

Digital integration also supports decision-making and planning. Farm management platforms can combine historical yield data, soil test results, and weather forecasts to recommend optimal fertiliser types, rates, and timing. These systems help farmers align fertiliser application with crop demand, reducing the risk of over-application and improving overall productivity. In many cases, recommendations are based on predictive models that simulate crop growth and nutrient uptake under different conditions.

Another key aspect is traceability and compliance. Digital systems can record when, where, and how fertilisers are applied, creating a detailed data trail. This is increasingly important for meeting regulatory requirements, sustainability certifications, and supply chain transparency expectations. For example, growers can demonstrate compliance with environmental regulations or nutrient management plans by providing digital records of fertiliser use.

For fertiliser producers and suppliers, digital agriculture integration enables closer engagement with customers. Companies can offer data-driven advisory services, customised fertiliser blends, and application recommendations based on field-specific data. This shifts the role of fertiliser suppliers from simply providing products to delivering integrated nutrient management solutions.

Digital agriculture also enhances efficiency and cost management. By applying fertilisers only where needed and in the correct amounts, farmers can reduce input costs while maintaining or improving yields. At the same time, reduced nutrient losses lower environmental risks and support sustainability goals. Over time, accumulated data allows for continuous improvement in fertiliser strategies and farm performance.

However, integrating digital agriculture comes with challenges. These include the cost of technology adoption, the need for technical skills and training, data management and privacy concerns, and ensuring compatibility between different digital systems. For small-scale producers, simplified or lower-cost digital tools, such as mobile apps, basic GPS systems, or shared advisory services, can provide an entry point into digital farming practices.

Both small and large-scale producers benefit from digital agriculture integration, but the way it affects their practices differs significantly. Small-scale producers tend to adopt simplified, low-cost tools and apply precision concepts manually, improving efficiency and sustainability without major capital investment. Large-scale producers, in contrast, implement fully automated, data-driven systems that optimise fertiliser use at a granular level across vast areas.

Example 1: Small-Scale Producer (Market Garden / Mixed Farming)

A small-scale vegetable grower operating on 5–10 hectares adopts basic digital agriculture tools, including a mobile soil testing app, a handheld GPS device, and access to satellite imagery through a low-cost subscription service. Previously, the farmer applied fertiliser uniformly across all fields, often leading to over-application in some areas and deficiencies in others.

With digital integration, the farmer begins mapping soil variability across the property. Soil tests reveal that certain sections of the field are already high in phosphorus but low in nitrogen, while others show the opposite pattern. Using this information, the farmer adjusts fertiliser practices manually—applying different blends or quantities to different sections of the field rather than treating the entire area the same. Although they do not use advanced Variable Rate Technology (VRT), they effectively implement "manual precision agriculture."

The impact on practices is significant. Fertiliser use decreases overall because inputs are no longer wasted on areas that do not require them. Crop performance improves in previously under-fertilised zones, leading to more consistent yields. The farmer also begins timing fertiliser applications more carefully by using weather forecast data to avoid heavy rainfall events, reducing nutrient runoff.

From a business perspective, the grower gains better control over costs and can market produce as being grown using data-informed, environmentally responsible practices, which appeals to local and sustainability-focused customers. However, the farmer must invest time in learning how to interpret data and integrate it into decision-making, highlighting the trade-off between simplicity and improved efficiency.

Example 2: Large-Scale Producer (Broadacre Cropping Operation)

A large-scale grain producer managing 5,000+ hectares integrates advanced digital agriculture systems across the entire operation. This includes GPS-guided tractors, drone and satellite imagery, in-field soil sensors, automated weather stations, and a central farm management platform. Fertiliser application is fully integrated with Variable Rate Technology (VRT), allowing machinery to adjust application rates in real time based on preloaded digital field maps.

Using several years of yield data, soil analysis, and remote sensing, the producer develops highly detailed nutrient maps for each field. These maps identify zones of high productivity, nutrient deficiency, and variability in soil moisture and organic matter. Fertiliser is then applied precisely according to these zones—for example, high-yield zones receive optimised nutrient rates to maximise output, while low-response areas receive reduced inputs to avoid waste.

The impact on practices is transformative. Fertiliser use becomes highly efficient, with significant reductions in input costs per hectare while maintaining or increasing total yield. Environmental outcomes improve as nutrient losses through leaching, runoff, and emissions are reduced. The producer also uses predictive modelling to plan fertiliser timing across the season, aligning applications with crop growth stages and expected weather conditions.

Digital systems also enhance compliance and reporting. Every fertiliser application is automatically recorded, including location, rate, and timing, allowing the producer to meet regulatory requirements and participate in sustainability certification programs. This traceability can provide access to premium markets or supply chains that require proof of sustainable practices.

However, the operation requires substantial investment in technology, data management systems, and skilled personnel to interpret and act on complex datasets. The scale of the operation makes these investments economically viable, but it also increases reliance on technology and data infrastructure.

In both cases, the core impact is the same: a shift from uniform fertiliser application to targeted, data-driven nutrient management. This results in improved input efficiency, better crop performance, reduced environmental impact, and stronger alignment with modern regulatory and sustainability expectations.

Biotechnology and Enhanced-Efficiency Fertilisers

Biotechnology and enhanced-efficiency fertilisers represent two complementary approaches aimed at improving how nutrients are supplied to crops, increasing productivity while reducing environmental impact. Together, they shift fertiliser use from a purely chemical input model toward a more biologically informed and efficiency-driven system.

Biotechnology in fertiliser systems focuses on harnessing biological processes, particularly microorganisms and plant–microbe interactions, to improve nutrient availability and uptake. Certain soil microbes play a critical role in nutrient cycling. For example, nitrogen-fixing bacteria (such as *Rhizobium* in legumes) convert atmospheric nitrogen into forms plants can use, while other microbes help solubilise phosphorus or enhance root growth. Biotechnology leverages these natural processes through products such as biofertilisers, microbial inoculants, and biostimulants, which are applied to seeds, soil, or crops to improve nutrient efficiency.

Biofertilisers, microbial inoculants, and biostimulants are biologically based inputs that enhance plant growth and nutrient use efficiency through natural processes rather than solely supplying nutrients in chemical form. While they are often grouped together, each plays a distinct role in plant nutrition and soil health, and each can be strategically integrated with fertiliser programs to improve performance and sustainability.

Biofertilisers are products that contain living microorganisms capable of directly increasing the availability of nutrients to plants. These organisms enhance nutrient supply by fixing atmospheric nitrogen, solubilising phosphorus, or mobilising other nutrients in the soil. For example, nitrogen-fixing bacteria such as *Rhizobium* form symbiotic relationships with legumes, converting atmospheric nitrogen into plant-available forms. Similarly, phosphate-solubilising bacteria can release phosphorus bound in soil minerals. Biofertilisers are typically applied to seeds, soil, or root zones.

When integrated with fertilisers, biofertilisers can reduce the need for synthetic inputs and improve nutrient efficiency. For instance, a farmer may apply a reduced rate of nitrogen fertiliser alongside a nitrogen-fixing biofertiliser, allowing biological processes to supplement nutrient supply. This not only lowers input costs but also reduces environmental impacts such as nitrogen leaching and greenhouse gas emissions.

Microbial inoculants are closely related to biofertilisers but are generally broader in function. They introduce beneficial microorganisms into the soil or plant environment to enhance overall soil biology, root development, and nutrient uptake. These may include bacteria, fungi (such as mycorrhizae), or consortia of multiple organisms. Mycorrhizal fungi, for example, form networks that extend beyond plant roots, increasing access to water and nutrients, particularly phosphorus.

In fertiliser programs, microbial inoculants are used to improve nutrient uptake efficiency rather than directly supply nutrients. A practical example is applying a standard fertiliser rate but combining it with a mycorrhizal inoculant, enabling plants to access more of the applied nutrients. This can lead to better crop performance without increasing fertiliser inputs. In some cases, fertiliser rates can be reduced over time as soil biological activity improves.

Biostimulants differ from both biofertilisers and microbial inoculants in that they do not necessarily supply nutrients or introduce living organisms. Instead, they are substances or compounds, such as seaweed extracts, humic acids, amino acids, or plant extracts, that stimulate plant physiological processes. Biostimulants enhance root growth, improve nutrient uptake, increase stress tolerance (e.g. drought or salinity), and support overall plant health.

When used alongside fertilisers, biostimulants help plants use nutrients more effectively. For example, applying a biostimulant during early growth stages can enhance root development, allowing the plant to access nutrients more efficiently from both soil reserves and applied fertilisers. This can improve yield outcomes and reduce the risk of nutrient losses, particularly under stressful environmental conditions.

In practice, integrating these biological products with fertilisers involves a coordinated approach. A farmer might use a base fertiliser program to supply essential nutrients, apply a microbial inoculant at planting to enhance root–soil interactions, and use a biostimulant during key growth stages to support plant development and stress resilience. The result is a more holistic nutrient management system, where chemical and biological inputs work together.

These biological solutions can reduce dependence on synthetic fertilisers by increasing the availability of nutrients already present in the soil. For instance, phosphate-solubilising bacteria can make otherwise unavailable phosphorus accessible to plants, while mycorrhizal fungi extend the root system's reach, improving water and nutrient uptake. In practice, this means that farmers can often achieve similar yields with lower fertiliser inputs, particularly when biological products are integrated into broader nutrient management strategies.

Biofertilisers, microbial inoculants, and biostimulants can be sourced from a range of natural, commercial, and industrial pathways. The source often depends on the scale of operation, regulatory requirements, and the level of consistency or performance required. Understanding where these inputs come from helps producers select appropriate, compliant, and effective products.

Biofertilisers are typically sourced from biological systems where beneficial microorganisms are isolated, cultured, and multiplied under controlled conditions. Commercially, they are produced by specialised agricultural biotechnology companies that grow strains such as nitrogen-fixing bacteria (*Rhizobium*, *Azotobacter*), phosphate-solubilising bacteria, or cyanobacteria in laboratories or fermentation facilities. These products are then formulated into liquids, powders, or granules for agricultural use.

At a smaller scale, biofertilisers can also be sourced or developed locally through composting systems, vermiculture (worm farming), and fermented organic inputs. For example, compost teas or liquid ferments made from plant materials and manure can contain active microbial communities that function similarly to biofertilisers. However, unlike commercial products, these may have variable composition and require careful management to ensure safety and effectiveness.

Microbial inoculants are most commonly sourced from commercial suppliers that specialise in soil biology and agricultural inputs. These products are developed through the isolation and selection of specific beneficial microorganisms, such as mycorrhizal fungi, *Bacillus* species, or *Trichoderma*. They are produced under controlled conditions to ensure viability, purity, and consistency, and are often packaged as seed coatings, soil amendments, or liquid applications.

Some microbial inoculants can also be sourced indirectly from natural ecosystems. Healthy soils, compost, and organic matter-rich environments naturally contain diverse microbial populations. Farmers may enhance these populations by incorporating high-quality compost, cover crops, or organic amendments, effectively "inoculating" their soils without purchasing commercial products. However, targeted inoculants are usually preferred when specific outcomes (e.g. improved phosphorus uptake) are required.

Biostimulants are sourced from a wide variety of natural and processed materials. Common sources include seaweed (particularly kelp species), plant extracts, humic and fulvic substances derived from decomposed organic matter, amino acids from protein hydrolysis, and microbial fermentation by-products. These materials are processed into liquid or soluble formulations that can be applied to crops through soil or foliar application.

Commercial biostimulants are typically manufactured by extracting, refining, and standardising these natural compounds to ensure consistent performance. For example, seaweed extracts are processed to retain plant growth hormones and bioactive compounds, while humic substances are extracted from leonardite or composted materials. At a small scale, some biostimulants can be produced on-farm, such as fermented plant extracts or compost-based liquids, although consistency and shelf life may be limited compared to commercial formulations.

Across all three categories, sourcing decisions are influenced by quality assurance, regulatory compliance, and intended use. Commercial products generally provide reliability, standardisation, and compliance with fertiliser regulations, making them suitable for larger-scale or market-facing operations. Locally produced or on-farm inputs offer cost advantages and support circular economy practices but may require additional testing and management to ensure safety and effectiveness.

In practice, many producers adopt a hybrid approach, using commercially sourced biofertilisers, inoculants, or biostimulants for targeted outcomes, while also enhancing soil biology through locally sourced organic materials. This combination allows for both consistency and sustainability, aligning biological inputs with broader fertiliser and nutrient management strategies.

Enhanced-efficiency fertilisers (EEFs), on the other hand, are engineered products designed to improve the timing and availability of nutrients from conventional fertilisers. Unlike standard fertilisers, which may release nutrients rapidly after application, EEFs are formulated to control, slow, or stabilise nutrient release, ensuring that nutrients are available when crops need them most.

There are several types of enhanced-efficiency fertilisers. Controlled-release fertilisers use coatings (such as polymers or sulphur) that gradually release nutrients over time, often influenced by soil temperature and moisture. Stabilised fertilisers include additives such as urease inhibitors or nitrification inhibitors, which slow down chemical processes in the soil that lead to nutrient loss. For example, urease inhibitors reduce ammonia volatilisation from urea, while nitrification inhibitors slow the conversion of ammonium to nitrate, reducing losses through leaching and denitrification.

Enhanced-efficiency fertilisers (EEFs) are designed to improve how nutrients are released, retained, and utilised by plants. Their primary purpose is to increase nutrient use efficiency while reducing environmental losses. EEFs are generally grouped into three main categories: controlled-release fertilisers, slow-release fertilisers, and stabilised fertilisers that contain inhibitors. Each type uses a different mechanism to regulate nutrient availability, but all aim to better align nutrient supply with plant demand.

Controlled-release fertilisers (CRFs) are manufactured with protective coatings that regulate how quickly nutrients are released into the soil. These coatings respond to environmental conditions such as soil temperature and moisture, allowing nutrients to diffuse gradually over time. Common examples include polymer-coated urea (PCU), sulphur-coated urea (SCU), and commercial products such as Osmocote Controlled Release Fertiliser and Polyon Fertiliser. The coating acts as a barrier that slows nutrient release, reducing leaching losses and ensuring a more consistent supply of nutrients over an extended period. This makes CRFs particularly suitable for horticulture, turf management, and situations where long-term nutrient availability is required.

Slow-release fertilisers (SRFs) differ from CRFs in that they rely on their chemical composition or low solubility to control nutrient release rather than a physical coating. Nutrients in these fertilisers become available gradually through microbial breakdown or slow dissolution in the soil. Examples include urea-formaldehyde (UF), isobutylidene diurea (IBDU), and crotonylidene diurea (CDU). The effectiveness of SRFs depends on soil conditions such as temperature, moisture, and microbial activity, as these factors influence the rate at which nutrients are released. As a result, their performance can vary across different environments and seasons.

Stabilised fertilisers incorporate chemical inhibitors that slow down the natural soil processes responsible for nutrient loss. These products are typically conventional fertilisers enhanced with additives such as urease inhibitors or nitrification inhibitors. For example, urea treated with NBPT (a urease inhibitor), such as Agrotain, reduces the conversion of urea to ammonia, thereby limiting volatilisation losses. Similarly, ammonium-based fertilisers containing nitrification inhibitors such as DMPP or DCD, including products like Entec Fertiliser, slow the conversion of ammonium to nitrate, reducing leaching and nitrous oxide emissions. Dual-inhibitor products such as SuperU Fertiliser combine both mechanisms to further enhance nitrogen retention in the soil.

In addition to these individual categories, some enhanced-efficiency fertilisers are formulated as integrated NPK blends that combine multiple technologies. These may include polymer-coated nutrients, added inhibitors, and micronutrients tailored to specific crops or growing conditions. Such products are commonly used in precision agriculture systems, where nutrient supply can be matched closely to crop requirements across different areas of a field.

Across all types of enhanced-efficiency fertilisers, the underlying objective is consistent: to match nutrient release with crop demand, reduce nutrient losses through leaching, volatilisation, and emissions, and improve overall fertiliser efficiency. By doing so, these products help optimise crop yields while minimising environmental impacts.

In practical terms, different agricultural sectors adopt EEFs in ways that suit their operational needs. A broadacre farmer may use urea with inhibitors to reduce nitrogen loss and improve efficiency across large fields. A horticulture grower may rely on products such as Osmocote to provide a steady nutrient supply for high-value crops. Similarly, turf managers often use coated fertilisers to achieve long-lasting, controlled nutrient release that supports consistent growth and reduces maintenance requirements.

The practical impact of EEFs is improved nutrient use efficiency (NUE). By matching nutrient release more closely with crop demand, these products reduce losses to the environment, lower greenhouse gas emissions (particularly nitrous oxide), and improve fertiliser effectiveness. This can lead to cost savings for farmers, as less fertiliser is required to achieve the same or better yields.

Mapping enhanced-efficiency fertilisers (EEFs) to specific industries depends on factors such as crop type, production system, nutrient demand patterns, and operational scale. Different EEF technologies, controlled-release, slow-release, and stabilised fertilisers, offer distinct advantages depending on how nutrients need to be managed in each context. The key objective across all industries is to align nutrient availability with crop demand while improving efficiency and minimising environmental losses.

In broadacre grain systems such as wheat, barley, and canola, production occurs over large areas and is highly sensitive to input costs and seasonal variability. In this setting, stabilised fertilisers containing inhibitors are generally the most suitable option. Nitrogen fertilisers such as urea are widely used but are prone to losses through volatilisation and leaching. The use of urease inhibitors (e.g. NBPT) and nitrification inhibitors (e.g. DMPP) helps retain nitrogen in plant-available forms for longer periods, which is particularly beneficial in dryland systems where rainfall timing is uncertain. Controlled-release fertilisers are less commonly used in these systems due to cost constraints, although they may have niche applications in precision agriculture. Overall, stabilised fertilisers provide the most practical balance between cost, efficiency, and scalability.

In vineyards, nutrient management is more precise and focused on optimising grape quality rather than simply maximising yield. Controlled-release fertilisers (CRFs) and enhanced-efficiency blends are therefore well suited to viticulture. CRFs provide a steady and controlled supply of nutrients throughout the growing season, aligning with vine uptake patterns and helping to avoid excessive vegetative growth that can compromise fruit quality. Fertigation systems are also commonly used, and these may incorporate stabilised fertilisers to improve nutrient retention in the root zone. In addition, tailored blends that include micronutrients can be integrated with precision agriculture technologies, enabling site-specific nutrient management across different areas of a vineyard.

Turf systems, including sports fields, golf courses, and landscaped areas, require consistent growth, uniform appearance, and minimal nutrient losses. In these highly managed environments, controlled-release fertilisers are the dominant EEF type. Polymer-coated fertilisers provide a predictable and extended nutrient release, reducing the need for frequent applications and supporting consistent turf performance. While slow-release fertilisers may also be used, controlled-release products are generally preferred due to their reliability and ability to minimise leaching and runoff.

Horticultural systems, which include vegetables, fruits, and nursery production, are typically high-value and demand precise nutrient management to maximise both yield and quality. In this sector, controlled-release and slow-release fertilisers are widely used, often in combination with fertigation systems. Controlled-release fertilisers are particularly important in nursery production and container-grown plants, where a consistent nutrient supply is critical. Slow-release fertilisers may be applied in field-grown crops where microbial activity supports gradual nutrient release. Stabilised fertilisers are also used in intensive systems to reduce nutrient losses, particularly in vegetable production. As a result, horticulture often benefits from a combination of EEF technologies tailored to specific crops and production methods.

Pasture systems, including grazing and dairy operations, involve large land areas and repeated fertiliser applications, especially nitrogen for grass growth. In this context, stabilised fertilisers are the most appropriate choice. Nitrogen losses can be significant due to volatilisation and leaching, particularly when using urea. The use of urease and nitrification inhibitors helps retain nitrogen in the soil, improving pasture productivity while reducing environmental impacts such as nitrate runoff and nitrous oxide emissions. Controlled-release fertilisers are less commonly used due to cost and the continuous nature of grazing systems, although they may be applied in more intensive or high-value pasture operations.

While all industries seek to improve nutrient efficiency, the selection of EEFs is influenced by scale, crop value, and management intensity. Large-scale, cost-sensitive systems such as grains and pasture tend to favour stabilised fertilisers due to their practicality and cost-effectiveness. In contrast, high-value and precision-driven systems such as horticulture, vineyards, and turf are more likely to adopt controlled-release and blended fertiliser technologies. This highlights that EEFs are not a one-size-fits-all solution, but rather a set of tools that must be matched to specific operational, economic, and environmental conditions.

When biotechnology and enhanced-efficiency fertilisers are used together, they create a more integrated nutrient management system. Biological inputs enhance the soil's natural ability to supply nutrients, while EEFs optimise the performance of applied fertilisers. For example, a farmer might apply a controlled-release nitrogen fertiliser alongside microbial inoculants that improve root development, resulting in both improved nutrient availability and uptake efficiency.

However, the effectiveness of these technologies can vary depending on soil type, climate, crop species, and management practices. Biotechnology products may require specific environmental conditions to perform well, while EEFs may involve higher upfront costs compared to conventional fertilisers. As a result, successful adoption often depends on proper understanding, correct application, and integration into broader farm management systems.

Biotechnology and enhanced-efficiency fertilisers represent a significant advancement in modern agriculture. They support more precise, efficient, and sustainable nutrient management by reducing waste, improving crop uptake, and minimising environmental impacts, aligning agricultural production with the goals of sustainability and resource efficiency.

Future Outlook: Sustainable Intensification

Sustainable intensification (SI) describes a pathway to raise agricultural productivity while reducing environmental footprints and preserving natural resources for future generations. In the fertiliser and nutrient-management domain, SI represents a shift from simply maximizing yields through higher inputs to optimizing nutrient use efficiency, resilience, and long-term sustainability within farming systems. This framing is widely echoed in scholarly discussions that link SI to more precise and efficient nutrient delivery, integration of biological processes, and a circular economy approach to nitrogen and phosphorus management [288-290]. Specifically, the emphasis is on producing more food from the same land area without converting additional ecosystems, a goal underscored by the combination of soil health maintenance, reduced pollution, and improved resource use efficiency that SI seeks to achieve in practice [291-293]. The general imperative, driven by population growth, climate change, and finite resources, grounds SI in a systems perspective that requires alignment among farmers, fertiliser producers, policymakers, and supply chains to incentivise practices that lower emissions and protect water quality while maintaining yields [288, 290, 294].

A central SI objective is enhancing nutrient use efficiency (NUE), defined as the crop uptake of nutrients relative to the amount applied. Achieving higher NUE involves applying fertilisers at the right rate, time, and place to minimise losses to leaching, volatilisation, and run-off. Precision agriculture, variable-rate application, and enhanced-efficiency fertilisers are core tools in this effort, enabling site-specific nutrient management that aligns inputs with crop demand and soil conditions. This focus on NUE is reinforced across multiple sources, which describe NUE improvements as a fundamental pathway to more sustainable fertiliser use alongside 4R Nutrient Stewardship principles (right source, right rate, right time, right place) to guide decision-making in fertiliser sourcing and application [289-292]. The transition toward SI does not universally entail reducing fertiliser volumes; rather, it stresses intelligent, targeted use to sustain yields while lowering environmental footprints [289, 291, 292].

Integrating biological and ecological processes into fertiliser management is another key element of SI. Biofertilisers, microbial inoculants, and biostimulants can support nutrient cycling, soil health, and crop resilience, reducing reliance on synthetic inputs over time. The regulatory and scientific landscape increasingly recognises biostimulants and microbial products as distinct from conventional fertilisers, with definitions enshrined in European Union regulations that emphasise improving nutrient use efficiency, stress tolerance, and nutrient availability in the rhizosphere independent of the product's own nutrient content (Mackiewicz-[288, 289, 291, 292]. Biostimulants and MBPs (microbial plant biostimulants) are highlighted as promising means to enhance NUE and crop tolerance to abiotic stress, thereby contributing to SI by fostering more robust soil-plant-microbe interactions and enabling greater nutrient efficiency from existing inputs [289, 291, 292, 295]. Within this biological integration, arbuscular mycorrhizal fungi (AMF) and plant growth-promoting rhizobacteria (PGPR) are repeatedly noted for their roles in improving nutrient acquisition and plant resilience, especially under nutrient-limited or stress-prone conditions [289, 293].

In parallel with biological integration, SI embraces transitions toward lower-emission and circular nutrient systems. This includes the development and deployment of low-carbon fertilisers and the use of recycled and organic nutrient sources, such as compost, biosolids, and recovered phosphorus. The rationale is to reduce dependence on finite mineral resources, lower the carbon intensity of fertiliser production, and close nutrient loops within the agricultural system. Several reviews underscore that circular-nutrient approaches, where waste streams (e.g., organic residues, wastewater concentrates) are converted into biofertilisers or biostimulants, can help reduce environmental footprints while sustaining productivity [68, 288, 294, 296]. The literature also discusses the role of microalgae- and algae-derived products as sources of biofertilisers and biostimulants, highlighting their potential to contribute to nutrient supply and soil health with a circular economy orientation, albeit with considerations around production efficiency and standardisation [294, 295, 297, 298].

Digital agriculture is identified as a major enabler of SI in fertiliser management. Sensor networks, remote sensing, soil moisture and nutrient sensors, and data-centric farm management platforms enable real-time decision-making, improved timing of applications, and better regulatory compliance. The value of digital tools lies in reducing waste, improving

NUE, and enabling continual improvement in nutrient strategies through data accumulation and learning across seasons and sites [289, 291, 299]. In addition, precision agriculture and VRF (variable-rate fertilisation) technologies demonstrate the potential to reduce input use while maintaining or increasing yields, with reported benefits in NUE and environmental impact reductions in various cropping contexts [289, 299].

From a systems perspective, achieving SI in fertiliser use requires coordination among multiple actors and policy instruments. Regulations, incentives, and market mechanisms increasingly shape farmer choices toward practices that reduce emissions, protect water quality, and promote soil health. Simultaneously, consumer demand for sustainably produced food creates market-driven pressure for cleaner fertiliser production and responsible nutrient management. The literature emphasises that policy and market alignment is essential to drive adoption of SI practices, particularly for small-scale producers who face higher barriers to adopting new technologies, while large-scale farming requires integration across extensive areas to realise efficiency gains [288, 290, 294].

There are notable challenges to SI implementation. The adoption of advanced technologies and novel inputs entails costs and requires technical knowledge and training. Soil variability, climate conditions, and crop requirements create complexity in decision-making and can impede adoption, especially for resource-constrained producers. Access to affordable technologies, extension services, and tailored support is critical for widespread adoption, particularly among smallholders, while larger operations must manage the integration of sophisticated systems across extensive land areas [288, 290, 291, 294]. A recurring theme across sources is that sustainable intensification does not automatically entail using less fertiliser in all cases; rather, it emphasizes smarter, more efficient, and more integrated management to reduce environmental harm while sustaining or enhancing productivity [289, 291, 292].

Looking to the future, the SI trajectory for fertiliser use is characterized by greater integration of precision nutrient management, biological innovations, circular nutrient streams, and digital decision support. The convergence of enhanced-efficiency fertilisers, bio-based products (biofertilisers, biostimulants, and microbial inoculants), and data-driven management systems is anticipated to support higher NUE, lower emissions, and more resilient farming systems. Some analyses argue for a nuanced path where fertiliser quantity may not always decrease; instead, improvements in the timing, placement, and efficiency of inputs can substantially lower environmental burdens while maintaining food security amid climate change and resource constraints [288-292, 295-297]. The overall outlook envisions a more integrated, efficient, and environmentally responsible fertiliser paradigm that aligns with broader sustainability agendas and consumer expectations, while recognising barriers to adoption and the need for continued research on the long-term outcomes and life-cycle implications of biostimulants, biofertilisers, and circular nutrient technologies [288-292, 295-297].

Appendices

Appendix A - Chemical Equations, Process Flow Diagrams and Fertiliser Calculations

Chemical equations describe the fundamental reactions that create fertilisers, while process flow diagrams show the real-world pathway from raw materials to finished products. Together, they provide a complete picture of how fertiliser production systems operate, bridging theory, engineering, and industrial practice.

Understanding fertiliser production requires both chemical insight (what reactions occur) and process understanding (how materials move through a system). This section outlines the key chemical equations involved in major fertiliser products and explains how these are represented in process flow diagrams (PFDs), which are used in industrial design, operations, and training.

Chemical Equations

Chemical equations are a way of representing chemical reactions using symbols and formulas. They show what substances are involved, how they interact, and what products are formed. In fertiliser production and agriculture, chemical equations are essential for understanding how nutrients are created, transformed, and made available to plants.

A chemical equation describes a reaction in a simple form: reactants → products. The reactants are the starting substances, while the products are the substances formed after the reaction. The arrow (→) means "reacts to form." For example, in ammonia production:

$$N_2 + 3H_2 \rightarrow 2NH_3$$

This means nitrogen gas reacts with hydrogen gas to produce ammonia.

Chemical equations are made up of several key components. Chemical formulas represent the substances involved, such as $N_2$for nitrogen gas, $H_2$for hydrogen gas, and $NH_3$for ammonia. Each formula shows the elements present and their proportions. Coefficients (numbers in front of formulas) indicate how many molecules or moles are involved in the reaction. For example:

$$2H_2 + O_2 \rightarrow 2H_2O$$

This shows that two molecules of hydrogen react with one molecule of oxygen to produce two molecules of water, ensuring the equation is balanced.

Subscripts (small numbers within formulas) show how many atoms are in each molecule. For example, H_2contains two hydrogen atoms, while CO_2contains one carbon atom and two oxygen atoms. Chemical equations may also include the state of matter, shown in brackets: (s) for solid, (l) for liquid, (g) for gas, and (aq) for aqueous (dissolved in water). For example:

$$NH_3(g) + H_2O(l) \rightarrow NH_4^+(aq) + OH^-(aq)$$

Additionally, reaction conditions such as temperature, pressure, or catalysts may be shown above or below the arrow, as in:

$$N_2 + 3H_2 \xrightarrow{\text{Fe catalyst, high pressure}} 2NH_3$$

A fundamental principle in chemical equations is the Law of Conservation of Mass, which states that matter cannot be created or destroyed. This means the number of atoms on the left side of the equation must equal the number on the right. For example, the unbalanced equation:

$$H_2 + O_2 \rightarrow H_2O$$

becomes balanced as:

$$2H_2 + O_2 \rightarrow 2H_2O$$

To read a chemical equation step-by-step, consider the fertiliser example:

$$NH_3 + HNO_3 \rightarrow NH_4NO_3$$

First, identify the reactants (ammonia and nitric acid). Next, identify the product (ammonium nitrate). Then interpret the reaction: ammonia reacts with nitric acid to form ammonium nitrate, a commonly used fertiliser. Finally, check that the equation is balanced by counting atoms on both sides.

Different types of chemical reactions are particularly relevant to fertilisers. In synthesis (combination) reactions, substances combine, as in:

$$N_2 + 3H_2 \rightarrow 2NH_3$$

In neutralisation reactions, an acid reacts with a base to form a salt:

$$NH_3 + HNO_3 \rightarrow NH_4NO_3$$

In decomposition reactions, a compound breaks down into simpler substances:

$$NH_2COONH_4 \rightarrow CO(NH_2)_2 + H_2O$$

Chemical equations are important in fertiliser science because they explain how fertilisers are manufactured (such as ammonia and urea), how nutrients transform in the soil, and how losses occur through processes like volatilisation and leaching. They also support improvements in efficiency and sustainability. For example, understanding nitrogen reactions can help reduce nitrous oxide emissions, while knowledge of solubility reactions can improve nutrient availability to plants.

When interpreting any chemical equation, it is useful to ask a few key questions: what is reacting, what is being produced, in what proportions, under what conditions, and why the reaction matters in practical terms—such as its impact on crop growth or the environment.

Key Chemical Equations in Fertiliser Production

a) Ammonia Production (Haber–Bosch Process)

Ammonia (NH_3) is the foundation of most nitrogen fertilisers.

Overall reaction:

$$N_2(g) + 3H_2(g) \rightleftharpoons 2NH_3(g)$$

Key conditions:

Temperature: ~400–500°C

Pressure: ~150–300 bar

Catalyst: Iron-based

Explanation: Nitrogen is sourced from air, and hydrogen is typically derived from natural gas (via steam methane reforming). The reaction is reversible and requires careful control of temperature and pressure to maximise yield.

b) Urea Production

Urea is one of the most widely used nitrogen fertilisers.

Step 1: Formation of ammonium carbamate

$$2NH_3 + CO_2 \rightarrow NH_2COONH_4$$

Step 2: Dehydration to urea

$$NH_2COONH_4 \rightarrow CO(NH_2)_2 + H_2O$$

Explanation: Ammonia reacts with carbon dioxide under high pressure to form an intermediate, which is then converted into urea.

c) Nitric Acid Production (Ostwald Process)

Used to produce ammonium nitrate fertilisers.

Step 1: Ammonia oxidation

$$4NH_3 + 5O_2 \rightarrow 4NO + 6H_2O$$

Step 2: Nitric oxide oxidation

$$2NO + O_2 \rightarrow 2NO_2$$

Step 3: Absorption in water

$$3NO_2 + H_2O \rightarrow 2HNO_3 + NO$$

d) Ammonium Nitrate Production

$$NH_3 + HNO_3 \rightarrow NH_4NO_3$$

Explanation: Ammonia neutralises nitric acid to produce ammonium nitrate, a high-nitrogen fertiliser.

e) Phosphate Fertiliser Production (Superphosphate)

Single superphosphate (SSP):

$$Ca_3(PO_4)_2 + 2H_2SO_4 \rightarrow Ca(H_2PO_4)_2 + 2CaSO_4$$

Explanation: Phosphate rock reacts with sulfuric acid to produce plant-available phosphorus.

f) Potash Fertilisers

Potassium fertilisers (e.g. KCl) are typically physically processed rather than chemically synthesised.

Example:

Mining → Crushing → Flotation → Purification → Granulation

Process Flow Diagrams (PFDs)

A Process Flow Diagram (PFD) is a visual representation of how a process operates, showing the movement of materials and the sequence of steps involved in converting raw inputs into finished products. In fertiliser production, PFDs are used to illustrate how materials such as natural gas, air, and minerals are transformed into products like ammonia, urea, or NPK fertilisers. They provide a simplified overview of complex industrial systems, making it easier to understand how different stages of production are connected.

PFDs present processes in a clear and structured format, allowing engineers, operators, and learners to understand how a system functions without needing to examine every mechanical detail. Their focus is on the overall process flow rather than detailed equipment design, which is typically shown in more complex diagrams such as Piping and Instrumentation Diagrams (P&IDs). This makes PFDs particularly useful for explaining concepts, training personnel, and analysing system performance at a high level.

A typical PFD includes several key elements. Process units—such as reactors, separators, heat exchangers, and storage tanks—are shown as labelled blocks or symbols, representing where physical or chemical changes occur. These units are connected by arrows or lines that indicate the direction of material flow through the system. Flow lines may also include additional information such as the type of material being transported (e.g. ammonia, steam, or water), as well as operating conditions like temperature, pressure, or flow rate.

In addition to process units and flow lines, PFDs often show input streams and output streams. Input streams represent raw materials entering the system, while output streams represent finished products, by-products, or waste materials leaving the process. Another important feature is the presence of recycle loops, where unreacted materials are returned to earlier stages to improve efficiency. For example, in ammonia production, unreacted nitrogen and hydrogen are recycled back into the reactor to maximise yield.

Reading a PFD involves following the process step-by-step, typically from left to right or from top to bottom. The first step is to identify the inputs, such as natural gas, air, water, or phosphate rock. From there, the arrows are followed through each stage of the process, with each piece of equipment representing a specific operation such as heating, reacting, separating, or cooling. As the flow is traced, it is important to observe how the material changes at each stage, including phase changes, chemical reactions, or pressure adjustments.

As you read the diagram, look for branching points where a stream splits into multiple paths, or converging points where different streams combine. Recycle streams should also be identified, as they indicate that the process is designed to reuse materials and improve efficiency. Finally, the outputs should be located at the end of the flow, showing the final products, by-products, or waste streams generated by the process.

A simplified ammonia production flow provides a clear example of how a PFD operates. In this process, natural gas and air enter the system, with natural gas being reformed to produce hydrogen and air processed to obtain nitrogen. These gases are then purified and compressed before entering the Haber–Bosch reactor, where ammonia is formed. After the reaction, the mixture is cooled and ammonia is separated from unreacted gases. The unreacted nitrogen and hydrogen are recycled back into the reactor, while the ammonia is sent to storage or further processing, such as urea production. This example highlights how a PFD shows both the sequence of operations and the flow of materials.

Although symbols can vary between diagrams, there are common conventions used in PFDs. Rectangles or vessel shapes typically represent reactors and processing units, while circles or ovals are used for pumps and compressors. Lines with arrows indicate material flow, and dashed lines may represent control or auxiliary systems. Clear labelling is essential, as it ensures that each unit and stream can be correctly identified and understood.

Process Flow Diagrams are important because they provide a big-picture view of how fertiliser production systems operate. They support process design and optimisation, help train operators and technicians, and assist in identifying inefficiencies or bottlenecks. They also contribute to understanding how materials and energy are used, which is critical for improving sustainability and ensuring safe operation.

When interpreting a PFD, it is helpful to ask key questions: what are the main inputs and outputs, what happens at each stage of the process, how materials move through the system, where reactions or separations occur, and whether there are recycle loops or waste streams. By answering these questions, the diagram becomes a practical tool for understanding both the structure and function of complex industrial processes.

a) Ammonia Production Flow (Simplified)

Process steps:

1. Air separation → nitrogen extraction
2. Natural gas reforming → hydrogen production
3. Gas purification (remove CO_2, CO, impurities)
4. Compression
5. Haber–Bosch reactor
6. Cooling and ammonia separation
7. Recycling unreacted gases

Flow concept:

Raw inputs → reaction → separation → recycle loop → product output

b) Urea Production Flow

Process steps:

1. Ammonia + CO_2 feed
2. High-pressure reactor (carbamate formation)
3. Dehydration section (urea formation)
4. Separation of unreacted gases
5. Recycling of NH_3 and CO_2
6. Evaporation and concentration
7. Granulation or prilling

c) Ammonium Nitrate Flow

Process steps:

1. Ammonia production
2. Nitric acid production (Ostwald process)
3. Neutralisation reactor
4. Concentration
5. Prilling or granulation
6. Cooling and storage

d) Phosphate Fertiliser Flow

Process steps:

1. Mining of phosphate rock
2. Crushing and grinding
3. Acidulation (reaction with sulfuric or phosphoric acid)
4. Curing and drying
5. Granulation
6. Screening and packaging

Process Flow Diagrams (PFDs) typically include several key components that represent the main functions within a process. Reactors are used where chemical reactions occur, transforming raw materials into new substances. Heat exchangers are included to control temperature, either by heating or cooling process streams to maintain optimal reaction conditions. Compressors are used to increase pressure, which is often necessary for reactions such as ammonia synthesis. Separators play a critical role in removing unreacted materials or isolating desired products from mixtures.

Another important feature in PFDs is the inclusion of recycle streams, which return unreacted or partially processed materials back into earlier stages of the process to improve efficiency and reduce waste. Storage tanks are also shown, providing capacity to hold intermediate or final products before further processing or distribution. Throughout the diagram, arrows indicate the direction of material flow, while labels provide information about key inputs, outputs, and operating conditions such as temperature and pressure.

Modern fertiliser plants are designed with a strong focus on integration and efficiency. Multiple processes are interconnected to maximise resource use and minimise waste. For example, heat recovery systems capture and reuse energy from hot process streams, significantly reducing overall energy consumption. Gas recycling loops, such as those used in the Haber process, increase production efficiency by reusing unreacted gases. Emission control systems are also incorporated to reduce the release of pollutants such as nitrogen oxides (NO_x), carbon dioxide (CO_2), and ammonia, supporting both environmental compliance and sustainability goals. In addition, process automation is widely used to improve consistency, optimise performance, and enhance operational safety.

From a practical perspective, there are clear differences between large-scale and small-scale systems. Large industrial fertiliser plants typically rely on continuous flow processes with integrated recycling loops, allowing for high efficiency and consistent production at scale. In contrast, small-scale or decentralised systems often use simpler batch processes, which are easier to manage but may be less efficient in terms of energy and resource use. Understanding both chemical equations and process flow diagrams is essential, as it links the underlying chemistry (what reactions occur), the engineering systems (how those reactions are carried out), and the operational aspects (how the process is controlled and optimised in practice).

Fertiliser Calculations

The following calculations can be used to accurately determine how much fertiliser should be applied to soil. Before applying any fertiliser, it is essential to assess the nutrient content of the soil through testing. Other important factors must also be considered, including soil type, soil depth, the current pasture or crop, and the paddock's history. These variables influence how nutrients behave in the soil and how effectively they will be used by plants.

Fertiliser recommendations are often expressed in terms of "units" of a nutrient, such as phosphorus. A unit is equivalent to one kilogram of that nutrient (for example, 1 unit of phosphorus equals 1 kg of phosphorus). However, this is not the same as one kilogram of fertiliser, because fertilisers contain different proportions of nutrients. These proportions are typically shown as percentages or expressed as N:P:K:S (Nitrogen: Phosphorus: Potassium: Sulphur). For example, single superphosphate is labelled 0:8.8:0:11, meaning that in 100 kg of fertiliser there are 8.8 kg of phosphorus and 11 kg of sulphur, with no nitrogen or potassium. Similarly, Starter 15 is labelled 14:12.7:0:11, meaning that 100 kg contains 14 kg of nitrogen, 12.7 kg of phosphorus, and 11 kg of sulphur.

This information allows users to calculate how much fertiliser is required to supply a specific nutrient. Calculation A is used to determine the required fertiliser rate:

Amount of fertiliser (kg/ha) = kg/ha nutrient ÷ % nutrient × 100.

For example, if 20 kg/ha of phosphorus is required and single superphosphate (8.8% P) is used, the calculation is:

$$20 \div 8.8 \times 100 = 227 \text{ kg/ha.}$$

This means 227 kg/ha of superphosphate must be applied to supply the required phosphorus.

The reverse calculation can be used to determine how much nutrient is being applied. Calculation B is:

Amount of nutrient (kg/ha) = fertiliser rate (kg/ha) × % nutrient ÷ 100.

For example, applying 125 kg/ha of single superphosphate results in:

$$125 \times 8.8 \div 100 = 11 \text{ kg/ha of phosphorus.}$$

Fertilisers can also be compared based on the cost of supplying a specific nutrient. Calculation C is used to determine the price per kilogram of nutrient:

Price per kg nutrient = price per tonne ÷ 10 ÷ % nutrient.

For example, if a paddock requires sulphur, single superphosphate at $262/t provides sulphur at $2.38 per kg, gypsum at $56/t provides sulphur at $0.39 per kg, and SF45 at $335/t provides sulphur at $0.80 per kg (noting that only part of the sulphur is immediately available). Similarly, if nitrogen is required, urea at $410/t provides nitrogen at $0.89 per kg, while DAP at $580/t provides nitrogen at $3.22 per kg.

When more than one nutrient is required, different fertiliser options can be compared. First, the required application rate for each fertiliser is calculated using Calculation A. Then, Calculation D is used to determine cost per hectare:

Cost per ha = price per tonne × rate (kg/ha) ÷ 1000.

For example, if 20 kg/ha of nitrogen and 20 kg/ha of phosphorus are required, one option is to use urea and single superphosphate. This results in a total cost of $77.31/ha. An alternative is to use DAP, which supplies both nutrients, resulting in a lower total cost of $63.80/ha. When using fertilisers that contain multiple nutrients, the application rate should be based on the nutrient present in the lowest proportion.

While cost is an important factor, fertiliser selection should not be based on price alone. Several practical considerations must also be taken into account. These include whether the crop or pasture will respond to the fertiliser, as some systems (such as clover-based pastures) may not require certain nutrients. Availability is also important, including whether the fertiliser can be sourced locally and at the required time, as well as any associated transport costs.

The availability of nutrients within the fertiliser should also be considered, particularly whether they are immediately available or released slowly over time. Handling requirements are another factor, including whether existing equipment can apply the required rates and whether the fertiliser is supplied in a manageable form. Timing and placement are critical, as fertilisers should be applied so that nutrients are available during key growth periods. Consideration must also be given to whether fertiliser can be safely applied with seed, whether it needs to be incorporated into the soil, and whether labour or contractor availability aligns with application timing.

Finally, potential side effects must be evaluated. Some fertilisers can contribute to soil acidification, contain salts or heavy metals, or pose risks of leaching and runoff. Environmental considerations such as erosion risk and contamination of waterways must be managed, including the use of buffer zones. There may also be animal health risks, such as nitrate toxicity from excessive nitrogen application. By considering these factors alongside calculations, fertiliser use can be both effective and sustainable.

Appendix B - Fertiliser Nutrient Conversion Tables

Fertiliser nutrient conversion tables are essential tools used to translate between different ways of expressing nutrient content. In fertiliser science and agriculture, nutrients are often reported in multiple forms, elemental (pure nutrient) form or oxide form (e.g. P_2O_5, K_2O). Understanding how to convert between these forms ensures accurate fertiliser selection, application rates, and nutrient management.

In practice, fertiliser labels, soil tests, and agronomic recommendations may use different units or chemical forms depending on regional standards or industry conventions. For example, phosphorus may be expressed as elemental phosphorus (P) or as phosphate (P_2O_5), while potassium may be expressed as potassium (K) or potash (K_2O). Conversion tables provide a simple way to move between these forms without needing to perform complex chemical calculations each time.

Fertiliser nutrient conversion tables provide a critical link between chemical composition and practical application. They ensure that nutrient calculations are accurate, regardless of how the fertiliser is labelled, supporting better decision-making, improved nutrient use efficiency, and more sustainable agricultural practices.

Different systems use different nutrient expressions:

- Elemental form → used in soil science and plant nutrition (e.g. P, K, S)
- Oxide form → commonly used in fertiliser labelling (e.g. P_2O_5, K_2O)

Because plants absorb nutrients in elemental form, but fertilisers are often labelled in oxide form, conversion is necessary to ensure correct application rates.

The following are standard conversion factors used in fertiliser calculations:

Phosphorus (P and P_2O_5)

- P_2O_5 → P: multiply by 0.44
- P → P_2O_5: multiply by 2.29

Potassium (K and K_2O)

- K_2O → K: multiply by 0.83
- K → K_2O: multiply by 1.20

Calcium (Ca and CaO)

- CaO → Ca: multiply by 0.71

- Ca → CaO: multiply by 1.40

Magnesium (Mg and MgO)

- MgO → Mg: multiply by 0.60
- Mg → MgO: multiply by 1.67

Sulphur (S and SO_3)

- SO_3 → S: multiply by 0.40
- S → SO_3: multiply by 2.50

Example Conversions

If a fertiliser label shows 20% P_2O_5, the actual phosphorus content can be calculated as:

$$20 \times 0.44 = 8.8\% \text{ P}$$

This explains why single superphosphate is often listed as approximately 8.8% phosphorus—it is derived from its phosphate (P_2O_5) content.

Similarly, if a fertiliser contains 60% K_2O, the elemental potassium content is:

$$60 \times 0.83 = 49.8\% \text{ K}$$

Converting Between Units (kg/ha)

Conversion tables are also used when calculating application rates.

For example, if a recommendation calls for 20 kg/ha of P, but the fertiliser is labelled in P_2O_5:

$$20 \times 2.29 = 45.8 \text{ kg/ha of } P_2O_5 \text{ required}$$

This ensures the correct amount of fertiliser is applied to meet crop nutrient needs.

Practical Use in Fertiliser Selection

Conversion tables are particularly useful when:

- Comparing fertilisers with different labelling systems
- Interpreting soil test results vs fertiliser labels
- Calculating nutrient application rates
- Evaluating nutrient efficiency and cost per unit nutrient

For example, one fertiliser may appear to have a higher phosphorus content, but after conversion, it may supply the same or less actual nutrient than another product.

Different countries use different conventions:

- Australia and New Zealand → often use elemental forms (P, K, S)
- United States → commonly uses oxide forms (P_2O_5, K_2O)
- Europe → may use a mix depending on regulation

Understanding conversion tables is therefore essential for working across different markets, standards, and product labels.

Sample Fertiliser Nutrient Conversion Table

The following table provides commonly used conversion factors between oxide forms (used in fertiliser labelling) and elemental forms (used in agronomy and plant nutrition).

1. Nutrient Conversion Factors

Nutrient	From (Oxide Form)	To (Elemental Form)	Multiply By	Reverse Conversion (Element → Oxide)
Phosphorus	P_2O_5	P	× 0.44	× 2.29
Potassium	K_2O	K	× 0.83	× 1.20
Calcium	CaO	Ca	× 0.71	× 1.40
Magnesium	MgO	Mg	× 0.60	× 1.67
Sulphur	SO_3	S	× 0.40	× 2.50

2. Practical Example Table

Fertiliser Label (Oxide Form)	Conversion	Elemental Equivalent
20% P_2O_5	20 × 0.44	8.8% P
60% K_2O	60 × 0.83	49.8% K
25% SO_3	25 × 0.40	10% S

3. Reverse Conversion Examples

Required Nutrient (Elemental)	Conversion	Oxide Equivalent Needed
20 kg/ha P	20 × 2.29	45.8 kg/ha P_2O_5

Required Nutrient (Elemental)	Conversion	Oxide Equivalent Needed
30 kg/ha K	30 × 1.20	36 kg/ha K_2O
10 kg/ha S	10 × 2.50	25 kg/ha SO_3

How to Use This Table

- Use oxide → elemental when interpreting fertiliser labels
- Use elemental → oxide when matching fertiliser to soil test recommendations
- Apply conversions when comparing fertilisers across different regions (e.g. USA vs Australia)

Appendix C - Safety Data and Handling Guidelines

The safe handling, storage, and use of fertilisers are critical to protecting human health, maintaining environmental integrity, and ensuring compliance with regulatory requirements. Fertilisers, particularly those containing nitrogen compounds, acids, or concentrated nutrients, can pose risks if not managed correctly. This section outlines key safety principles, the role of Safety Data Sheets (SDS), and practical guidelines for handling fertilisers in both small- and large-scale operations.

1. Understanding Safety Data Sheets (SDS)

A Safety Data Sheet (SDS) is a standardised document that provides detailed information about a chemical product, including fertilisers. It is an essential reference for anyone involved in handling, storing, or transporting fertilisers.

An SDS typically includes:

- **Identification** – product name, supplier details, and recommended use
- **Hazard identification** – classification of risks (e.g. irritant, oxidiser)
- **Composition** – chemical ingredients and concentrations
- **First aid measures** – actions to take in case of exposure
- **Fire-fighting measures** – suitable extinguishing methods
- **Accidental release measures** – spill response procedures
- **Handling and storage** – safe practices and conditions
- **Exposure controls and PPE** – required protective equipment
- **Physical and chemical properties** – appearance, solubility, reactivity
- **Stability and reactivity** – conditions to avoid
- **Toxicological and environmental information**

Understanding and following SDS guidance is a fundamental requirement under workplace health and safety (WHS) regulations.

Sample Safety Data Sheet (SDS)
Product: Granular NPK Fertiliser (15-15-15)
Prepared for Training and Example Purposes Only

1. Identification

- **Product Name:** NPK Fertiliser 15-15-15
- **Recommended Use:** Agricultural fertiliser for crop and soil nutrient supplementation
- **Supplier:** Example Agri Supplies Pty Ltd

- **Emergency Contact:** 000 (Australia Emergency Services)

2. Hazard Identification

- **Classification:**
  - Not classified as hazardous under normal use (may vary by jurisdiction)
  - Dust may cause irritation
- **Hazards:**
  - Eye irritation
  - Skin irritation (prolonged contact)
  - Respiratory irritation from dust inhalation
- **Precautionary Statements:**
  - **Avoid breathing dust**
  - **Wear protective gloves and eye protection**
  - **Wash hands after handling**

3. Composition / Information on Ingredients

Component	Concentration (%)
Nitrogen (N)	15%
Phosphorus (P_2O_5)	15%
Potassium (K_2O)	15%
Inert materials	Balance

4. First Aid Measures

- Inhalation: Move to fresh air. Seek medical advice if symptoms persist
- Skin Contact: Wash with soap and water
- Eye Contact: Rinse cautiously with water for several minutes
- Ingestion: Rinse mouth, do not induce vomiting, seek medical advice if unwell

5. Fire-Fighting Measures

- **Suitable Extinguishing Media:** Water spray, foam, dry chemical
- **Specific Hazards:** May release nitrogen oxides (NO_x) when heated
- **Protective Equipment:** Firefighters should wear self-contained breathing apparatus

6. Accidental Release Measures

- Avoid dust generation
- Collect material using dry methods (e.g. sweeping or vacuuming)
- Prevent entry into waterways or drains

- Dispose of according to local regulations

7. Handling and Storage
Handling:

- Avoid inhalation of dust
- Use appropriate PPE
- Do not eat, drink, or smoke while handling

Storage:

- Store in a cool, dry, well-ventilated area
- Keep away from moisture
- Store away from incompatible substances (e.g. strong acids, fuels)

8. Exposure Controls / Personal Protection
Engineering Controls: Adequate ventilation
PPE:

- Gloves
- Safety glasses
- Dust mask (if airborne particles present)

9. Physical and Chemical Properties

- Appearance: Granular solid
- Colour: Grey/brown
- Odour: Slight or none
- Solubility: Partially soluble in water
- pH: Neutral to slightly acidic

10. Stability and Reactivity

- Stability: Stable under normal conditions
- Conditions to Avoid: Excessive heat, moisture
- Incompatible Materials: Strong acids, fuels, combustible materials
- Hazardous Decomposition: Nitrogen oxides (NO_x), ammonia

11. Toxicological Information

- Low toxicity under normal use
- May cause irritation to eyes, skin, and respiratory tract
- Prolonged exposure to dust may cause discomfort

12. Ecological Information

- Nutrient-rich material may cause eutrophication if released into waterways
- Avoid over-application or runoff
- Moderate environmental impact if mismanaged

13. Disposal Considerations

- Dispose of in accordance with local regulations
- Do not discharge into drains or waterways
- Reuse where possible (e.g. apply to soil at recommended rates)

14. Transport Information

- Not classified as dangerous goods (for this formulation)
- Ensure secure packaging during transport
- Avoid contamination or spillage

15. Regulatory Information

- Complies with general fertiliser regulations (jurisdiction dependent)
- Subject to workplace health and safety (WHS) requirements
- Refer to local environmental and agricultural guidelines

16. Other Information
This SDS is provided as an example for educational purposes and does not replace manufacturer-specific documentation. Always refer to the official SDS supplied with the product.

2. Key Hazards Associated with Fertilisers

Fertilisers can present a range of hazards depending on their composition:

- **Chemical hazards** – some fertilisers (e.g. ammonium nitrate) are oxidising agents and can contribute to fire or explosion under certain conditions
- **Health hazards** – dust, fumes, or direct contact may cause skin irritation, eye damage, or respiratory issues
- **Environmental hazards** – spills or over-application can lead to water contamination, eutrophication, or soil imbalance
- **Reactivity risks** – mixing incompatible substances (e.g. fertilisers with fuels or certain chemicals) can be dangerous

Understanding the specific risks of each product is essential before handling.

Chemical fertilisers present a range of hazards that must be understood and actively managed to ensure safe handling, storage, and use. Each hazard type relates to different risks—chemical, human health, environmental, and reactive—and requires specific mitigation strategies.

Chemical hazards arise because some fertilisers, particularly nitrogen-based products such as ammonium nitrate, act as strong oxidising agents. While they are not fuels themselves, they can significantly intensify combustion and, under certain conditions (such as contamination, confinement, or high temperatures), contribute to fire or even explosion. This risk is especially relevant in bulk storage or transport situations. To mitigate these hazards, fertilisers should be stored in cool, well-ventilated areas away from heat sources and combustible materials. Segregation from fuels, organic materials, and other incompatible substances is essential. Facilities should implement fire prevention measures, temperature monitoring, and emergency response plans, particularly for large-scale storage.

Health hazards are associated with exposure to fertiliser dust, fumes, or direct contact with the product. Fine particles can irritate the respiratory system if inhaled, while contact with skin or eyes may cause irritation or, in some cases, chemical burns depending on the formulation. Long-term or repeated exposure can increase the risk of more serious health effects. Mitigation involves the use of appropriate personal protective equipment (PPE), such as gloves, safety glasses, and dust masks or respirators where necessary. Good hygiene practices—such as washing hands after handling and avoiding eating or drinking during use—are also important. Ensuring adequate ventilation in enclosed areas and using dust suppression methods can further reduce exposure risks.

Environmental hazards occur when fertilisers are misapplied, spilled, or allowed to enter natural ecosystems. Excess nutrients, particularly nitrogen and phosphorus, can run off into waterways, leading to eutrophication, algal blooms, and oxygen depletion, which harm aquatic life. Over-application can also disrupt soil nutrient balance and degrade soil health over time. Mitigation strategies focus on responsible nutrient management, including applying fertilisers at the correct rate, time, and place (aligned with crop needs), and avoiding application before heavy rainfall. Buffer zones should be maintained near waterways, and storage areas should be designed to prevent runoff or leaching. Spill containment systems and proper clean-up procedures are also critical to minimising environmental impact.

Reactivity risks involve the potential for dangerous reactions when fertilisers are mixed with incompatible substances. For example, combining oxidising fertilisers with fuels, oils, or certain chemicals can create conditions for rapid combustion or chemical reactions. Contamination during storage or handling can significantly increase these risks. To mitigate reactivity hazards, fertilisers should be stored and handled separately from incompatible materials, with clear labelling and segregation protocols in place. Equipment used for handling should be clean and free from residues of other chemicals. Staff should be trained to recognise incompatible substances and follow strict handling procedures to prevent accidental mixing.

Overall, effective hazard management relies on a combination of proper storage, safe handling practices, appropriate PPE, environmental controls, and staff training. By understanding the nature of these hazards and implementing targeted mitigation measures, fertiliser use can be both safe and sustainable across different scales of operation.

3. Personal Protective Equipment (PPE)

Appropriate PPE should always be used when handling fertilisers, particularly in bulk or concentrated forms. This may include:

- Gloves (chemical-resistant)
- Safety glasses or goggles
- Dust masks or respirators (for powdered fertilisers)
- Protective clothing and footwear

The type of PPE required should be guided by the SDS and workplace risk assessments.

4. Safe Handling Practices

Safe handling procedures help minimise risks during fertiliser use:

- Avoid inhaling dust or vapours
- Prevent direct contact with skin and eyes
- Use appropriate equipment for lifting and application
- Do not eat, drink, or smoke while handling fertilisers
- Wash hands and exposed skin after use

Care should also be taken to avoid cross-contamination, particularly when handling different fertiliser types or chemicals.

Safe Handling of Fertilisers Policy and Procedure (Sample)

1. Policy Statement

This policy establishes the requirements for the safe handling of fertilisers to minimise risks to workers, the environment, and operational processes. All personnel involved in the storage, handling, transport, or application of fertilisers must follow these procedures in accordance with workplace health and safety (WHS) legislation, Safety Data Sheets (SDS), and organisational safety standards.

The organisation is committed to providing a safe working environment by implementing appropriate controls to prevent exposure to fertiliser dust, vapours, and direct contact hazards, while also preventing cross-contamination between products.

2. Scope

This policy applies to:

- All employees, contractors, and visitors handling fertilisers
- All fertiliser types (granular, liquid, organic, and synthetic)
- All operational areas including storage, mixing, transport, and application

3. Responsibilities

Management:

- Ensure appropriate training, supervision, and resources are provided
- Maintain access to current SDS for all fertilisers
- Provide suitable PPE and handling equipment

Workers:

- Follow all safe handling procedures
- Use PPE as required
- Report hazards, incidents, or unsafe conditions

4. Safe Handling Procedures

To minimise risks during fertiliser use, the following procedures must be followed:

4.1 Control of Dust and Vapours

- Avoid inhaling fertiliser dust or vapours at all times
- Use dust suppression methods where applicable (e.g. dampening materials, enclosed systems)
- Ensure adequate ventilation in enclosed or confined spaces
- Wear appropriate respiratory protection when handling dusty materials

4.2 Prevention of Skin and Eye Contact

- Avoid direct contact with fertilisers
- Wear appropriate PPE, including gloves and safety glasses
- Do not handle fertilisers with bare hands
- In case of contact, wash affected areas immediately

4.3 Use of Appropriate Equipment

- Use suitable mechanical aids (e.g. loaders, spreaders, conveyors) for lifting and application
- Avoid manual handling of heavy loads where possible
- Ensure equipment is maintained, clean, and fit for purpose
- Follow manufacturer instructions for all application equipment

4.4 Personal Hygiene Requirements

- Do not eat, drink, or smoke while handling fertilisers
- Wash hands thoroughly after handling fertilisers and before breaks
- Wash exposed skin after completing tasks
- Remove and clean contaminated clothing before reuse

4.5 Prevention of Cross-Contamination

- Keep different fertiliser types separated during storage and handling
- Clean equipment between handling different products
- Avoid mixing fertilisers unless specifically required and approved
- Clearly label all containers and storage areas

5. Storage and Handling Controls

- Store fertilisers in designated, well-ventilated areas
- Keep storage areas clean and free from contaminants
- Segregate incompatible materials (e.g. fertilisers and fuels)
- Ensure spill containment measures are in place

6. Training and Competency

All personnel must:

- Receive training in safe fertiliser handling procedures
- Understand SDS information and hazard controls
- Be competent in the use of PPE and equipment
- Participate in refresher training as required

7. Incident and Hazard Reporting

- Report all incidents, spills, or exposure events immediately
- Follow emergency procedures as outlined in SDS
- Participate in incident investigations where required

8. Monitoring and Review

This policy will be:

- Reviewed regularly to ensure compliance with current regulations
- Updated in response to incidents, audits, or operational changes
- Monitored through workplace inspections and safety audits

5. Storage Guidelines

Proper storage is essential to maintain product integrity and prevent hazards:

- Store fertilisers in a cool, dry, well-ventilated area
- Keep away from incompatible materials (e.g. fuels, acids, organic materials in some cases)
- Protect from moisture to prevent caking or chemical changes
- Ensure secure storage to prevent unauthorised access
- Follow segregation requirements for hazardous fertilisers (e.g. ammonium nitrate)

Bulk storage facilities should also include spill containment and fire prevention measures.

Fertiliser Storage Policy and Procedure

1. Policy Statement

This policy establishes the requirements for the safe and effective storage of fertilisers to maintain product quality, prevent hazards, and ensure compliance with workplace health and safety (WHS) and environmental regulations. Proper storage practices are essential to minimise risks associated with contamination, fire, chemical reactions, and environmental harm.

The organisation is committed to ensuring that all fertilisers are stored in conditions that preserve their integrity and reduce risks to people, property, and the environment.

2. Scope

This policy applies to:

- All employees, contractors, and visitors involved in fertiliser storage
- All fertiliser types (solid, liquid, organic, and synthetic)
- All storage locations, including sheds, warehouses, and bulk storage facilities

3. Responsibilities

Management:

- Provide appropriate storage facilities and infrastructure
- Ensure compliance with relevant regulations and standards
- Maintain access to Safety Data Sheets (SDS)
- Implement fire prevention and spill control systems

Workers:

- Follow all storage procedures and safety requirements
- Report hazards, damage, or unsafe storage conditions
- Maintain cleanliness and organisation of storage areas

4. Storage Requirements

4.1 Storage Environment

- Store fertilisers in a cool, dry, and well-ventilated area
- Protect fertilisers from direct sunlight and excessive heat
- Ensure adequate airflow to prevent accumulation of fumes or moisture

4.2 Moisture Control

- Keep fertilisers protected from rain, humidity, and water exposure
- Use sealed containers, covered storage, or pallets to prevent ground moisture contact
- Regularly inspect for signs of caking, degradation, or contamination

4.3 Segregation and Compatibility

- Store fertilisers away from incompatible materials, including:
    - Fuels and oils
    - Strong acids or alkalis
    - Organic materials (where applicable)
- Follow specific segregation requirements for hazardous fertilisers (e.g. ammonium nitrate)
- Clearly label storage zones and materials

4.4 Security and Access Control

- Ensure storage areas are secure and restrict access to authorised personnel only
- Lock storage facilities where required
- Maintain inventory control and access records where applicable

5. Bulk Storage Requirements

For large-scale or bulk fertiliser storage, additional controls must be implemented:

- Install spill containment systems (e.g. bunding, drainage control) to prevent environmental contamination
- Implement fire prevention measures, including:
    - Separation distances between storage areas
    - Fire detection and suppression systems
    - Removal of combustible materials
- Ensure structural integrity of storage facilities (e.g. silos, bins, sheds)
- Provide clear signage indicating hazards and emergency procedures

6. Handling Within Storage Areas

- Use appropriate equipment for moving and stacking fertilisers
- Avoid damaging packaging or containers
- Keep aisles and access points clear
- Clean up spills immediately to prevent contamination or hazards

7. Inspection and Maintenance

- Conduct regular inspections of storage areas to identify:
    - Moisture ingress
    - Product degradation
    - Contamination or mixing of materials
    - Safety hazards
- Maintain records of inspections and corrective actions

- Ensure all storage infrastructure is maintained and fit for purpose

8. Emergency Preparedness

- Maintain spill kits and fire-fighting equipment in accessible locations
- Ensure personnel are trained in emergency response procedures
- Follow SDS guidance for specific fertiliser hazards
- Report and respond to incidents promptly

9. Training and Competency

- All personnel involved in fertiliser storage must:
    - Be trained in safe storage practices and hazard awareness
    - Understand segregation and compatibility requirements
    - Be familiar with emergency procedures and SDS information

10. Monitoring and Review

This policy will be:

- Reviewed regularly to ensure compliance with current regulations
- Updated in response to incidents, audits, or operational changes
- Monitored through inspections and safety audits

6. Transport and Handling

Fertilisers must be transported in accordance with relevant regulations, particularly if classified as hazardous goods.

Key considerations include:

- Proper labelling and documentation
- Secure packaging to prevent leaks or spills
- Compliance with dangerous goods transport regulations (where applicable)
- Safe loading and unloading procedures

For large-scale operations, transport may involve bulk carriers, rail, or shipping, requiring additional safety controls and monitoring.

Fertiliser Transport, Labelling, and Handling Policy and Procedure (Sample)

1. Policy Statement

This policy establishes the requirements for the safe labelling, packaging, transport, loading, and unloading of fertilisers. The organisation is committed to ensuring that all fertiliser products are handled and transported in a manner that protects workers, the public, and the environment, while complying with all relevant workplace health and safety (WHS), environmental, and dangerous goods transport regulations.

2. Scope

This policy applies to:

- All employees, contractors, and transport operators involved in fertiliser handling and transport
- All fertiliser types, including granular, liquid, organic, and hazardous fertilisers
- All stages of transport, including preparation, loading, transit, and unloading

3. Responsibilities

Management:

- Ensure compliance with applicable transport and dangerous goods regulations
- Provide appropriate packaging, labelling systems, and equipment
- Ensure training and competency of personnel

Workers and Operators:

- Follow all procedures for labelling, packaging, and transport
- Use equipment safely during loading and unloading
- Report any incidents, damage, or non-compliance

4. Labelling and Documentation

4.1 Product Labelling

- All fertiliser products must be clearly labelled with:
    - Product name and type
    - Nutrient composition (e.g. N:P:K:S)
    - Hazard warnings (if applicable)
    - Manufacturer or supplier details
- Labels must be legible, durable, and securely attached

4.2 Documentation

- Maintain accurate transport documentation, including:
    - Delivery dockets or consignment notes
    - Safety Data Sheets (SDS)

  - Dangerous goods declarations (where required)
- Ensure documentation is accessible during transport and handling

5. Packaging Requirements

- Fertilisers must be packaged to prevent leaks, spills, or contamination
- Approved packaging types include:
  - Sealed bags (e.g. 25 kg or bulk bags)
  - Bulk containers (e.g. silos, tankers)
- Inspect packaging prior to transport to ensure:
  - No damage, tears, or leaks
  - Secure closures and seals
- Do not transport fertilisers in damaged or compromised packaging

6. Dangerous Goods Compliance

- Identify whether the fertiliser is classified as dangerous goods (e.g. ammonium nitrate-based products)
- Where applicable:
  - Comply with national and international dangerous goods transport regulations
  - Use appropriate hazard labels and placards
  - Ensure vehicles meet regulatory requirements
  - Ensure drivers are appropriately licensed and trained
- Follow segregation rules for incompatible substances during transport

7. Loading Procedures

- Conduct a pre-loading inspection of equipment and packaging
- Ensure loading areas are stable, level, and free from hazards
- Use appropriate equipment (e.g. forklifts, loaders) for handling fertilisers
- Secure loads to prevent movement during transport
- Avoid overloading vehicles or containers
- Maintain separation between incompatible materials

8. Unloading Procedures

- Inspect the unloading area for safety and accessibility
- Use appropriate equipment and safe handling techniques
- Avoid dropping or damaging packaging
- Contain and clean up any spills immediately
- Ensure fertilisers are placed in designated storage areas

9. Spill and Incident Management

- In the event of a spill or leak:

- Stop the source of the spill if safe to do so
- Contain the material to prevent environmental contamination
- Use appropriate cleanup methods
- Report the incident according to organisational procedures
- Refer to SDS for specific emergency response actions

10. Training and Competency

All personnel involved in transport and handling must:

- Be trained in safe handling, loading, and unloading procedures
- Understand labelling and documentation requirements
- Be aware of dangerous goods regulations (if applicable)
- Demonstrate competency in the use of handling equipment

11. Monitoring and Review

This policy will be:

- Reviewed regularly to ensure compliance with current regulations
- Updated following incidents, audits, or operational changes
- Monitored through inspections and transport audits

7. Spill and Emergency Response

In the event of a spill or accidental release:

- Contain the material to prevent spread
- Avoid contamination of waterways or drains
- Use appropriate cleanup methods (e.g. dry collection for solids)
- Dispose of waste according to local regulations
- Refer to the SDS for specific emergency procedures

Emergency preparedness, including access to spill kits and trained personnel, is essential.

Fertiliser Spill Response and Emergency Management Policy and Procedure

1. Policy Statement

This policy establishes the procedures for responding to fertiliser spills or accidental releases to minimise risks to people, property, and the environment. The organisation is committed to ensuring that all incidents are managed quickly, safely, and in accordance with workplace health and safety (WHS), environmental regulations, and Safety Data Sheet (SDS) requirements.

Effective spill response relies on preparedness, trained personnel, and access to appropriate equipment.

2. Scope

This policy applies to:

- All employees, contractors, and visitors involved in fertiliser handling
- All fertiliser types (solid, liquid, organic, and synthetic)
- All operational areas including storage, transport, and application sites

3. Responsibilities

Management:

- Provide spill response equipment (e.g. spill kits)
- Ensure personnel are trained in spill response procedures
- Maintain access to current SDS for all fertilisers
- Ensure compliance with environmental and safety regulations

Workers:

- Follow spill response procedures
- Report all spills and incidents immediately
- Use appropriate PPE and equipment
- Participate in training and emergency drills

4. Emergency Preparedness

- Spill kits must be readily available in all areas where fertilisers are stored or handled
- Spill kits should include absorbent materials, containment barriers, and cleanup tools
- Personnel must be trained in:
    - Spill response procedures
    - Use of spill kits
    - Emergency communication protocols
- Emergency contact details must be clearly displayed

5. Spill Response Procedures

5.1 Initial Response

- Stop work immediately and assess the situation
- Identify the type and quantity of fertiliser involved
- Refer to the SDS for specific hazards and response actions
- Isolate the area and restrict access if necessary

5.2 Containment

- Contain the spill to prevent spread
- Use barriers such as absorbent materials, sand, or bunding
- Prevent fertiliser from entering waterways, drains, or sensitive areas

5.3 Cleanup

- Use appropriate cleanup methods based on fertiliser type:
    - Dry fertilisers: collect using sweeping, vacuuming, or mechanical recovery
    - Liquid fertilisers: absorb using appropriate materials (e.g. absorbent pads)
- Avoid creating dust or spreading contamination
- Place collected material into suitable containers for disposal or reuse (if safe)

5.4 Disposal

- Dispose of contaminated materials in accordance with local regulations
- Do not wash fertiliser into drains or waterways
- Where possible, reuse uncontaminated material safely

5.5 Decontamination

- Clean affected surfaces and equipment
- Remove and safely clean or dispose of contaminated PPE
- Wash hands and exposed skin thoroughly after cleanup

6. Incident Reporting

- All spills and releases must be reported immediately to supervisors
- Record details including:
    - Location and time
    - Type and quantity of fertiliser
    - Cause of the spill
    - Actions taken
- Notify regulatory authorities if required

7. Environmental Protection Measures

- Establish buffer zones around waterways and drainage systems
- Implement drainage controls in storage and handling areas
- Monitor for potential environmental impacts following a spill
- Take corrective action to prevent recurrence

8. Training and Competency

All personnel must:

- Be trained in spill response procedures and SDS interpretation
- Demonstrate competency in the use of spill kits
- Participate in regular refresher training and emergency drills

9. Monitoring and Review

This policy will be:

- Reviewed regularly to ensure compliance with current regulations
- Updated following incidents or identified improvements
- Monitored through inspections and incident audits

8. Environmental Protection Measures

Safe fertiliser handling also involves protecting the environment:

- Avoid over-application or application before heavy rainfall
- Maintain buffer zones near waterways
- Store fertilisers away from drainage areas
- Manage runoff and erosion risks

These practices help prevent nutrient losses and reduce environmental impacts such as eutrophication.

Environmental Protection in Fertiliser Use Policy and Procedure

1. Policy Statement

This policy establishes requirements for environmentally responsible fertiliser use to minimise nutrient losses, protect soil and water quality, and support sustainable agricultural practices. The organisation is committed to preventing environmental harm, including runoff, leaching, and eutrophication, through effective planning, application, and management of fertilisers.

2. Scope

This policy applies to:

- All employees, contractors, and operators applying or managing fertilisers
- All fertiliser types (organic, synthetic, liquid, and granular)
- All land-based operations, including cropping, pasture, and horticulture systems

3. Responsibilities

Management:

- Ensure environmental protection measures are implemented
- Provide training and resources for sustainable fertiliser use
- Monitor compliance with environmental regulations

Workers and Operators:

- Follow application and environmental protection procedures
- Monitor site conditions and report risks
- Apply fertilisers in accordance with approved plans and guidelines

4. Environmental Protection Procedures

4.1 Application Timing and Rate Control

- Apply fertilisers at rates aligned with crop or pasture nutrient requirements
- Avoid over-application to prevent nutrient buildup and losses
- Do not apply fertilisers before heavy rainfall or irrigation events
- Schedule applications during periods of active plant growth to maximise uptake

4.2 Buffer Zones and Waterway Protection

- Establish and maintain buffer zones between fertilised areas and:
    - Rivers, streams, and lakes
    - Drainage channels and wetlands
- Do not apply fertilisers within designated buffer zones
- Use vegetation strips or barriers to reduce runoff into waterways

4.3 Storage Location and Drainage Control

- Store fertilisers away from drainage lines, stormwater systems, and waterways

- Ensure storage areas are elevated or protected from flooding
- Implement containment measures to prevent accidental runoff or leakage

4.4 Runoff and Erosion Management

- Implement soil conservation practices such as:
    - Ground cover maintenance
    - Contour farming or strip cropping
    - Reduced tillage practices
- Avoid fertiliser application on steep slopes or highly erodible soils
- Use appropriate irrigation practices to prevent excessive runoff
- Monitor weather and soil conditions to reduce erosion risk

5. Planning and Monitoring

- Develop nutrient management plans based on soil testing and crop requirements
- Monitor soil nutrient levels and adjust fertiliser applications accordingly
- Regularly inspect fields for signs of runoff, erosion, or nutrient loss
- Maintain records of fertiliser applications, including timing, rates, and conditions

6. Incident Management

- In the event of runoff, contamination, or environmental impact:
    - Identify and stop the source
    - Implement corrective actions (e.g. containment, remediation)
    - Report incidents in accordance with organisational procedures
- Review practices to prevent recurrence

7. Training and Competency

All personnel involved in fertiliser application must:

- Be trained in environmentally responsible fertiliser use
- Understand nutrient management principles
- Be aware of site-specific environmental risks
- Participate in ongoing training and updates

8. Monitoring and Review

This policy will be:

- Reviewed regularly to ensure compliance with environmental regulations
- Updated based on monitoring results, incidents, or best practice developments
- Audited through environmental inspections and performance reviews

9. Differences Between Small-Scale and Large-Scale Operations

Safety requirements apply to all operations, but implementation varies by scale.

- Small-scale producers and farmers typically use simpler storage systems, manual handling, and basic PPE, but must still follow SDS guidance and safe practices
- Large-scale manufacturers and distributors implement formal safety management systems, including risk assessments, training programs, monitoring systems, and regulatory compliance frameworks

Regardless of scale, safety responsibility remains critical.

Appendix D – Sample Business Plan for Small Scale Production

Sample Business Plan: Small-Scale Fertiliser Production

1. Executive Summary

This business plan outlines the establishment of a small-scale fertiliser production facility focused on producing blended NPK fertilisers and/or organic fertilisers for local agricultural markets. The business aims to supply cost-effective, high-quality fertiliser products tailored to regional soil and crop needs.

The operation will target small to medium-sized farms, horticulture enterprises, and landscaping sectors. By leveraging local raw materials, efficient blending technologies, and sustainable practices, the business will position itself as a reliable and environmentally responsible supplier.

2. Business Description

The business will operate as a fertiliser blending and/or production facility producing:

- NPK blended fertilisers (custom formulations)
- Organic fertilisers (compost-based or bio-enhanced)
- Specialty fertilisers (e.g. slow-release or micronutrient blends)

Business objectives:

- Establish production within 6–12 months
- Achieve break-even within 2–3 years
- Build strong relationships with local farmers and distributors
- Expand into value-added products (e.g. enhanced-efficiency fertilisers)

3. Market Analysis

The fertiliser industry is shaped by growing agricultural demand, the need to restore soil nutrients, and an increasing focus on sustainability, with expansion driven by rising food production requirements and the adoption of precision agriculture. Key target markets include broadacre farmers, horticulture producers, turf and landscaping businesses, and agricultural retailers, all of whom require affordable products, consistent nutrient quality, tailored formulations, and environmentally responsible solutions. While competition comes from large manufacturers, regional distributors, and import suppliers, smaller producers can gain a competitive advantage through localised production, customised blends, faster delivery, and reduced transport costs.

3.1 Industry Overview

The fertiliser industry is driven by agricultural demand, soil health management, and increasing focus on sustainability. Growth is influenced by:

- Rising food production requirements
- Soil nutrient depletion
- Adoption of precision agriculture

3.2 Target Market

- Broadacre farmers (grains, pasture)
- Horticulture growers (vegetables, fruit)
- Turf and landscaping businesses
- Retail agricultural suppliers

3.3 Customer Needs

- Affordable fertilisers
- Reliable nutrient content
- Custom blends for specific crops/soils
- Environmentally sustainable products

3.4 Competitive Analysis

Competitors include:

- Large fertiliser manufacturers
- Regional distributors
- Import suppliers

Competitive advantage:

- Localised production
- Custom formulations
- Faster delivery
- Lower transport costs

4. Products and Services

The business will offer a range of fertiliser products, including standard NPK blends, urea-based nitrogen fertilisers, and phosphate and potassium formulations, alongside value-added options such as controlled-release and stabilised fertilisers, organic compost blends, and biofertilisers. In addition to products, the business will provide services such as soil testing, custom fertiliser blending, and agronomic advice to support targeted nutrient management. Production will rely on key raw materials including urea, ammonium phosphate (DAP/MAP), potash, organic inputs such as manure and compost, and various additives like micronutrients and inhibitors to enhance fertiliser performance.

4.1 Core Products

- Standard NPK blends (e.g. 15:15:15, 20:10:10)
- Urea-based nitrogen fertilisers
- Phosphate and potassium blends

4.2 Value-Added Products

- Controlled-release fertilisers
- Stabilised fertilisers (with inhibitors)
- Organic fertilisers and compost blends
- Biofertilisers and microbial additives

4.3 Services

- Soil testing and nutrient recommendations
- Custom fertiliser blending
- Agronomic advice

5. Production Plan

The production plan involves sourcing and storing key raw materials such as urea, ammonium phosphate (DAP/MAP), potash, organic inputs, and performance-enhancing additives, which are then processed through a structured sequence of weighing, batching, mechanical blending, optional granulation, drying, cooling, screening, and quality control before final packaging and storage. The operation requires essential equipment including fertiliser blenders, conveyors, weighing systems, bagging machines, and storage bins or silos, with a typical small-scale production capacity ranging from approximately 1 to 5 tonnes per hour.

5.1 Raw Materials

- Urea (nitrogen source)
- Ammonium phosphate (DAP/MAP)
- Potash (KCl or K_2SO_4)
- Organic inputs (manure, compost)
- Additives (micronutrients, inhibitors)

5.2 Production Process

- Raw material sourcing and storage
- Weighing and batching
- Blending (mechanical mixer)
- Granulation (if applicable)
- Drying and cooling
- Screening and quality control
- Packaging and storage

5.3 Equipment Required

- Fertiliser blender (horizontal/vertical mixer)
- Conveyors and feeders
- Weighing systems
- Bagging and sealing machines
- Storage bins/silos

5.4 Capacity

- Example: 1–5 tonnes per hour (small-scale plant)

6. Regulatory and Compliance Requirements

The business must comply with:

- Fertiliser regulations (labelling, nutrient content)
- Workplace Health and Safety (WHS) laws
- Environmental protection regulations
- Dangerous goods storage (if applicable)

Key requirements:

- Accurate product labelling (N:P:K content)
- Safety Data Sheets (SDS)
- Storage and handling compliance
- Environmental management plans

7. Operations Plan

The operations plan requires establishing the business at a suitable industrial or rural location with good transport access and sufficient space for production and storage. Staffing will include plant operators, a production supervisor, sales and marketing personnel, and administrative support to ensure efficient day-to-day operations. A reliable supply chain is essential, involving secure sourcing of raw materials, a local distribution network, and effective logistics to deliver fertiliser products to farms and other customers.

7.1 Location

- Industrial or rural site with transport access
- Adequate space for storage and production

7.2 Staffing

- Plant operator(s)
- Production supervisor
- Sales/marketing representative
- Administrative support

7.3 Supply Chain

- Secure suppliers for raw materials
- Local distribution network
- Logistics for delivery to farms

8. Marketing and Sales Strategy

The marketing and sales strategy focuses on positioning the business as a local, reliable, and sustainable fertiliser provider, with an emphasis on customised solutions and agronomic support. Sales will be driven through multiple channels, including direct engagement with farmers, partnerships with agricultural retailers, and online or phone-based ordering. Promotion will involve practical field demonstrations, participation in industry events and agricultural shows, and targeted digital and local advertising to build brand awareness and customer trust.

8.1 Branding

- Emphasise local, reliable, and sustainable production
- Highlight customisation and agronomic support

8.2 Sales Channels

- Direct sales to farmers
- Partnerships with agricultural retailers
- Online and phone ordering

8.3 Promotion

- Field demonstrations

- Industry events and agricultural shows
- Digital marketing and local advertising

9. Financial Plan

The financial plan outlines estimated startup costs ranging from approximately $210,000 to $580,000, covering equipment, facility setup, initial raw materials, and licensing requirements. Ongoing operating costs include raw materials, labour, utilities, maintenance, and transport. Revenue will be generated through bulk and bagged fertiliser sales, custom blending services, and agronomic consulting. Profitability will depend on input costs and pricing strategies, with higher margins achievable through value-added products such as enhanced-efficiency fertilisers and biofertilisers.

9.1 Startup Costs (Indicative)

- Equipment: $100,000 – $300,000
- Facility setup: $50,000 – $150,000
- Initial raw materials: $50,000 – $100,000
- Licensing and compliance: $10,000 – $30,000

9.2 Operating Costs

- Raw materials
- Labour
- Utilities (electricity, fuel)
- Maintenance
- Transport

9.3 Revenue Streams

- Fertiliser sales (bulk and bagged)
- Custom blending services
- Agronomic consulting

9.4 Profitability

- Margins depend on raw material costs and pricing strategy
- Value-added products (EEFs, biofertilisers) offer higher margins

10. Risk Assessment

The key risks to the business include fluctuations in raw material prices, changes in regulatory requirements, environmental compliance challenges, and strong market competition. These risks can be managed through strategies such as diversifying suppliers to reduce dependency, maintaining robust compliance systems, focusing on niche or customised product offerings to differentiate in the market, and implementing strong quality assurance processes to ensure consistent product standards and operational reliability.

Key Risks:

- Raw material price volatility
- Regulatory changes
- Environmental compliance risks
- Market competition

Mitigation Strategies:

- Diversify suppliers
- Maintain compliance systems
- Focus on niche/custom products
- Implement quality assurance processes

11. Sustainability Strategy

- Use recycled and organic inputs where possible
- Promote efficient fertiliser use (4R principles)
- Minimise emissions and waste
- Develop eco-friendly product lines

12. Growth and Expansion

- Future opportunities include:
- Expansion into enhanced-efficiency fertilisers
- Integration with digital agriculture platforms
- Export opportunities
- Scaling production capacity

13. Implementation Timeline (Example)

Phase	Timeline
Feasibility & planning	1–3 months
Site setup & equipment	3–6 months
Regulatory approvals	2–4 months
Pilot production	1–2 months
Full operations	6–12 months

References

1. Timilsena, Y.P., et al., *Enhanced Efficiency Fertilisers: A Review of Formulation and Nutrient Release Patterns*. Journal of the Science of Food and Agriculture, 2014. **95**(6): p. 1131-1142.
2. Scheer, C., et al., *Effect of Enhanced Efficiency Fertilisers on Nitrous Oxide Emissions in a Sub-Tropical Cereal Cropping System*. Soil Research, 2016. **54**(5): p. 544-551.
3. Darzi, R., et al., *Evaluating How Enhanced Efficiency Nitrogen Fertilizers Improve Agricultural Sustainability: Greenhouse Multi-Phase Tracking System*. Agriculture, 2023. **13**(7): p. 1384.
4. Singh, K., et al., *Optimizing Nutrient Delivery in Agronomic Crops: A Review of Enhanced Efficiency Fertilizers*. Journal of Plant Nutrition and Soil Science, 2025. **188**(4): p. 570-584.
5. Carpenter, H.C.H., *Percy Carlyle Gilchrist, 1851 - 1935*. Obituary Notices of Fellows of the Royal Society, 1936. **2**(5): p. 19-24.
6. *Dr. Albert A. Gray*. Nature, 1936. **137**(3457): p. 178-178.
7. Henderson, C.H., *Note on the Modifications of the Bessemer Steel Process*. Journal of the Franklin Institute, 1887. **123**(6): p. 454-462.
8. Carpenter, H.C.H., *The Future of French Metallurgy*. Nature, 1917. **98**(2469): p. 493-494.
9. Beaven, E.S., *Dr. J. M. H. Munro*. Nature, 1937. **139**(3506): p. 60-60.
10. Arrobas, M., et al., *Nitrogen-Use Efficiency and Economic Efficiency of Slow-Release N Fertilisers Applied to Irrigated Turfs in a Mediterranean Environment*. Nutrient Cycling in Agroecosystems, 2010. **89**(3): p. 329-339.
11. Mahmoud, A.W.M. and S.S. Taha, *Main Sulphur Content in Essential Oil of Eruca Sativa as Affected by Nano Iron and Nano Zinc Mixed With Organic Manure*. Agriculture (Pol Nohospodárstvo), 2018. **64**(2): p. 65-79.
12. Ahmad, N. and T. Muhammad, *Fertiliser, Plant Nutrient Management, and Self-Reliance In Agriculture*. The Pakistan Development Review, 1998. **37**(4II): p. 217-233.
13. Masso, C., et al., *Dilemma of Nitrogen Management for Future Food Security in Sub-Saharan Africa – A Review*. Soil Research, 2017. **55**(6): p. 425-434.
14. Hajdys, D., *The European Green Deal and the Opportunities and Risks of Organic Farming in Poland*. Journal of Finance and Financial Law, 2024. **2**(42): p. 171-190.
15. Salim, I., et al., *Assessing the Environmental Sustainability of Glucose From Wheat as a Fermentation Feedstock*. Journal of Environmental Management, 2019. **247**: p. 323-332.
16. Magnone, D., et al., *The Impact of Phosphorus on Projected Sub-Saharan Africa Food Security Futures*. Nature Communications, 2022. **13**(1).
17. Uçak, H., Y. Arı, and E. Yelgen, *The Volatility Connectedness Among Fertilisers and Agricultural Crop Prices: Evidence From Selected Main Agricultural Products*. Agricultural Economics (Zemědělská Ekonomika), 2022. **68**(9): p. 348-360.

18. Bor, Ö. and N. Dağıstan, *The Impact Of fluctuating International Fertiliser Prices and Exchange Rates On domestic Fertiliser Prices In Türkiye*. Agricultural Economics (Zemědělská Ekonomika), 2024. **70**(1): p. 12-23.
19. Brienza, C., et al., *Technical Performance and Mass- And Energy Balances of Five Large-Scale Anearobic Digesters Applying Nutrient Recovery and Reuse : A Report Within the H2020 Project SYSTEMIC*. 2021.
20. Peterková, D., *Current Market Situation of Urea for Agricultural Use and Other Nitrogen Fertilisers*.
21. Ngo, V.M., et al., *A Big Data Smart Agricultural System: Recommending Optimum Fertilisers for Crops*. International Journal of Information Technology, 2023. **15**(1): p. 249-265.
22. Viana, L., et al., *Life Cycle Assessment of Oat Flake Production With Two End-of-Life Options for Agro-Industrial Residue Management*. Sustainability, 2023. **15**(6): p. 5124.
23. Garcia, J.N., et al., *Production of a Map of Greenhouse Gas Emissions and Energy Use From Australian Agriculture*. 2013.
24. Magee, K., et al., *Valorisation of Organic Waste by-Products Using Black Soldier Fly (Hermetia Illucens) as a Bio-Convertor*. Sustainability, 2021. **13**(15): p. 8345.
25. Pieralli, S., et al., *Implications of the <scp>EU</Scp>'s Carbon Border Adjustment Mechanism for Fertiliser and Food Markets*. Eurochoices, 2025. **24**(2): p. 23-30.
26. Majumder, R., *Balancing Food Security and Environmental Safety: Rethinking Modern Agricultural Practices*. Environmental and Experimental Biology, 2024. **21**(4): p. 101-110.
27. Curtis, E., J. Virzi, and C. Welsh, *Chokepoint: How the War with Iran Threatens Global Food Security*. 2026, Center for Strategic & International Studies.
28. Cánovas, C.R., et al., *Valorization of Wastes From the Fertilizer Industry: Current Status and Future Trends*. Journal of Cleaner Production, 2018. **174**: p. 678-690.
29. Papaslioti, E.M., et al., *Effects of Redox Oscillations on the Phosphogypsum Waste in an Estuarine Salt-Marsh System*. Chemosphere, 2020. **242**: p. 125174.
30. Illera, V., et al., *Immobilization of the Heavy Metals Cd, Cu and Pb in an Acid Soil Amended With Gypsum - and Lime -rich Industrial By -products*. European Journal of Soil Science, 2003. **55**(1): p. 135-145.
31. Gezerman, A.O., *Effect of Phosphogypsum Use as a Waste Recycling on GHG Emissions by Mineral Carbonisation Method*. International Journal of Chemistry and Technology, 2022. **6**(2): p. 102-107.
32. Krainiuk, O., et al., *Geoecological Analysis of Threats of Using Phosphogypsum in Construction of Roads*. Journal of Geology Geography and Geoecology, 2023. **32**(1): p. 79-88.
33. Arhouni, F.E., et al., *Evaluation of the Physicochemical, Heavy Metal and Radiological Contamination From Phosphogypsum Discharges of the Phosphoric Acid Production Unit on the Coast of El Jadida Province in Morocco*. Journal of Radioanalytical and Nuclear Chemistry, 2023. **332**(10): p. 4019-4028.
34. Lou, X.-Y., et al., *From Grave to Cradle: Treatment, Resource Recycling, and Valorization of Phosphogypsum Wastes*. Environmental Science & Technology Letters, 2024. **11**(9): p. 908-919.

35. Macías, F., et al., *An Anomalous Metal-Rich Phosphogypsum: Characterization and Classification According to International Regulations*. Journal of Hazardous Materials, 2017. **331**: p. 99-108.
36. Жантасов, K.T., et al., *Ecologically Friendly, Slow-Release Granular Fertilizers With Phosphogypsum*. Polish Journal of Environmental Studies, 2022. **31**(3): p. 2935-2942.
37. Chen, C., et al., *Study on Controlling the Dissolution of Calcium Dihydrogen Phosphate Using Ternary Phase Diagrams to Improve the Production Process of Phosphoric Acid*. Industrial & Engineering Chemistry Research, 2022.
38. Saint Humanic Acid, *Potassium humate is used in fertilizers*. 2026, Saint Humanic Acid,.
39. Ercoli, L., et al., *As Durum Wheat Productivity Is Affected by Nitrogen Fertilisation Management in Central Italy*. European Journal of Agronomy, 2013. **44**: p. 38-45.
40. Prisa, D. and G. Menci, *Innovative Fertilizers With Added Plant Extracts in the Cultivation of Valeriana Officinalis and Raphanus Sativus and in the Control of Botrytis and Powdery Mildew*. GSC Biological and Pharmaceutical Sciences, 2024. **27**(2): p. 127-136.
41. Hamidi, N.H., et al., *Soil Nitrogen Sorption Using Charcoal and Wood Ash*. Agronomy, 2021. **11**(9): p. 1801.
42. Martinez, C., et al., *Integrated Field Assessment of Nitrogen Release Dynamics and Crop Recovery of Band-Applied Controlled-Release Fertilisers*. Plant and Soil, 2021. **466**(1-2): p. 257-273.
43. Toda, M., F. Walder, and M.G.A.v.d. Heijden, *Organic Management and Soil Health Promote Nutrient Use Efficiency*. Journal of Sustainable Agriculture and Environment, 2023. **2**(3): p. 215-224.
44. Feng, X., et al., *Slow-Release Nitrogen Fertiliser Suitable for One-Time Fertilisation of Spring Maize in Northeast China*. Plant Soil and Environment, 2021. **67**(3): p. 164-172.
45. Snyder, C.S., *Enhanced Nitrogen Fertiliser Technologies Support the '4R' Concept to Optimise Crop Production and Minimise Environmental Losses*. Soil Research, 2017. **55**(6): p. 463-472.
46. Wilkinson, S., A.K. Weston, and D.J. Marks, *Stabilising Urea Amine Nitrogen Increases Potato Tuber Yield by Increasing Chlorophyll Content, Reducing Shoot Growth Rate and Increasing Biomass Partitioning to Roots and Tubers*. Potato Research, 2019. **63**(2): p. 217-239.
47. Hailegnaw, N.S., et al., *Co-Application of High Temperature Biochar With 3,4-Dimethylpyrazole-Phosphate Treated Ammonium Sulphate Improves Nitrogen Use Efficiency in Maize*. Scientific Reports, 2021. **11**(1).
48. Demirci, T., T.T. Gelen, and N. Bölgen, *A Biopolymer Coating Strategy for the Slow Release of Urea From Urea Fertiliser*. Kemija U Industriji, 2024. **73**(11-12): p. 467-476.
49. Hua, K., et al., *Enhancing Rice Yield, Quality, and Resource Utilisation With Slow-Release Fertiliser in Alternate Wetting and Drying Irrigation*. Plant Soil and Environment, 2024. **70**(5): p. 253-262.
50. Toleikienė, M., et al., *Processing of Legume Green Manures Slowdowns C Release, Reduces N Losses and Increases N Synchronisation Index for Two Years*. Sustainability, 2024. **16**(5): p. 2152.

51. Edmeades, D.C., *The Evaluation of a Controlled Release Nitrogen Fertiliser.* Journal of New Zealand Grasslands, 2015. **77**: p. 147-152.
52. Sáez-Tovar, J.A., et al., *Comparative Study of Different Fertilisation Strategies on Lettuce Crop Production Including Remote Sensing Precision Agriculture Techniques: Effects on Soil Properties, Crop Yield and Greenhouse Gas Emissions.* Soil Use and Management, 2025. **41**(4).
53. Gebremichael, A.W., et al., *Ammonium-Based Compound Fertilisers Mitigate Nitrous Oxide Emissions in Temperate Grassland.* Agronomy, 2021. **11**(9): p. 1712.
54. Bryant, R.H., J.A. Greig, and M.C. Mangwe, *Can Additives or Controlled Release Coating Improve the Nitrogen Use Efficiency of Urea Fertiliser.* Journal of New Zealand Grasslands, 2024: p. 161-168.
55. Bordoloi, N., K.K. Baruah, and T.K. Maji, *Nitrous Oxide Emission From a Transplanted Rice Field in Alluvial Soil as Influenced by Management of Nitrogen Fertiliser.* Soil Use and Management, 2016. **32**(4): p. 573-582.
56. Yadav, S.P.S., et al., *Advancing Effective Methods for Mitigating Greenhouse Gas Emissions From Rice (<i>Oryza Sativa</I> L.) Fields.* Journal of Sustainable Agriculture and Environment, 2024. **3**(4).
57. Phillips, I., et al., *Combination of Inorganic Nitrogen and Organic Soil Amendment Improves Nitrogen Use Efficiency While Reducing Nitrogen Runoff.* Nitrogen, 2022. **3**(1): p. 58-73.
58. Ward, C.R., D.R. Chadwick, and P.W. Hill, *Potential to Improve Nitrogen Use Efficiency (NUE) by Use of Perennial Mobile Green Manures.* Nutrient Cycling in Agroecosystems, 2022. **125**(1): p. 43-62.
59. Meng, C., et al., *Slow-Release Fertilisers Increased Microflora and Nitrogen Use Efficiency and Thus Promoted Peanut Growth and Yield.* Plant Soil and Environment, 2024. **70**(2): p. 61-71.
60. Bhardwaj, D., et al., *Biofertilizers Function as Key Player in Sustainable Agriculture by Improving Soil Fertility, Plant Tolerance and Crop Productivity.* Microbial Cell Factories, 2014. **13**(1).
61. Raymond, N.S., et al., *Phosphate-solubilising Microorganisms for Improved Crop Productivity: A Critical Assessment.* New Phytologist, 2020. **229**(3): p. 1268-1277.
62. Alori, E.T., B.R. Glick, and O.O. Babalola, *Microbial Phosphorus Solubilization and Its Potential for Use in Sustainable Agriculture.* Frontiers in Microbiology, 2017. **8**.
63. Gutiérrez-Martín, M., et al., *Single Superphosphate-P Fractions Influence Bacterial Communities in Wheat Rhizosphere Soil Under Microcosm Conditions.* European Journal of Soil Science, 2025. **76**(2).
64. Keni, M.F., *Improving the Growth Performance of Oil Palm Seedlings by Mixtures of Organic and Chemical Fertilisers.* Journal of Oil Palm Research, 2022.
65. Jain, S., J. Jain, and J. Singh, *The Rhizosphere Microbiome: Microbial Communities and Plant Health.* 2020: p. 175-190.
66. Saini, R. and J.P. Sharma, *Climate Resilient Microbes in Sustainable Crop Production.* 2019: p. 264-283.
67. Chavez, M.Y. and M. Uchanski, *Insect Left-Over Substrate as Plant Fertiliser.* Journal of Insects as Food and Feed, 2021. **7**(5): p. 683-694.
68. Malusà, E. and N. Vassilev, *A Contribution to Set a Legal Framework for Biofertilisers.* Applied Microbiology and Biotechnology, 2014. **98**(15): p. 6599-6607.

69. Beura, K., et al., *Combined Effect of Enriched Compost and Microbial Inoculants on Soil Nutrients and Phosphorus Uptake by Rice*. International Journal of Current Microbiology and Applied Sciences, 2019. **8**(02): p. 176-181.
70. Bhattacharjya, S., et al., *Evaluation of Microbial Solubilisation of Nano Rock Phosphate*. International Journal of Current Microbiology and Applied Sciences, 2019. **8**(01): p. 1055-1069.
71. Sundram, S., *Integrated Balanced Fertiliser Management in Soil Health Rejuvenation for a Sustainable Oil Palm Cultivation: A Review*. Journal of Oil Palm Research, 2019.
72. Gao, Y. and A.C. Serrenho, *Greenhouse Gas Emissions From Nitrogen Fertilizers Could Be Reduced by Up to One-Fifth of Current Levels by 2050 With Combined Interventions*. Nature Food, 2023.
73. Momot, R.V. and A.Y. Nebesnaya, *Research of the Main Segments of the Fertilizer Market*. Bio Web of Conferences, 2024. **103**: p. 00073.
74. Smith, C., A.K. Hill, and L. Torrente-Murciano, *Current and Future Role of Haber–Bosch Ammonia in a Carbon-Free Energy Landscape*. Energy & Environmental Science, 2020. **13**(2): p. 331-344.
75. Tomšík, K., et al., *Global Trends in Fertilizer Commerce: A Dual Analysis of General and Nitrogen Fertilizer Markets*. 2024.
76. Mallast, J., et al., *A Three-Year Data Set of Gaseous Field Emissions From Crop Sequence at Three Sites in Germany*. Scientific Data, 2022. **9**(1).
77. Abu, R., K. Patchigolla, and N.J. Simms, *A Review on Qualitative Assessment of Natural Gas Utilisation Options for Eliminating Routine Nigerian Gas Flaring*. Gases, 2023. **3**(1): p. 1-24.
78. Balaram, V., *Potential Future Alternative Resources for Rare Earth Elements: Opportunities and Challenges*. 2023.
79. McKelvey, V., D.L. Everhart, and R.M. Garrels, *Origin of Uranium Deposits*. 1955.
80. Alqudah, M., et al., *Paleoenvironmental Study of the Late Cretaceous–Eocene Tethyan Sea Associated With Phosphorite Deposits in Jordan*. Applied Sciences, 2023. **13**(3): p. 1568.
81. Abzalov, M., et al., *Geology and Metallogeny of Jordanian Uranium Deposits*. Applied Earth Science Transactions of the Institutions of Mining and Metallurgy Section B, 2015. **124**(2): p. 63-77.
82. Walan, P., et al., *Phosphate Rock Production and Depletion: Regional Disaggregated Modeling and Global Implications*. Resources Conservation and Recycling, 2014. **93**: p. 178-187.
83. Bamiki, R.E., et al., *Phosphate Rocks: A Review of Sedimentary and Igneous Occurrences in Morocco*. Minerals, 2021. **11**(10): p. 1137.
84. El-Anwar, E.A.A. and S.H.A.E. Rahim, *Mineralogy, Geochemistry and Origin of the Phosphorites at Um El-Huwtat Mine, Quseir, Central Eastern Desert, Egypt*. Carbonates and Evaporites, 2022. **37**(1).
85. Ahmad, F., et al., *Geochemistry and Stable Isotopes of the Upper Campanian–lower Maastrichtian Phosphorite-bearing Sequence, Central Jordan: Implications for Their Age, Origin, and Diagenesis*. Geological Journal, 2019. **55**(6): p. 4453-4468.
86. Álvaro, J.J., et al., *Ediacaran–Cambrian Phosphorites From the Western Margins of Gondwana and Baltica*. Sedimentology, 2015. **63**(2): p. 350-377.

87. Brownlie, W.J., et al., *Global Food Security Threatened by Potassium Neglect.* Nature Food, 2024. **5**(2): p. 111-115.
88. Ciceri, D., D.A.C. Manning, and A. Allanore, *Historical and Technical Developments of Potassium Resources.* The Science of the Total Environment, 2015. **502**: p. 590-601.
89. Oecd, *The Supply of Critical Raw Materials Endangered by Russia's War on Ukraine.* 2022.
90. Bullen, J., et al., *The Potential for Current Sodium and Potassium Production to Support a Global Switch to the Use of Potassium-Enriched Salt: A Desktop Research Study.* Public Health Nutrition, 2024. **27**(1).
91. *Mineral Commodity Summaries 2023.* 2023.
92. Hellegers, P., *Food Security Vulnerability Due to Trade Dependencies on Russia and Ukraine.* Food Security, 2022. **14**(6): p. 1503-1510.
93. Jagtap, S., et al., *The Russia-Ukraine Conflict: Its Implications for the Global Food Supply Chains.* Foods, 2022. **11**(14): p. 2098.
94. None, N., *Europe and Central Asia Economic Update, Spring 2022: War in the Region*. 2022.
95. Sohag, K., et al., *Food Inflation and Geopolitical Risks: Analyzing European Regions Amid the Russia-Ukraine War.* British Food Journal, 2022. **125**(7): p. 2368-2391.
96. Costola, M. and M. Lorusso, *Spillovers Among Energy Commodities and the Russian Stock Market.* Journal of Commodity Markets, 2022. **28**: p. 100249.
97. Chepeliev, M., M. Maliszewska, and M.F.S.e. Pereira, *The War in Ukraine, Food Security and the Role for Europe.* Eurochoices, 2023. **22**(1): p. 4-13.
98. Tampubolon, M., *Russia's Invasion of Ukraine and Its Impact on Global Geopolitics.* European Scientific Journal Esj, 2022. **18**(20): p. 48.
99. Klymenko, K. and M. Savostianenko, *Cooperation of Ukraine With International Financial Partners in Martial Law.* Fìnansi Ukraïni, 2022. **2022**(5): p. 43-60.
100. Kamal-Chaoui, L., et al., *The Implications for OECD Regions of the War in Ukraine.* 2022.
101. Kornher, L., T. Baležentis, and F.G. Santeramo, *<scp>EU</Scp> Food Price Inflation Amid Global Market Turbulences During the <scp>COVID</Scp>-19 Pandemic and the Russia–Ukraine War.* Applied Economic Perspectives and Policy, 2024. **46**(4): p. 1563-1584.
102. Tonelli, D., et al., *Cost-Competitive Decentralized Ammonia Fertilizer Production Can Increase Food Security.* Nature Food, 2024. **5**(6): p. 469-479.
103. Tonelli, D., et al., *Cost-Competitiveness of Distributed Ammonia Production For the Global Fertilizer Industry.* 2023.
104. Rouwenhorst, K.H.R., et al., *Ammonia, 4. Green Ammonia Production*. 2020: p. 1-20.
105. Jain, M., R. Muthalathu, and X. Wu, *Electrified Ammonia Production as a Commodity and Energy Storage Medium to Connect the Food, Energy, and Trade Sectors.* Iscience, 2022. **25**(8): p. 104724.
106. Fernández, C.A., et al., *Achieving Decentralized, Electrified, and Decarbonized Ammonia Production.* Environmental Science & Technology, 2024. **58**(16): p. 6964-6977.

107. Rezaie, F., et al., *Low-Temperature Electrochemical Ammonia Synthesis: Measurement Reliability and Comparison to Haber–Bosch in Terms of Energy Efficiency.* Energy Technology, 2023. **11**(10).
108. Varanasi, S.A., C.I. Fernández, and M.C. Hatzell, *Energy Management and Economic Considerations of Intermittent Photovoltaic-Driven Electrochemical Ammonia Production*. Energy & Fuels, 2023. **37**(19): p. 15222-15230.
109. Lim, J., et al., *Ammonia and Nitric Acid Demands for Fertilizer Use in 2050.* Acs Energy Letters, 2021. **6**(10): p. 3676-3685.
110. Verleysen, K., A. Parente, and F. Contino, *How Sensitive Is a Dynamic Ammonia Synthesis Process? Global Sensitivity Analysis of a Dynamic Haber-Bosch Process (For Flexible Seasonal Energy Storage).* Energy, 2021. **232**: p. 121016.
111. Shin, C., et al., *Recovery of Clean Water and Ammonia From Domestic Wastewater: Impacts on Embodied Energy and Greenhouse Gas Emissions.* Environmental Science & Technology, 2022. **56**(12): p. 8712-8721.
112. Fertilizer Industrial Services, *Urea technology*. 2026, Fertilizer Industrial Services,.
113. Tanzeem, M. and M. Al-Thubaiti, *Blue Ammonia an Attractive Pathway to Decarbonize Conventional Ammonia Plants.* World Journal of Advanced Engineering Technology and Sciences, 2023. **9**(1): p. 119-126.
114. Mikulčić, H., et al., *Flexible Carbon Capture and Utilization Technologies in Future Energy Systems and the Utilization Pathways of Captured CO2.* Renewable and Sustainable Energy Reviews, 2019. **114**: p. 109338.
115. Quadrelli, E.A., et al., *Carbon Dioxide Recycling: Emerging Large-Scale Technologies With Industrial Potential.* Chemsuschem, 2011. **4**(9): p. 1194-1215.
116. Leonzio, G., et al., *Scenario Analysis of Carbon Capture, Utilization (Particularly Producing Methane and Methanol), and Storage (CCUS) Systems.* Industrial & Engineering Chemistry Research, 2020. **59**(15): p. 6961-6976.
117. Siddiqui, O., et al., *Performance Investigation of a New Renewable Energy-based Carbon Dioxide Capturing System With Aqueous Ammonia*. International Journal of Energy Research, 2019. **44**(3): p. 2252-2263.
118. Solomon, M.B., et al., *Decarbonization in Australia and India: Bilateral Opportunities and Challenges for the Net Zero Transformation.* Acs Engineering Au, 2024. **4**(3): p. 295-311.
119. Shirmohammadi, R., et al., *CO2 Utilization via Integration of an Industrial Post-Combustion Capture Process With a Urea Plant: Process Modelling and Sensitivity Analysis.* Processes, 2020. **8**(9): p. 1144.
120. Zhang, Z., et al., *Recent Advances in Carbon Dioxide Utilization.* Renewable and Sustainable Energy Reviews, 2020. **125**: p. 109799.
121. Zhou, W., L. Pan, and X. Mao, *Optimization and Comparative Analysis of Different CCUS Systems in China: The Case of Shanxi Province.* Sustainability, 2023. **15**(18): p. 13455.
122. Dalei, N.N. and J.M. Joshi, *Potential Matching of Carbon Capture Storage and Utilization (CCSU) as Enhanced Oil Recovery in Perspective to Indian Oil Refineries.* Clean Technologies and Environmental Policy, 2022. **24**(9): p. 2701-2717.
123. Leonzio, G., I.D.L. Bogle, and P.U. Foscolo, *Life Cycle Assessment of a Carbon Capture Utilization and Storage Supply Chain in Italy and Germany: Comparison Between Carbon Dioxide Storage and Utilization Systems.* Sustainable Energy Technologies and Assessments, 2023. **55**: p. 102743.

124. Liu, J.Y., et al., *Development Status and Energy Penalty of Carbon Dioxide Capture Technologies*. Advanced Materials Research, 2013. **864-867**: p. 1598-1601.
125. Clenet, D., et al., *A spray freeze dried micropellet based formulation proof-of-concept for a yellow fever vaccine candidate*. European Journal of Pharmaceutics and Biopharmaceutics, 2019. **142**.
126. Berger, Á., *Evaluation of the Experience of International Accidents Related to the Storage and Handling of Ammonium Nitrate Fertilisers*. Hadmérnök, 2022. **17**(2): p. 129-145.
127. Oluwoye, I., et al., *Atmospheric Emission of NO From Mining Explosives: A Critical Review*. Atmospheric Environment, 2017. **167**: p. 81-96.
128. Baraza, X., A. Pey, and J. Giménez, *The Self-Sustaining Decomposition of Ammonium Nitrate Fertiliser: Case Study, Escombreras Valley, Spain*. Journal of Hazardous Materials, 2020. **387**: p. 121674.
129. Pandey, D.K. and D.P. Bose, *Development of Green Propellants for Short Range Solid Rocket : An Overview*. International Journal of Engineering Research And, 2016. **V5**(09).
130. Boeck, L.R. and P.W. Mahan, *Loss Prevention Learnings From Beirut and Similar Ammonium Nitrate Explosions*. Process Safety Progress, 2021. **41**(2): p. 276-282.
131. Kumar, M.A., et al., *Development of Green Energetic Oxidizers for Solid Propellant Applications: Present Status*. Propellants Explosives Pyrotechnics, 2024. **49**(12).
132. Al-Hajj, S., et al., *Beirut Ammonium Nitrate Blast: Analysis, Review, and Recommendations*. Frontiers in Public Health, 2021. **9**.
133. Hadden, R.M., F.X. Jervis, and G. Rein, *Investigation of the Fertilizer Fire Aboard the Ostedijk*. Fire Safety Science, 2008. **9**: p. 1091-1101.
134. Davis, S., T. DeBold, and C. Marsegan, *Investigation Findings and Lessons Learned in the West Fertilizer Explosion*. Journal of Fire Sciences, 2017. **35**(5): p. 379-395.
135. Babrauskas, V., *The Ammonium Nitrate Explosion at West, Texas: A Disaster That Could Have Been Avoided*. Fire and Materials, 2017. **42**(2): p. 164-172.
136. Guo, J., et al., *Thermal Hazard Analysis of Two Non-Ideal Explosives Based on Ammonium Perchlorate/Ammonium Nitrate and Aluminium Powder*. Molecules, 2024. **29**(11): p. 2680.
137. Fabin, M., et al., *Following the Decomposition of Hydrogen Peroxide in on-Site Mixture Explosives: Study of the Effect of the Auxiliary Oxidising Agent and Binder*. Molecules, 2023. **28**(16): p. 5957.
138. Baraza, X., et al., *Lessons Learned From the Barracas Accident: Ammonium Nitrate Explosion During Road Transport*. Process Safety Progress, 2022. **41**(3): p. 519-530.
139. Pasman, H.J., et al., *Beirut Ammonium Nitrate Explosion: Are Not We Really Learning Anything?* Process Safety Progress, 2020. **39**(4).
140. Australian Trade and Investment Commission, *Perdaman builds A$6 billion urea plant in Western Australia*. 2026, Australian Government.
141. Perdaman, *Australia's Largest Urea Manufacturing Plant*. 2026, Perdaman.
142. Multotec, *Typical Phosphates Beneficiation Flow Sheet*. 2026, Multotec.
143. Mew, M., G. Steiner, and B. Geißler, *Phosphorus Supply Chain—Scientific, Technical, and Economic Foundations: A Transdisciplinary Orientation*. Sustainability, 2018. **10**(4): p. 1087.
144. McKelvey, V., *Phosphate Deposits*. 1967.

145. Ryszko, U., P. Rusek, and D. Kołodyńska, *Quality of Phosphate Rocks From Various Deposits Used in Wet Phosphoric Acid and P-Fertilizer Production*. Materials, 2023. **16**(2): p. 793.
146. Geißler, B., et al., *Striving Toward a Circular Economy for Phosphorus: The Role of Phosphate Rock Mining*. Minerals, 2018. **8**(9): p. 395.
147. Valco, *Phosphoric acid – Manufacturing process for Phosphoric acid*. 2026, Valco.
148. Jihad, S.N. and K.A. Saeed, *A Review of Solidification/Stabilization of Heavy Metal Contaminated Soil*. Journal of Engineering and Sustainable Development, 2023. **27**(2): p. 227-244.
149. Men, J., et al., *Recycling Phosphogypsum in Road Construction Materials and Associated Environmental Considerations: A Review*. Heliyon, 2022. **8**(11): p. e11518.
150. Camacho, A., et al., *Alpha Emitter Intrinsic Concentration in Copper Required for Nuclear Spectrometry Application*. Journal of Nuclear Physics Material Sciences Radiation and Applications, 2017. **5**(1): p. 213-222.
151. Lutskiy, D., et al., *Complex Processing of Phosphogypsum - A Way of Recycling of Dumps With Reception of Commodity Production of Wide Application*. Journal of Ecological Engineering, 2018. **19**(2): p. 221-225.
152. Li, W., et al., *Sustainable Utilization of Phosphogypsum in Multi-Solid Waste Recycled Aggregates: Environmental Impact and Economic Viability*. Sustainability, 2024. **16**(3): p. 1161.
153. Anamika, B. and D. Giri, *Utilisation of Phosphogypsum Along With Other Additives in Geo- Engineering- A Review*. Materiales De Construcción, 2022. **72**(347): p. e288.
154. Lacy, H. and K.A. Barnes, *Tailings Storage Facilities –Decommissioning Planning Is Vital for Successful Closure*. 2006.
155. Mukhopadhyay, R., et al., *Soil Salinity Under Climate Change: Challenges for Sustainable Agriculture and Food Security*. Journal of Environmental Management, 2021. **280**: p. 111736.
156. Lersow, M. and P.T. Schmidt, *The Wismut Remediation Project*. 2006: p. 181-190.
157. Huang, L., et al., *Ecological Engineering to Accelerate Mineral Weathering and Transformation Underpins Sustainable Tailings Rehabilitation*. 2019.
158. Araujo, F.S.M., et al., *Recycling and Reuse of Mine Tailings: A Review of Advancements and Their Implications*. 2022.
159. Chernysh, Y., et al., *Soil Contamination by Heavy Metals and Radionuclides and Related Bioremediation Techniques: A Review*. Soil Systems, 2024. **8**(2): p. 36.
160. Mukaba, J.-L., et al., *Rare Earths' Recovery From Phosphogypsum: An Overview on Direct and Indirect Leaching Techniques*. Minerals, 2021. **11**(10): p. 1051.
161. Silva, G.O.A.d., G. Southam, and E.J. Gagen, *Accelerating Soil Aggregate Formation: A Review on Microbial Processes as the Critical Step in a Post-Mining Rehabilitation Context*. Soil Research, 2022. **61**(3): p. 209-223.
162. Levchenko, E.N., I.G. Spiridonov, and D. Klyucharev, *Environmental Pollution Problems in the Mining Regions of Russia*. 2019: p. 453-456.
163. Li, Y., S. Jia, and J. Liu, *Solidification, Remediation and Long-Term Stability of Heavy Metal Contaminated Soil Under the Background of Sustainable Development*. Scientific Reports, 2022. **12**(1).
164. Yan, S., et al., *Gypsum-Enhanced Red Mud Composites: A Study on Strength, Durability, and Leaching Characteristics*. Buildings, 2024. **14**(7): p. 1979.

165. Ji, X., et al., *Green Grouting Material Based on Phosphogypsum–Slag Geopolymer: Feasibility and Performance Evaluation for Trenchless Road Repair.* Materials, 2025. **18**(21): p. 4901.
166. Erro, J., et al., *Incorporation of humic-derived active molecules into compound NPK granulated fertilizers: Main technical difficulties and potential solutions.* Chemical and Biological Technologies in Agriculture, 2016. **3**.
167. Engineers Guide, *Diammonium phosphate (DAP) fertilizer manufacturing flow diagram*. 2026, Engineers Guide,.
168. Incitec Pivot Limited, *MAP and DAP*. 2026, Incitec Pivot Limited.
169. Capitaine, S.L. and C. Carlson, *Processing Phosphates for Use in the Fertilizer Industry*. 2026, FEECO International, Inc.
170. FerTech Inform, *Ammonium phosphates*. 2022, FerTech Inform.
171. IMARC Group, *Monoammonium Phosphate (MAP) Production Cost Analysis Report 2026: Industry Trends, Plant Setup, Machinery, Raw Materials, Investment Opportunities, Cost and Revenue*. 2026.
172. Chaabouni, A., et al., *Study the Nature and the Effects of the Impurities of Phosphate Rock in the Plants of Production of Phosphoric Acid.* Journal of Advances in Chemistry, 2008. **7**(2): p. 1296-1299.
173. McCoy, M., *A New Route to Hydrogen Fluoride.* C&en Global Enterprise, 2020. **98**(32): p. 24-25.
174. Guendouzi, M.E., M. Skafi, and A. Rifai, *Hexafluorosilicate Salts in Wet Phosphoric Acid Processes: Properties of X₂SiF₆–H₂O With X = Na⁺, K⁺, or NH₄⁺ in Aqueous Solutions at 353.15 K.* Journal of Chemical & Engineering Data, 2016. **61**(5): p. 1728-1734.
175. Schlesinger, W.H., E.M. Klein, and A. Vengosh, *Global Biogeochemical Cycle of Fluorine.* Global Biogeochemical Cycles, 2020. **34**(12).
176. Tomaszewska, M., *Concentration and Purification of Fluosilicic Acid by Membrane Distillation.* Industrial & Engineering Chemistry Research, 2000. **39**(8): p. 3038-3041.
177. Zhang, W., et al., *Millimeter-Sized K₂MnF₆:Si⁴⁺, NH₄⁺ Red-Luminescent Crystals With High Absorption Efficiency (AE_{max} of 93.5%) and External Quantum Efficiency (EQE_{max} of 68.9%) Grown by Cooling-Induced Crystallization.* Inorganic Chemistry, 2023. **63**(1): p. 833-841.
178. He, S.H., *Study on Damage and Solution of Acidizing Technology in Sandstone Reservoir.* Advanced Materials Research, 2013. **631-632**: p. 729-732.
179. Ennaciri, Y., M. Bettach, and H.E. Alaoui-Belghiti, *Phosphogypsum Conversion Into Calcium Fluoride and Sodium Sulfate.* Annales De Chimie Science Des Matériaux, 2020. **44**(6): p. 407-412.
180. Fuge, R., *Fluorine in the Environment, a Review of Its Sources and Geochemistry.* Applied Geochemistry, 2019. **100**: p. 393-406.
181. Ji, C.-H., et al., *Reconciling Environmental Safety and Economic Feasibility: A Review of Soil Fluorine Management Strategies in the Republic of Korea*. 2024.
182. Box, S.E. and P.M. Cossette, *Mineral Resource Inventory of North Dakota*. 2021.

183. Border, S. and L. Sawyer, *Evaporites and Brines – Geological, Hydrological and Chemical Aspects of Resource Estimation*. Applied Earth Science Transactions of the Institutions of Mining and Metallurgy Section B, 2014. **123**(2): p. 95-106.
184. Yager, D.B., *Potash—A Vital Agricultural Nutrient Sourced From Geologic Deposits*. 2016.
185. Brokaw, A.L., *Geology and Hydrology of the Carlsbad Potash Area, Eddy and Lea Counties, New Mexico*. 1972.
186. Hite, R.J., *Subsurface Geology of a Potential Waste Emplacement Site, Salt Valley Anticline, Grand County, Utah*. 1977.
187. John, C.B., et al., *Potash Ore Reserves in the Proposed Waste Isolation Pilot Plant Area, Eddy County, Southeastern New Mexico*. 1978.
188. Kleine-Kleffmann, U., *The Discovery of the First Potash Mine and the Development of the Potash Industry Since 1861*. Journal of Plant Nutrition and Soil Science, 2023. **186**(6): p. 615-622.
189. Tourtelot, E.B. and J.D. Vine, *Copper Deposits in Sedimentary and Volcanogenic Rocks*. 1976.
190. Lavrenko, S.A. and D.I. Shishlyannikov, *Performance Evaluation of Heading-and-Winning Machines in the Conditions of Potash Mines*. Applied Sciences, 2021. **11**(8): p. 3444.
191. Rubiolo, D.G., *Potash Exploration in North and South America*. Revista Geonomos, 2013.
192. Pedley, A., J. Neubert, and S.v.d. Klauw, *Potash Deposits in Africa*. Episodes, 2016. **39**(2): p. 447-457.
193. Qin, X., et al., *Origin and Evolution of Saline Spring Water in North and Central Laos Based on Hydrochemistry and Stable Isotopes (ΔD, Δ18O, Δ11B, and Δ37Cl)*. Water, 2021. **13**(24): p. 3568.
194. Williams, D.A., *Mineral Resources of the U. S. S. R.** Nature, 1942. **150**(3810): p. 539-541.
195. Raup, O.B. and R.J. Hite, *Bromine Geochemistry of Chloride Rocks of the Middle Pennsylvanian Paradox Formation of the Hermosa Group, Paradox Basin, Utah and Colorado*. 1996.
196. Xu, Y., Y. Cao, and C. Liu, *Whether the Middle Eocene Salt-Forming Brine in the Kuqa Basin Reached the Potash-Forming Stage: Quantitative Evidence From Halite Fluid Inclusions*. Geofluids, 2021. **2021**: p. 1-12.
197. Soulliere, S.J., G.K. Lee, and C.M. Martin, *Mineral Resources of the Horseshoe Canyon North Wilderness Study Area, Emery and Wayne Counties, Utah*. 1988.
198. Cooper, J.B., *Road Log From Carlsbad, New Mexico, to Project Gnome Site*. 1962.
199. Rime, V., et al., *Nature and Significance of Late Pleistocene to Holocene Thick Evaporite Deposits of the Danakil Depression, Afar, Ethiopia*. Sedimentology, 2024. **72**(2): p. 475-506.
200. García, J.M.L., et al., *Origin and Evolution of the Halo-Volcanic Complex of Dallol: Proto-Volcanism in Northern Afar (Ethiopia)*. Frontiers in Earth Science, 2020. **7**.
201. Schofield, N., et al., *Mobilizing Salt: Magma-Salt Interactions*. Geology, 2014. **42**(7): p. 599-602.
202. Otálora, F.n., et al., *Mineralochemical Mechanism for the Formation of Salt Volcanoes: The Case of Mount Dallol (Afar Triangle, Ethiopia)*. Acs Earth and Space Chemistry, 2022. **6**(12): p. 2767-2778.

203. Bastow, I.D., et al., *The Development of Late-Stage Continental Breakup: Seismic Reflection and Borehole Evidence From the Danakil Depression, Ethiopia.* Tectonics, 2018. **37**(9): p. 2848-2862.
204. Kebede, B.A., et al., *Continuous Subsidence of Dallol Volcano Caused by Magmatic, Hydrothermal, and Salt Dissolution Processes: Insights From InSAR Observations.* Geochemistry Geophysics Geosystems, 2025. **26**(4).
205. Mège, D., et al., *Tectonic and Hydrothermal Activity at the Yellow Lake Fissure in Response to the 2004 Dallol Dyke Intrusion Event in Afar.* Frontiers in Earth Science, 2023. **11**.
206. Cavalazzi, B., et al., *The Dallol Geothermal Area, Northern Afar (Ethiopia)—An Exceptional Planetary Field Analog on Earth.* Astrobiology, 2019. **19**(4): p. 553-578.
207. Sanz, J.L., et al., *Methanogenesis at High Temperature, High Ionic Strength and Low pH in the Volcanic Area of Dallol, Ethiopia.* Microorganisms, 2021. **9**(6): p. 1231.
208. Viltres, R., et al., *Kinematics and Deformation of the Southern Red Sea Region From GPS Observations.* Geophysical Journal International, 2020. **221**(3): p. 2143-2154.
209. Hurman, G., et al., *Quantitative Analysis of Faulting in the Danakil Depression Rift of Afar: The Importance of Faulting in the Final Stages of Magma-Rich Rifting.* Tectonics, 2023. **42**(6).
210. Hutchinson, R.W. and G.G. Engels, *A Discussion on the Structure and Evolution of the Red Sea and the Nature of the Red Sea, Gulf of Aden and Ethiopia Rift Junction - Tectonic Significance of Regional Geology and Evaporite Lithfacies in Northeastern Ethiopia.* Philosophical Transactions of the Royal Society of London Series a Mathematical and Physical Sciences, 1970. **267**(1181): p. 313-329.
211. Sahle, Y. and A. Beyin, *Archaeological Reconnaissance of the Late Pleistocene Red Sea Coast in the Danakil.* Antiquity, 2017. **91**(359).
212. Kotopoulou, E., et al., *A Polyextreme Hydrothermal System Controlled by Iron: The Case of Dallol at the Afar Triangle.* Acs Earth and Space Chemistry, 2018. **3**(1): p. 90-99.
213. Melaku, A.A., W. Getaneh, and B. Atnafu, *Genesis of the Enkafela Mn Deposit: A Record of Submarine Hydrothermal Activity in the Afar Depression, Northeast Ethiopia.* Applied Earth Science Transactions of the Institutions of Mining and Metallurgy, 2022. **131**(1): p. 2-14.
214. Jaramillo-Vogel, D., et al., *Pleistocene Sea-floor Fibrous Crusts and Spherulites in the Danakil Depression (Afar, Ethiopia).* Sedimentology, 2018. **66**(2): p. 480-512.
215. Humbert, R., *Potassium in Relation to Food Production.* Hortscience, 1969. **4**(1): p. 35-36.
216. Islam, M.S., et al., *Effect of Potassium on Yield, Quality and Shelf Life of Tomato.* Bangladesh Journal of Agricultural Research, 2020. **44**(4): p. 659-668.
217. Das, U., R. Bhattacharyya, and K.K. Baruah, *Bunch Yield of 'Grand Naine' (AAA) Banana as Influenced by the Varied Components of Precision Farming Systems.* International Journal of Current Microbiology and Applied Sciences, 2018. **7**(05): p. 501-506.
218. Mustaffa, M.M. and V. Kumar, *Banana Production and Productivity Enhancement Through Spatial, Water and Nutrient Management.* Journal of Horticultural Sciences, 2012. **7**(1): p. 1-28.

219. Johnson, V.J. and A. Mirza, *Role of Macro and Micronutrients in the Growth and Development of Plants*. International Journal of Current Microbiology and Applied Sciences, 2020. **9**(11): p. 576-587.
220. Abu, M., *Effect of Compound and Single-Based Fertilizerson Shelf Life of Exotic/Minor Vegetable Crops Cultivated for the Fresh Market*. Horticulture International Journal, 2021. **5**(2): p. 90-95.
221. Arah, I.K., et al., *Preharvest and Postharvest Factors Affecting the Quality and Shelf Life of Harvested Tomatoes: A Mini Review*. International Journal of Agronomy, 2015. **2015**: p. 1-6.
222. Ilić, Z.S., et al., *Summer Production of Lettuce Under Shading Conditions*. 2017.
223. Ilin, Ž., et al., *GAP Recommendations for Permanent and Temporary Protection System - Early Potato Production*. 2017.
224. Herzog, E., A.J. Keutgen, and J. Baláš, *Biostimulants Affect Quality and Yield of Tomato Fruits From Open-Field Production*. 2017.
225. Villette, S., et al., *Simple imaging system to measure velocity and improve the quality of fertilizer spreading in agriculture*. J. Electronic Imaging, 2008. **17**: p. 031109.
226. Ehsani, R., A.W. Schumann, and M. Salyani, *Variable Rate Technology for Florida Citrus*. Edis, 2009. **2009**(2).
227. Finger, R., et al., *Precision Farming at the Nexus of Agricultural Production and the Environment*. Annual Review of Resource Economics, 2019. **11**(1): p. 313-335.
228. Afroj, M., M.M.H. Kazal, and M.M. Rahman, *Precision Agriculture in the World and Its Prospect in Bangladesh*. Research in Agriculture Livestock and Fisheries, 2016. **3**(1): p. 1-14.
229. Vizzari, M., F.S. Santaga, and P. Benincasa, *Sentinel 2-Based Nitrogen VRT Fertilization in Wheat: Comparison Between Traditional and Simple Precision Practices*. Agronomy, 2019. **9**(6): p. 278.
230. Santaga, F.S., et al., *Simplified and Advanced Sentinel-2-Based Precision Nitrogen Management of Wheat*. Agronomy, 2021. **11**(6): p. 1156.
231. Rovira-Más, F., et al., *Open-Format Prescription Maps for Variable Rate Spraying in Orchard Farming*. Journal of the Asabe, 2024. **67**(2): p. 243-257.
232. Ammoniaci, M., et al., *State of the Art of Monitoring Technologies and Data Processing for Precision Viticulture*. Agriculture, 2021. **11**(3): p. 201.
233. Baio, F.H.R., et al., *A New Proposal for Soybean Plant Stand: Variation Based on the Law of the Minimum*. Plants, 2024. **13**(22): p. 3193.
234. Puntel, L.A., et al., *How Digital Is Agriculture in a Subset of Countries From South America? Adoption and Limitations*. Australian Journal of Agricultural Research, 2022. **74**(6): p. 555-572.
235. Thompson, N.M., N.D. DeLay, and J. Mintert, *Understanding the Farm Data Lifecycle: Collection, Use, and Impact of Farm Data on U.S. Commercial Corn and Soybean Farms*. Precision Agriculture, 2021. **22**(6): p. 1685-1710.
236. Khanna, M., *Digital Transformation of the Agricultural Sector: Pathways, Drivers and Policy Implications*. Applied Economic Perspectives and Policy, 2020. **43**(4): p. 1221-1242.
237. Pawase, P.P., et al., *Variable Rate Fertilizer Application Technology for Nutrient Management: A Review*. International Journal of Agricultural and Biological Engineering, 2023. **16**(4): p. 11-19.

238. Serrano, J., et al., *Technological Upgrade of a Vicon RS-EDW Spreader: Development of a Microcontroller for Variable Rate Application.* Agriengineering, 2024. **6**(2): p. 1436-1449.
239. Masi, M., et al., *Precision Farming: Barriers of Variable Rate Technology Adoption in Italy.* Land, 2023. **12**(5): p. 1084.
240. Mitchell, S., A. Weersink, and B. Erickson, *Adoption of Precision Agriculture Technologies in Ontario Crop Production.* Canadian Journal of Plant Science, 2018. **98**(6): p. 1384-1388.
241. Zhang, S., et al., *Development of a Data Acquisition and Processing System for Precision Agriculture.* 2007.
242. Ali, A., M. Hassan, and H.P. Kaul, *Broad Scope of Site-Specific Crop Management and Specific Role of Remote Sensing Technologies Within It—A Review.* Journal of Agronomy and Crop Science, 2024. **210**(4).
243. Schimmelpfennig, D. and J. Lowenberg-DeBoer, *Farm Types and Precision Agriculture Adoption: Crops, Regions, Soil Variability, and Farm Size.* SSRN Electronic Journal, 2020.
244. Latherow, T., et al., *Optimising Precision Agriculture Choices for Arable Farmers in Germany and the <scp>UK</Scp>: The <scp>LINKDAPA</Scp> Approach.* Eurochoices, 2024. **23**(3): p. 29-35.
245. Grizzetti, B., et al., *Nitrogen as a Threat to European Water Quality.* 2011: p. 379-404.
246. Guan, D., et al., *Viewpoints of Dominant Environmental Factors Influencing Algal Blooms.* Applied Mechanics and Materials, 2011. **66-68**: p. 155-159.
247. Paerl, H.W., et al., *It Takes Two to Tango: When and Where Dual Nutrient (N &Amp; P) Reductions Are Needed to Protect Lakes and Downstream Ecosystems.* Environmental Science & Technology, 2016. **50**(20): p. 10805-10813.
248. Silvarrey, A., V. Sposito, and R. Faggian, *Climate Change and Cyanobacteria Harmful Algae Blooms: Adaptation Practices for Developing Countries.* Marine & Freshwater Research, 2021. **72**(12): p. 1722-1734.
249. Thieu, V., et al., *Nitrogen Cycling in a Hypothetical Scenario of Generalised Organic Agriculture in the Seine, Somme and Scheldt Watersheds.* Regional Environmental Change, 2010. **11**(2): p. 359-370.
250. Newton, A.R. and R. Melaram, *Harmful Algal Blooms in Agricultural Irrigation: Risks, Benefits, and Management.* Frontiers in Water, 2023. **5**.
251. Kim, J.H., et al., *Machine Learning-Based Early Warning Level Prediction for Cyanobacterial Blooms Using Environmental Variable Selection and Data Resampling.* Toxics, 2023. **11**(12): p. 955.
252. Brooks, B.W., et al., *Are Harmful Algal Blooms Becoming the Greatest Inland Water Quality Threat to Public Health and Aquatic Ecosystems?* Environmental Toxicology and Chemistry, 2016. **35**(1): p. 6-13.
253. Koreivienė, J., et al., *EU Project of Life Programme "Algae Service for Life" Creates Tools for Ecological Service to Mitigate Cyanobacteria and Macroalgae Blooms in Freshwater Ecosystems.* Botanica, 2019. **25**(1): p. 65-73.
254. Bowmer, K.H., *Ecosystem Effects From Nutrient and Pesticide Pollutants: Catchment Care as a Solution.* Resources, 2013. **2**(3): p. 439-456.

255. Sultana, S., et al., *Cyanobacterial Blooms in Earthen Aquaculture Ponds and Their Impact on Fisheries and Human Health in Bangladesh.* Aquaculture Research, 2022. **53**(15): p. 5129-5141.

256. Cowan, N., et al., *Nitrous Oxide Emission Factors of Mineral Fertilisers in the UK and Ireland: A Bayesian Analysis of 20 years of Experimental Data.* Environment International, 2020. **135**: p. 105366.

257. Roche, L., et al., *Impact of Fertiliser Nitrogen Formulation, and N Stabilisers on Nitrous Oxide Emissions in Spring Barley.* Agriculture Ecosystems & Environment, 2016. **233**: p. 229-237.

258. Król, D., et al., *Nitrogen Fertilisers With Urease Inhibitors Reduce Nitrous Oxide and Ammonia Losses, While Retaining Yield in Temperate Grassland.* The Science of the Total Environment, 2020. **725**: p. 138329.

259. Mateo-Marín, N., et al., *Feasibility of Stabilised Nitrogen Fertilisers Decreasing Greenhouse Gas Emissions Under Optimal Management in Sprinkler Irrigated Conditions.* Agriculture Ecosystems & Environment, 2020. **290**: p. 106725.

260. Dougherty, W.J., et al., *Nitrification (DMPP) and Urease (NBPT) Inhibitors Had No Effect on Pasture Yield, Nitrous Oxide Emissions, or Nitrate Leaching Under Irrigation in a Hot-Dry Climate.* Soil Research, 2016. **54**(5): p. 675-683.

261. Wallace, A.J., et al., *Nitrogen Use Efficiency of 15N Urea Applied to Wheat Based on Fertiliser Timing and Use of Inhibitors.* Nutrient Cycling in Agroecosystems, 2019. **116**(1): p. 41-56.

262. Islam, Z.F., P.V. Cherepanov, and H.W. Hu, *Engineering Biodegradable Coatings for Sustainable Fertilisers.* Microbiology Australia, 2023. **44**(1): p. 9-12.

263. Aarons, S.R., J.F. Angus, and C.J.P. Gourley, *Efficient Use of Reactive Nitrogen for Productive Agroecosystems.* Soil Research, 2017. **55**(6): p. 413-416.

264. Mazzetto, A.M., et al., *Region-Specific Emission Factors for Brazil Increase the Estimate of Nitrous Oxide Emissions From Nitrogen Fertiliser Application by 21%.* Atmospheric Environment, 2020. **230**: p. 117506.

265. Carbonell-Bojollo, R.M., et al., *Soil Management, Irrigation and Fertilisation Strategies for N2O Emissions Mitigation in Mediterranean Agricultural Systems.* Agronomy, 2022. **12**(6): p. 1349.

266. Skocaj, D., Y. Everingham, and B.L. Schroeder, *Nitrogen Management Guidelines for Sugarcane Production in Australia: Can These Be Modified for Wet Tropical Conditions Using Seasonal Climate Forecasting?* Springer Science Reviews, 2013. **1**(1-2): p. 51-71.

267. Affendi, N.M.N., N. Mansor, and S.S. Samiri, *Addition of Chemical and Natural Urease Inhibitors in Reducing Ammonia and Nitrous Oxide Losses.* Journal of Soil Science and Plant Nutrition, 2019. **20**(1): p. 253-258.

268. Deng, B., et al., *Urease Inhibitor and Biochar Independently Affected N₂O Emissions From Camellia Oleifera Soils.* Plant Soil and Environment, 2022. **68**(9): p. 424-430.

269. Leahy, S.C., et al., *Mitigating Greenhouse Gas Emissions From New Zealand Pasture-Based Livestock Farm Systems.* Journal of New Zealand Grasslands, 2019: p. 101-110.

270. Bruulsema, T., H. Peterson, and L.I. Prochnow, *The Science of 4R Nutrient Stewardship for Phosphorus Management Across Latitudes.* Journal of Environmental Quality, 2019. **48**(5): p. 1295-1299.

271. Beek, C.v., et al., *The Fertile Grounds Initiative: A Concerted Action for Integrated Soil Fertility Management*. Advances in Plants & Agriculture Research, 2017. **6**(3): p. 69-76.
272. Davidson, E.A., et al., *N-Related Greenhouse Gases in North America: Innovations for a Sustainable Future*. Current Opinion in Environmental Sustainability, 2014. **9-10**: p. 1-8.
273. Nigon, L.L., *Integrating Enhanced-Efficiency Fertilizers in 4R Nutrient Management*. Crops & Soils, 2023. **57**(1): p. 32-37.
274. Jarvie, H.P., et al., *<i>Phosphorus Mirabilis</I>: Illuminating the Past and Future of Phosphorus Stewardship*. Journal of Environmental Quality, 2019. **48**(5): p. 1127-1132.
275. Pearson, S., et al., *Decarbonising Our Food Systems: Contextualising Digitalisation for Net Zero*. Frontiers in Sustainable Food Systems, 2023. **7**.
276. Luman, R., *Blockchain Driven Supply Chain Transparency in Saf Production: Enhancing Traceability and Regulatory Compliance*. International Journal of Advanced Research in Computer Science, 2024. **15**(5): p. 89-98.
277. Kivarina, M., *Digital Agriculture: The Experience of Russian Regions*. Russian Journal of Management, 2024. **12**(2): p. 211-220.
278. Jiang, S., et al., *A Tertiary Review on Blockchain and Sustainability With Focus on Sustainable Development Goals*. Ieee Access, 2022. **10**: p. 114975-115006.
279. Nurhayati, N., *Digital Innovation in Halal Logistics: An Indonesian Case Study*. Agripreneur Jurnal Pertanian Agribisnis, 2023. **12**(2): p. 23-37.
280. Li, J., A. Maiti, and J. Fei, *Features and Scope of Regulatory Technologies: Challenges and Opportunities With Industrial Internet of Things*. Future Internet, 2023. **15**(8): p. 256.
281. Mallouppas, G., C. Ioannou, and E.A. Yfantis, *A Review of the Latest Trends in the Use of Green Ammonia as an Energy Carrier in Maritime Industry*. Energies, 2022. **15**(4): p. 1453.
282. Nayak-Luke, R., et al., *Assessing the Viability of Decarbonising India's Nitrogenous Fertiliser Consumption*. Journal of Cleaner Production, 2022. **366**: p. 132462.
283. Hamukoshi, S.S., et al., *An Overview of the Socio-Economic Impacts of the Green Hydrogen Value Chain in Southern Africa*. Journal of Energy in Southern Africa, 2022. **33**(3): p. 12-21.
284. Negro, V., M. Noussan, and D. Chiaramonti, *The Potential Role of Ammonia for Hydrogen Storage and Transport: A Critical Review of Challenges and Opportunities*. Energies, 2023. **16**(17): p. 6192.
285. Oladele, K., et al., *Blockchain Technology Adoption in Developing Countries and Its Effect on Their Achievement of Sustainability Goals*. Journal of Southwest Jiaotong University, 2023. **58**(6).
286. Carmona-Martínez, A.A., et al., *Charting the Course: Navigating Decarbonisation Pathways in Greece, Germany, the Netherlands, and Spain's Industrial Sectors*. Sustainability, 2024. **16**(14): p. 6176.
287. Diéguez, M.S., et al., *High Technical and Temporal Resolution Integrated Energy System Modelling of Industrial Decarbonisation*. Advances in Applied Energy, 2022. **7**: p. 100105.
288. Fernández, J.A., et al., *Current Trends in Organic Vegetable Crop Production: Practices and Techniques*. Horticulturae, 2022. **8**(10): p. 893.

289. Kulhánek, M., et al., *Plant Nutrition—New Methods Based on the Lessons of History: A Review.* Plants, 2023. **12**(24): p. 4150.

290. Bene, C.D., et al., *Barriers and Opportunities for Sustainable Farming Practices and Crop Diversification Strategies in Mediterranean Cereal-Based Systems.* Frontiers in Environmental Science, 2022. **10**.

291. Mackiewicz-Walec, E. and M. Olszewska, *Biostimulants in the Production of Forage Grasses and Turfgrasses.* Agriculture, 2023. **13**(9): p. 1796.

292. Costa, D., et al., *Impact of Biostimulant Use in Agricultural Crops (Strawberries, Leafy Greens and Mushrooms) Under Different Horticultural Cropping Systems: A Systematic Review.* Annals of Applied Biology, 2025. **187**(3): p. 304-317.

293. Rouphael, Y., et al., *Arbuscular Mycorrhizal Fungi Act as Biostimulants in Horticultural Crops.* Scientia Horticulturae, 2015. **196**: p. 91-108.

294. Ronga, D., et al., *Microalgal Biostimulants and Biofertilisers in Crop Productions.* Agronomy, 2019. **9**(4): p. 192.

295. Gonçalves, A.L., *The Use of Microalgae and Cyanobacteria in the Improvement of Agricultural Practices: A Review on Their Biofertilising, Biostimulating and Biopesticide Roles.* Applied Sciences, 2021. **11**(2): p. 871.

296. Morillas-España, A., et al., *Microalgae Based Wastewater Treatment Coupled to the Production of High Value Agricultural Products: Current Needs and Challenges.* Chemosphere, 2022. **291**: p. 132968.

297. Prisa, D. and D. Spagnuolo, *Plant Production With Microalgal Biostimulants.* Horticulturae, 2023. **9**(7): p. 829.

298. Köninger, J., et al., *Manure Management and Soil Biodiversity: Towards More Sustainable Food Systems in the EU.* Agricultural Systems, 2021. **194**: p. 103251.

299. Fasusi, O.A. and O.O. Babalola, *The Multifaceted Plant-Beneficial Rhizobacteria Toward Agricultural Sustainability.* Plant Protection Science, 2021. **57**(2): p. 95-111.

Index

A

Acid soils, 59
Acidification, 11, 419
Acidulation, 108, 117, 118, 128, 129, 234, 236, 239, 251, 253, 254, 255, 257, 261, 274, 416
Additives, 5, 10, 74, 102, 141, 190, 205, 241, 303, 312, 319, 320, 321, 322, 326, 400, 401, 448, 449, 457, 463
Adsorption, 152, 154, 166, 242
Aeration, 38, 40, 41, 42, 43, 52, 54, 67, 76
Agribusiness, 2
Agriculture, 3, 2, 3, 4, 6, 7, 8, 11, 12, 13, 14, 15, 20, 34, 39, 40, 42, 46, 47, 50, 67, 75, 79, 82, 83, 86, 87, 89, 90, 93, 98, 100, 101, 107, 123, 129, 165, 170, 188, 231, 241, 243, 262, 277, 280, 293, 307, 308, 312, 319, 330, 333, 334, 347, 350, 358, 359, 360, 362, 365, 368, 370, 371, 375, 376, 379, 381, 383, 385, 392, 404, 409, 420, 454, 455, 464, 467, 468, 469, 470, 472
Agronomy, 15, 17, 69, 312, 422, 456, 457, 467, 468, 470, 472
Alkaline soils, 44, 52, 59, 62, 69, 71, 72, 73, 191, 232, 267, 355
Amino acids, 20, 24, 26, 106, 397, 399
Ammonia, 3, 4, 2, 10, 13, 16, 27, 33, 34, 46, 51, 83, 87, 88, 96, 98, 100, 102, 103, 104, 107, 109, 110, 111, 112, 113, 127, 130, 131, 132, 133, 134, 135, 136, 137, 138, 139, 140, 141, 142, 143, 144, 145, 146, 147, 148, 149, 150, 151, 152, 153, 154, 155, 156, 157, 158, 159, 160, 161, 162, 163, 164, 165, 166, 167, 168, 169, 170, 172, 173, 174, 175, 179, 180, 181, 182, 184, 185, 186, 187, 188, 189, 190, 191, 192, 193, 194, 196, 197, 198, 199, 200, 205, 206, 211, 212, 213, 214, 215, 216, 217, 218, 232, 262, 263, 264, 266, 267, 268, 269, 270, 271, 272, 273, 320, 321, 355, 363, 365, 366, 367, 368, 369, 370, 382, 386, 400, 401, 409, 410, 411, 412, 413, 414, 415, 416, 417, 427, 459, 460, 461, 469, 470, 471
Ammonia synthesis, 4, 109, 111, 130, 134, 136, 139, 140, 141, 142, 143, 144, 145, 146, 147, 148, 149, 150, 151, 152, 153, 154, 155, 156, 157, 158, 159, 160, 163, 165, 166, 184, 200, 211, 365, 366, 416, 460
Ammonium, 4, 8, 13, 47, 51, 61, 63, 64, 73, 75, 77, 81, 84, 85, 87, 88, 90, 93, 102, 104, 110, 131, 136, 137, 143, 145, 156, 169, 170, 172, 173, 176, 180, 181, 186, 187, 188, 189, 190, 191, 192, 193, 194, 195, 196, 197, 200, 201, 202, 204, 205, 206, 207, 208, 209,210, 211, 212, 213, 214, 215, 232, 242, 243, 262, 263, 267, 268, 269, 273, 298, 302, 303, 304, 305, 317, 320, 323, 325, 327, 344, 345, 353, 354, 355, 359, 364, 371, 378, 385, 386, 400, 401, 410, 412, 415, 428, 433, 434, 437, 448, 449, 457, 462, 464
Ammonium nitrate, 4, 8, 13, 61, 81, 85, 87, 93, 110, 131, 136, 138, 143, 145, 156, 169, 170, 187, 188, 189, 190, 191, 192, 193, 194, 195, 196, 197, 200, 201, 202, 204, 205, 206, 207, 208, 209, 210, 211, 212, 213, 214, 215, 317, 325, 345, 355, 378, 385, 386, 410, 412, 415,428, 433, 434, 437, 462
Ammonium sulphate, 4, 87, 110, 136, 138, 145, 170, 187, 188, 191, 192, 193, 194, 195, 205, 206, 242, 243, 298, 302, 304, 305, 317, 325, 327, 344, 364, 457
ATP (adenosine triphosphate), 23
Availability of nutrients, 7, 44, 71, 397, 398, 399, 418

B

Biofertilisers, 3, 13, 77, 80, 98, 100, 101, 106, 396, 397, 398, 399, 405, 406, 448, 451, 452, 458, 472
Biogeochemical cycles, 465
Biomass, 5, 8, 53, 76, 82, 183, 318, 347, 349, 350, 351, 457
Biotechnology, 5, 3, 375, 396, 398, 403, 404, 458
Blended fertilisers, 3, 74, 81, 82, 86, 89, 90, 315, 320, 321, 446
Boron, 27, 55, 60, 83, 323, 345
Bulk blending, 5, 311, 313, 314, 315, 327

C

Calcium, 14, 25, 28, 29, 44, 48, 54, 55, 57, 58, 62, 65, 68, 69, 70, 71, 72, 73, 83, 115, 118, 119, 120, 122, 129, 141, 232, 233, 236, 241, 242, 243, 244, 245, 251, 301, 303, 323, 324, 325, 327, 342, 345, 346, 363, 369, 421, 422, 456, 465
Calvin cycle, 22, 23
Carbon dioxide emissions, 110, 149, 153
Carbon sequestration, 53
Carnallite, 4, 87, 125, 280, 282, 283, 284, 285, 286, 287, 288, 289, 290, 291, 292, 299, 301
Catalysts, 4, 109, 137, 140, 143, 147, 148, 150, 151, 153, 154, 155, 156, 157, 158, 160, 161, 162, 164, 183, 184, 185, 194, 200, 410
Cation exchange capacity, 52
CEC, 52
Cell division, 24, 25
Cell wall, 20, 25, 27, 279, 306
Chlorine, 27
Chlorophyll, 6, 20, 21, 22, 25, 27, 29, 47, 53, 55, 58, 82, 98, 313, 347, 349, 350, 457
Chloroplast, 21
Chlorosis, 20, 25, 26, 27, 55, 58, 59, 60, 69, 72
Circular economy, 5, 32, 35, 231, 243, 358, 363, 364, 371, 372, 374, 379, 381, 383, 384, 392, 399, 404, 405, 463
Clay soils, 40, 52
Climate change, 30, 34, 113, 138, 358, 359, 360, 362, 404, 406, 463, 469
Coatings, 9, 10, 32, 62, 64, 74, 83, 90, 91, 92, 104, 212, 317, 319, 320, 322, 326, 399, 400, 470
Compost, 49, 78, 81, 84, 91, 99, 101, 364, 373, 374, 384, 390, 391, 398, 399, 405, 446, 448, 449, 458
Compound fertilisers, 64, 81, 88, 89, 90, 119, 131, 294, 313, 457
Controlled-release fertilisers, 83, 91, 92, 320, 353, 355, 400, 402, 403, 448, 456
Copper, 27, 83, 148, 150, 154, 157, 160, 463, 465
Crop establishment, 56
Crop nutrition, 14, 88, 319
Crop productivity, 7, 14, 60, 62, 68, 69, 84, 138, 458
Crop quality, 6, 7, 9, 11, 26, 55, 57, 58, 82, 277, 279, 294, 307, 316, 318, 324, 347, 353
Crop rotation, 11, 12, 14, 373
Crop yield, 5, 7, 9, 13, 20, 22, 29, 30, 31, 33, 35, 43, 44, 53, 78, 85, 133, 170, 277, 279, 293, 295, 314, 357, 362, 366, 401, 457
Crystallisation, 87, 126, 127, 191, 192, 193, 194, 236, 281, 285, 287, 289, 292, 294, 298, 299, 300, 301, 302, 303, 304, 305
Custom fertilisers, 323
Cyanobacteria, 398, 469, 472

D

Decision-support systems, 348
Decomposition, 14, 41, 47, 48, 53, 68, 91, 122, 175, 176, 180, 181, 182, 194, 206, 210, 211, 238, 242, 278, 290, 292, 298, 302, 303, 304, 305, 306, 361, 411, 427, 462
Deficiency symptoms, 3, 24, 37, 54, 55, 56, 59, 60, 61, 69, 71, 72
Denitrification, 9, 10, 47, 50, 62, 101, 102, 320, 350, 351, 352, 353, 359, 400
Digital agriculture, 5, 3, 348, 375, 393, 394, 395, 405, 453, 471
Dissolution, 76, 84, 85, 97, 130, 286, 290, 299, 303, 305, 400, 456, 466
Drainage, 38, 40, 41, 43, 52, 67, 68, 69, 70, 262, 275, 359, 361, 369, 434, 441, 442, 443

E

EEF, 402, 403
Efficiency, 4, 5, 6, 8, 9, 11, 13, 17, 22, 29, 30, 32, 35, 51, 62, 64, 65, 67, 71, 76, 77, 79, 80, 86, 93, 94, 96, 97, 98, 101, 102, 104, 105, 110, 115, 116, 117, 120, 124, 129, 131, 133, 134, 135, 137, 143, 144, 147, 150, 151, 153, 154, 155, 156, 158, 159, 160, 161, 162, 163, 164, 168, 169, 172, 174, 176, 179, 180, 181, 182, 186, 188, 191, 192, 193, 197, 198, 199, 200, 205, 225, 226, 227, 228, 234, 235, 236, 237, 238, 239, 240, 244, 246, 247, 248, 249, 250, 251, 253, 255, 274, 275, 280, 284, 289, 291, 292, 299, 303, 309, 312, 314, 315, 316, 317, 319, 320, 323, 324, 334, 341, 346, 348, 350, 351, 353, 354, 358, 359, 360, 362, 364, 365, 366, 368, 369, 370, 371, 373, 374, 376, 379, 394, 395, 396, 397, 399, 400, 401, 402, 403, 404, 405, 406, 411, 414, 416, 417, 422, 454, 457, 458, 465, 470, 471
Energy consumption, 4, 134, 138, 144, 154, 161, 162, 168, 179, 180, 181, 182, 239, 246, 248, 249, 250, 291, 292, 301, 416

Energy efficiency, 162, 164, 170, 173, 181, 182, 193, 199, 238, 239, 460
Energy transfer, 4, 6, 7, 22, 23, 25, 29, 48, 55, 56, 82, 225, 279, 293, 295, 313, 350
Enhanced-Efficiency Fertilisers, 5, 3, 51, 321, 360, 366, 371, 375, 385, 396, 400, 401, 402, 403, 404, 406, 446, 451, 453
Environmental impact, 10, 11, 13, 34, 35, 40, 50, 51, 53, 62, 74, 80, 81, 83, 92, 100, 101, 106, 117, 121, 138, 149, 153, 173, 200, 308, 312, 316, 317, 320, 321, 324, 337, 348, 353, 363, 364, 365, 368, 369, 370, 371, 376, 379, 381, 382, 386, 389, 393, 396, 397, 401, 403, 404,406, 427, 429, 441, 442, 444, 463
Environmental regulation, 124, 128, 131, 200, 254, 385, 393, 433, 439, 442, 444
Environmental stewardship, 3, 11, 32, 375
Enzyme activation, 6, 24, 25, 27, 49, 52, 82
Enzymes, 6, 23, 24, 26, 47, 98, 277, 279, 293, 295
Erosion, 40, 48, 51, 66, 361, 362, 419, 442, 443
Eutrophication, 5, 11, 32, 34, 48, 92, 321, 357, 360, 361, 362, 363, 371, 377, 427, 428, 429, 442

F

Fertigation, 5, 63, 80, 83, 93, 94, 95, 97, 101, 106, 295, 307, 308, 309, 317, 318, 319, 323, 325, 328, 329, 330, 332, 341, 342, 343, 344, 345, 346, 352, 353, 355, 363, 369, 370, 373, 402, 403
Fertiliser application, 5, 15, 20, 28, 35, 38, 40, 44, 68, 73, 97, 233, 330, 332, 337, 341, 343, 351, 352, 360, 382, 393, 395, 396, 403, 443, 444, 470
Fertiliser blending, 5, 311, 312, 446, 448
Fertiliser classification, 84
Fertiliser demand, 3, 30, 31, 32, 370
Fertiliser efficiency, 43, 46, 54, 77, 121, 320, 322, 330, 337, 356, 360, 401
Fertiliser formulation, 5, 35, 51, 64, 82, 120, 316, 369
Fertiliser granules, 10, 84, 204, 205, 313, 319, 320, 322
Fertiliser industry, 5, 14, 17, 34, 35, 111, 138, 224, 233, 386, 446, 447
Fertiliser losses, 364
Fertiliser manufacturing, 28, 29, 33, 50, 51, 54, 60, 73, 74, 108, 111, 112, 117, 120, 128, 132, 159, 161, 171, 185, 192, 196, 200, 253, 260, 369
Fertiliser production, 1, 2, 3, 4, 2, 3, 4, 5, 13, 14, 16, 17, 28, 29, 30, 32, 33, 34, 35, 46, 50, 51, 61, 74, 107, 108, 109, 110, 111, 112, 113, 114, 115, 117, 118, 119, 123, 124, 125, 128, 129, 130, 131, 132, 135, 137, 148, 149, 155, 157, 158, 165, 166, 169, 170, 171, 183, 184, 193, 200, 201, 205, 211, 212, 213, 214, 215, 223, 224, 225, 227, 229, 230, 231, 238, 239, 240, 248, 255, 268, 274, 277, 278, 280, 311, 319, 321, 358, 363, 365, 366, 367, 368, 370, 372, 379, 380, 382, 386, 405, 406, 409, 411, 413, 414, 446
Fertiliser rates, 393, 397
Fertiliser regulation, 5, 377, 378, 379, 380, 381, 382, 383, 384, 385, 386, 399, 427, 449
Fertiliser storage, 209, 214, 433, 434, 435
Fertiliser technology, 9, 10, 64, 65, 80, 104
Fertiliser transport, 165, 436
Fertiliser types, 40, 84, 107, 232, 311, 330, 352, 356, 393, 430, 431, 433, 436, 439, 442
Food security, 3, 2, 5, 7, 8, 11, 14, 30, 31, 32, 35, 36, 50, 64, 74, 108, 133, 277, 375, 406, 454, 455, 459, 460, 463
Fruit quality, 9, 57, 307, 309, 324, 352, 402
Future innovations, 5, 375

G

Geopolitics, 229, 231, 460
Global agriculture, 3, 4, 114, 125, 138, 145
Granulation, 4, 5, 87, 88, 93, 118, 170, 172, 177, 178, 181, 186, 188, 190, 201, 203, 204, 215, 217, 218, 220, 221, 222, 252, 254, 256, 257, 258, 263, 264, 265, 267, 268, 269, 270, 271, 272, 273, 292, 305, 306, 311, 313, 315, 322, 413, 415, 416, 448, 449
Green ammonia, 5, 13, 34, 35, 110, 113, 133, 134, 138, 144, 163, 165, 166, 183, 184, 185, 358, 365, 366, 367, 368, 369, 370, 373, 381, 385, 389, 460, 471
Green revolution, 13, 17
Greenhouse gases, 360, 470
Gypsum, 14, 17, 71, 73, 87, 118, 119, 120, 232, 233, 234, 235, 236, 237, 238, 240, 241, 244, 245, 246, 248, 250, 251, 252, 253, 256, 281, 285, 298, 303, 304, 327, 418, 455, 464

H

Haber–Bosch process, 4, 13, 16, 87, 110, 130, 134, 136, 137, 138, 139, 141, 145, 149, 151, 157, 164, 165, 172, 359, 363, 365, 386, 389, 411
Handling risks, 211, 273
Harvest removal, 8
Heavy metals, 35, 53, 123, 124, 130, 223, 274, 275, 363, 364, 372, 377, 378, 384, 419, 455, 464
Humic substances, 51, 74, 76, 399
Humus, 14, 17, 53, 66
Hydrogen production, 4, 110, 130, 136, 138, 142, 143, 144, 147, 149, 150, 152, 153, 155, 157, 160, 161, 162, 163, 164, 165, 166, 172, 173, 175, 366, 369, 415
Hydrolysis, 10, 360, 399

I

Industrial processes, 2, 5, 46, 51, 81, 84, 107, 119, 123, 131, 191, 242, 278, 415
Industry standards, 380
Inhibitors, 9, 10, 32, 54, 62, 73, 74, 101, 102, 103, 320, 355, 359, 365, 366, 367, 369, 370, 400, 401, 402, 403, 448, 449, 470
Inorganic fertilisers, 3, 13, 79, 81, 84, 85, 86
Integrated nutrient management, 8, 11, 106, 360, 394, 403
Ion exchange, 302, 303
Iron, 15, 26, 27, 28, 30, 48, 54, 55, 59, 62, 67, 69, 71, 72, 73, 83, 96, 105, 110, 115, 116, 117, 136, 137, 141, 143, 148, 150, 151, 154, 157, 158, 160, 163, 166, 167, 274, 275, 323, 345, 411, 454, 467
Irrigation, 13, 20, 54, 70, 72, 83, 85, 93, 94, 95, 97, 101, 103, 308, 309, 317, 323, 325, 326, 328, 330, 332, 341, 342, 343, 346, 352, 354, 355, 359, 361, 362, 366, 367, 373, 374, 443, 457, 469, 470
Irrigation systems, 54, 83, 93, 94, 95, 97, 101, 323, 328, 330, 342

L

Land degradation, 30, 124
Leaching, 5, 9, 10, 34, 40, 43, 46, 47, 49, 50, 52, 62, 63, 67, 71, 74, 76, 83, 84, 85, 92, 93, 95, 101, 102, 104, 241, 242, 243, 289, 292, 299, 301, 305, 309, 317, 319, 320, 324, 333, 344, 348, 349, 350, 351, 353, 354, 355, 361, 362, 364, 373, 374, 384, 391, 395,397, 400, 401, 402, 403, 404, 411, 419, 429, 442, 464
Light-dependent reactions, 21, 23
Lime, 12, 69, 72, 73, 90, 389, 455
Liquid fertilisers, 83, 93, 96, 97, 106, 335, 373, 374, 390, 440
Logistics, 132, 166, 217, 218, 387, 389, 390, 391, 450, 471

M

Macronutrients, 4, 5, 6, 7, 18, 20, 28, 54, 79, 81, 82, 125, 277, 373
Magnesium, 22, 25, 28, 29, 44, 54, 55, 58, 59, 62, 69, 70, 83, 284, 285, 287, 290, 291, 292, 299, 300, 301, 323, 324, 325, 327, 345, 371, 421, 423
Manganese, 27, 54, 69, 71, 72, 83, 96
Market dynamics, 5, 31, 308, 376, 386, 388, 389, 390, 391, 392
Microbial activity, 41, 42, 52, 53, 65, 67, 71, 72, 75, 76, 77, 84, 91, 100, 106, 320, 322, 353, 373, 400, 403
Microbial decomposition, 81
Micronutrients, 3, 5, 18, 26, 28, 30, 44, 52, 54, 62, 69, 71, 72, 73, 74, 75, 80, 82, 83, 96, 97, 100, 105, 321, 323, 325, 327, 345, 346, 401, 402, 448, 449, 467
Mineral fertilisers, 84, 101, 469
Mineralisation, 47, 75, 76, 84
Moisture content, 87, 117, 204, 205, 229, 238, 265, 292, 299, 301, 312, 322
Molybdenum, 27, 83
Monocropping, 8

N

Natural gas, 3, 30, 33, 34, 54, 87, 107, 109, 110, 111, 112, 113, 121, 127, 128, 130, 131, 132, 136, 138, 142, 144, 145, 147, 149, 153, 159, 162, 164, 166, 167, 365, 366, 369, 386, 388, 389, 391, 411, 413, 414, 415, 459
Nickel, 27, 145, 147, 148, 151, 152, 153, 154, 157, 158, 163
Nitrate, 10, 12, 17, 27, 32, 46, 47, 51, 61, 63, 64, 73, 75, 77, 84, 85, 92, 100, 102, 104, 188, 190, 191, 195, 200, 210, 213, 232, 279, 307, 308, 309, 320, 321, 323, 325, 327, 328, 345, 346, 351, 353, 354, 355, 359, 361, 362, 369, 378, 400, 401, 403, 410, 419, 470

Nitrate leaching, 10, 32, 64, 77, 92, 100, 104, 321, 362, 369, 470
Nitric acid, 4, 16, 17, 88, 131, 170, 188, 189, 190, 193, 194, 195, 196, 197, 198, 199, 200, 211, 212, 232, 233, 410, 412, 416, 460
Nitrification, 9, 10, 46, 62, 74, 75, 101, 102, 191, 320, 355, 359, 366, 367, 368, 369, 400, 401, 402, 403, 470
Nitrogen, 3, 4, 6, 7, 9, 10, 12, 13, 14, 15, 16, 17, 20, 22, 24, 26, 27, 28, 29, 30, 31, 32, 33, 34, 41, 44, 46, 47, 49, 50, 51, 52, 54, 55, 56, 61, 63, 65, 69, 73, 74, 75, 77, 81, 82, 84, 85, 86, 87, 88, 89, 91, 92, 93, 96, 98, 99, 100, 101, 102, 103, 104, 105, 107, 109, 110, 111, 114, 123, 125, 127, 128, 130, 131, 132, 136, 137, 138, 139, 140, 141, 143, 144, 145, 146, 147, 149, 151, 152, 153, 156, 157, 158, 159, 162, 165, 166, 169, 170, 171, 172, 173, 183, 184, 187, 188, 189, 191, 192, 193, 194, 196, 197, 198, 200, 205, 211, 212, 213, 214, 215, 232, 262, 267, 279, 312, 313, 314, 315, 316, 317, 318, 319, 320, 321, 323, 324, 325, 326, 327, 328, 333, 336, 339, 344, 345, 346, 347, 349, 350, 351, 352, 353, 354, 355, 359, 360, 361, 363, 365, 366, 368, 369, 371, 373, 377, 378, 380, 382, 385, 386, 388, 391, 394, 396, 397, 398, 401, 402, 403, 404, 409, 411, 412, 414, 415, 416, 417, 418, 419, 424, 425, 427, 428, 429, 448, 449, 454, 455, 456, 457, 458, 459, 468, 469, 470
Nitrogen cycle, 15, 46, 47, 50
Nitrogen fixation, 16, 41, 46, 98, 99, 100, 165
Nitrogen volatilisation, 339
Nitrous oxide, 11, 32, 34, 64, 77, 104, 357, 359, 366, 401, 403, 411, 454, 457, 469, 470
Nucleic acids, 20, 24
NUE, 9, 10, 11, 12, 29, 35, 50, 54, 61, 62, 64, 65, 74, 76, 91, 104, 169, 353, 362, 366, 367, 368, 401, 404, 405, 406, 457
Nutrient availability, 3, 5, 7, 10, 15, 28, 36, 37, 39, 40, 41, 42, 44, 46, 51, 52, 53, 60, 64, 68, 69, 71, 72, 73, 74, 75, 77, 79, 93, 100, 104, 191, 233, 323, 330, 343, 346, 352, 353, 354, 373, 396, 400, 402, 403, 405, 411
Nutrient balance, 5, 27, 29, 70, 429
Nutrient cycling, 11, 37, 41, 43, 45, 53, 62, 66, 68, 75, 77, 79, 100, 101, 361, 372, 373, 396, 405, 454, 458, 470
Nutrient deficiency, 24, 60, 61, 86, 318, 324, 347, 395
Nutrient demand, 17, 82, 323, 325, 326, 341, 350, 351, 402
Nutrient losses, 46, 53, 62, 63, 74, 77, 81, 83, 101, 317, 319, 321, 324, 333, 337, 345, 348, 350, 354, 356, 362, 369, 372, 373, 374, 377, 380, 381, 382, 389, 394, 395, 398, 401, 402, 403, 442
Nutrient mobility, 29, 46, 330, 344
Nutrient ratios, 312, 316, 317, 323, 325
Nutrient recovery, 32, 371, 372, 455
Nutrient replenishment, 8
Nutrient solution, 341, 342, 346, 369, 384
Nutrient uptake, 5, 18, 28, 29, 40, 41, 43, 46, 63, 67, 69, 72, 74, 75, 83, 93, 100, 104, 106, 307, 327, 341, 350, 352, 354, 369, 393, 397, 398
Nutrient use efficiency, 3, 9, 13, 29, 31, 32, 35, 37, 40, 50, 53, 54, 62, 64, 68, 72, 74, 76, 80, 83, 91, 92, 95, 105, 312, 316, 324, 326, 328, 337, 347, 352, 353, 362, 369, 393, 396, 400, 401, 404, 405, 420, 456

O

Organic amendments, 11, 64, 75, 76, 78, 101, 399
Organic fertilisers, 81, 84, 86, 91, 378, 391, 446, 448
Osmotic regulation, 27
Ostwald process, 4, 17, 188, 189, 194, 195, 196, 199, 200, 412, 416
Oxidation, 189, 196, 197, 198, 199, 412

P

pH, 3, 28, 36, 44, 53, 55, 59, 61, 62, 63, 69, 71, 72, 73, 74, 76, 77, 78, 85, 96, 101, 105, 106, 117, 190, 191, 192, 193, 194, 240, 268, 273, 309, 323, 324, 327, 343, 349, 353, 355, 369, 426, 466
Phloem, 18, 19, 20, 29
Phosphate rock, 3, 4, 13, 15, 34, 51, 54, 64, 87, 88, 106, 107, 114, 115, 116, 117, 118, 119, 120, 121, 122, 123, 124, 126, 127, 128, 129, 130, 131, 132, 223, 225, 226, 227, 229, 230, 231, 232, 233, 234, 235, 236, 239, 244, 245, 246, 247, 249, 251, 252, 253, 255, 257, 259, 260, 261, 262, 274, 275, 372, 386, 389, 413, 414, 416, 459, 463, 464
Phosphates, 29, 122, 138, 228, 232, 243, 274, 275, 342, 346, 462, 464

Phosphogypsum, 4, 34, 119, 121, 124, 129, 130, 223, 238, 240, 241, 245, 249, 274, 303, 363, 455, 456, 463, 464, 465
Phosphoric acid, 34, 51, 88, 108, 118, 119, 120, 128, 129, 130, 131, 132, 138, 223, 232, 233, 234, 235, 236, 237, 238, 239, 244, 245, 246, 247, 248, 249, 250, 252, 253, 254, 255, 262, 263, 266, 267, 268, 269, 270, 271, 273, 274, 275, 325, 345, 363, 416, 456, 463, 464
Phosphorus, 3, 4, 6, 7, 9, 12, 15, 17, 20, 22, 23, 24, 28, 29, 30, 31, 33, 34, 44, 46, 47, 48, 49, 50, 51, 52, 54, 55, 56, 61, 63, 65, 69, 71, 73, 74, 75, 76, 77, 81, 82, 84, 85, 86, 87, 88, 89, 93, 100, 105, 107, 114, 115, 116, 117, 118, 119, 120, 121, 122, 123, 124, 125, 127, 128, 129, 130, 131, 138, 225, 231, 232, 233, 251, 252, 253, 262, 267, 279, 312, 313, 314, 315, 316, 317, 318, 319, 321, 323, 324, 325, 327, 328, 337, 338, 345, 346, 349, 350, 351, 352, 353, 361, 362, 364, 371, 377, 381, 386, 394, 396, 397, 398, 399, 404, 405, 413, 417, 418, 420, 421, 422, 425, 429, 455, 458, 463, 470, 471
Phosphorus cycle, 47, 48, 51
Photosynthesis, 4, 5, 7, 18, 19, 20, 21, 22, 23, 25, 27, 29, 33, 52, 53, 55, 56, 58, 59, 82, 279, 293, 295, 350
Plant anatomy, 4, 18, 20, 28
Plant growth, 2, 4, 5, 7, 10, 13, 15, 18, 20, 22, 23, 27, 28, 29, 30, 33, 37, 38, 41, 43, 44, 45, 46, 47, 51, 53, 54, 55, 61, 63, 65, 66, 67, 68, 69, 70, 71, 79, 81, 83, 84, 98, 100, 106, 125, 200, 225, 242, 277, 279, 294, 295, 312, 313, 316, 318, 322, 371, 396, 399, 405, 443
Plant metabolism, 26
Plant nutrition, 3, 2, 4, 6, 13, 14, 15, 17, 22, 28, 36, 37, 52, 124, 396, 420, 422, 454, 465, 470, 471
Plant physiology, 4, 12, 277, 279, 306
Pollination, 27, 60
Pollutants, 200, 416, 469
Polymer coatings, 320
Porosity, 45, 52
Potash, 3, 4, 13, 33, 34, 54, 107, 125, 126, 127, 128, 131, 132, 277, 278, 279, 280, 281, 282, 283, 284, 285, 286, 288, 289, 290, 291, 292, 293, 294, 298, 299, 300, 302, 304, 307, 308, 309, 313, 314, 316, 317, 323, 345, 363, 386, 388, 389, 391, 413, 420, 448, 449, 465, 466
Potash mining, 3, 107, 125, 282
Potassium, 3, 4, 6, 7, 9, 12, 15, 20, 22, 24, 25, 28, 29, 30, 31, 33, 44, 46, 48, 49, 50, 51, 52, 53, 54, 55, 57, 61, 70, 72, 73, 81, 82, 83, 84, 85, 86, 87, 88, 89, 93, 105, 107, 125, 126, 127, 128, 131, 141, 232, 277, 278, 279, 280, 281, 282, 284, 285, 287, 288, 289, 290, 291, 292, 293, 294, 295, 296, 297, 298, 299, 300, 301, 302, 303, 304, 305, 306, 307, 308, 309, 312, 313, 314, 315, 316, 317, 318, 319, 321, 323, 324, 325, 326, 327, 328, 339, 345, 346, 349, 350, 351, 352, 353, 354, 369, 371, 377, 386, 388, 391, 413, 417, 420, 421, 422, 425, 448, 456, 459, 467
Potassium chloride, 51, 73, 81, 85, 86, 87, 88, 89, 93, 125, 127, 279, 290, 291, 292, 293, 294, 295, 296, 297, 298, 302, 303, 304, 305
Potassium cycle, 48, 49, 51
Potassium sulphate, 51, 73, 88, 89, 127, 279, 294, 295, 296, 297, 298, 299, 300, 301, 302, 303, 304, 305
Precision agriculture, 5, 6, 9, 13, 30, 31, 32, 35, 97, 315, 319, 322, 324, 326, 335, 343, 346, 347, 348, 349, 350, 353, 360, 363, 366, 371, 394, 401, 402, 404, 406, 447, 457, 468, 469
Prilling, 4, 87, 93, 170, 172, 177, 178, 181, 186, 188, 190, 201, 202, 203, 215, 217, 218, 220, 222, 415, 416
Processing, 3, 4, 9, 13, 15, 33, 35, 105, 107, 109, 115, 116, 118, 119, 121, 122, 123, 124, 125, 126, 127, 128, 130, 131, 145, 163, 172, 190, 211, 223, 226, 227, 229, 230, 231, 237, 238, 240, 241, 243, 251, 254, 269, 270, 274, 275, 278, 280, 281, 283, 284, 285, 287,288, 290, 291, 292, 293, 298, 299, 300, 301, 303, 304, 305, 306, 307, 309, 315, 363, 364, 372, 414, 416, 457, 463, 464, 468
Product quality, 117, 156, 158, 193, 202, 203, 205, 215, 216, 220, 221, 225, 227, 238, 246, 250, 254, 268, 279, 295, 319, 328, 350, 377, 384, 391, 433
Production systems, 6, 11, 56, 68, 93, 147, 165, 291, 307, 312, 319, 330, 341, 346, 360, 369, 389
Protein synthesis, 27, 295, 307, 309
Proteins, 6, 20, 23, 26, 47, 82, 85, 98, 279

Q

Quality control, 5, 77, 87, 90, 117, 241, 312, 321, 322, 383, 385, 448, 449

R

Raw materials, 3, 17, 34, 54, 87, 88, 106, 107, 109, 114, 131, 132, 253, 260, 261, 269, 272, 273, 278, 293, 304, 305, 315, 317, 326, 377, 384, 386, 389, 409, 414, 416, 446, 448, 449, 450, 451, 459, 464
Reactor design, 4, 155, 156, 157, 158, 159, 161
Recycling nutrients, 35, 54, 133
Release kinetics, 64
Renewable energy, 34, 110, 113, 133, 138, 144, 149, 163, 165, 169, 365, 366, 367, 368, 369, 370, 461
Resource efficiency, 106, 173, 224, 243, 250, 357, 358, 363, 371, 381, 392, 404
Respiration, 22, 23, 27, 41, 45, 359
Root development, 4, 6, 7, 23, 40, 41, 43, 46, 48, 52, 53, 55, 63, 68, 72, 82, 100, 101, 106, 225, 280, 308, 313, 316, 318, 325, 350, 351, 397, 403
Root hairs, 18, 19, 29
Runoff, 10, 11, 13, 31, 34, 40, 48, 50, 51, 63, 64, 71, 74, 85, 92, 93, 95, 317, 324, 333, 348, 353, 355, 361, 362, 364, 374, 380, 384, 393, 395, 402, 403, 419, 427, 429, 442, 443, 444, 457

S

Salinity, 3, 36, 53, 69, 70, 71, 72, 73, 74, 106, 285, 306, 309, 323, 397, 463
Secondary nutrients, 5, 18, 25, 28, 54, 79, 82, 323, 325, 327, 345
Seed development, 9, 27
SMR, 4, 109, 136, 142, 145, 147, 148, 149, 159, 166, 175
Soil aeration, 45
Soil biology, 3, 75, 76, 77, 397, 398, 399
Soil chemistry, 14, 37, 61, 69, 73, 85, 323, 325
Soil compaction, 45
Soil composition, 3, 38, 44, 45, 79
Soil degradation, 31, 32, 86, 317, 377
Soil fertility, 3, 6, 8, 11, 12, 13, 14, 37, 38, 41, 46, 50, 51, 53, 54, 55, 60, 61, 62, 63, 64, 65, 66, 67, 79, 84, 86, 100, 101, 107, 200, 277, 393, 458, 470
Soil health, 3, 8, 11, 13, 51, 54, 62, 64, 68, 74, 75, 79, 84, 86, 98, 100, 242, 277, 360, 364, 370, 371, 372, 373, 374, 378, 379, 390, 391, 396, 404, 405, 406, 429, 447, 456, 458
Soil horizons, 67
Soil management, 28, 35, 44, 60, 74, 86, 279, 470
Soil microorganisms, 18, 41, 71, 76, 84, 91
Soil moisture, 93, 106, 321, 347, 353, 359, 395, 405
Soil organic matter, 11, 53, 65, 75, 77, 84
Soil phases, 44, 46
Soil profile, 3, 65, 66, 67, 68, 69, 76, 354, 361
Soil structure, 41, 42, 43, 52, 53, 66, 70, 73, 75, 76, 81, 84, 101, 241, 362, 371, 373, 391
Soil testing, 63, 316, 323, 328, 394, 443, 448
Soil texture, 11, 36, 39, 40, 42, 68, 71
Soil texture triangle, 39, 40
Solubility, 51, 61, 69, 71, 73, 74, 77, 91, 104, 105, 127, 233, 281, 285, 289, 295, 299, 300, 303, 304, 317, 325, 342, 344, 400, 411, 424, 426
Solution mining, 4, 126, 127, 277, 281, 283, 285, 286, 287, 288, 289, 291, 294, 304
Stabilised fertilisers, 10, 35, 64, 317, 320, 365, 366, 367, 368, 369, 400, 401, 402, 403, 448
Steam methane reforming, 4, 34, 109, 136, 138, 142, 144, 145, 147, 148, 149, 150, 155, 156, 157, 159, 164, 166, 172, 411
Stomata, 19, 21
Storage systems, 206, 251, 298, 322, 444
Sulphur, 3, 10, 25, 26, 28, 33, 54, 55, 69, 73, 75, 83, 91, 108, 109, 118, 119, 120, 128, 130, 131, 136, 142, 147, 152, 153, 157, 164, 166, 191, 193, 194, 232, 241, 251, 253, 294, 295, 307, 308, 309, 316, 320, 323, 324, 325, 327, 328, 400, 417, 418, 421, 423, 454
Supply chains, 4, 17, 31, 54, 107, 108, 112, 113, 124, 128, 131, 132, 180, 229, 230, 253, 254, 363, 364, 366, 381, 383, 386, 388, 389, 390, 391, 392, 395, 404, 460
Surface application, 96, 339
Sustainability, 3, 5, 3, 5, 11, 12, 13, 14, 31, 32, 34, 35, 74, 78, 86, 98, 108, 124, 134, 144, 149, 169, 200, 223, 304, 309, 312, 324, 357, 359, 360, 362, 363, 364, 369, 370, 372, 374, 375, 379, 380, 381, 382, 383, 384, 386, 389, 390, 391, 392, 393, 394, 395, 396, 399,404, 406, 411, 414, 417, 446, 447, 452, 454, 455, 457, 461, 463, 470, 471, 472
Sustainable agriculture, 9, 35, 51, 62, 67, 78, 92, 321, 372, 382, 390, 456, 457, 458, 463
Sustainable intensification, 5, 64, 101, 368, 370, 376, 404, 406
Synthetic fertilisers, 11, 53, 84, 101, 364, 370, 371, 372, 373, 391, 398

T

Technological innovation, 134, 375
Timing of application, 406
Trace elements, 4, 26, 83, 274, 275, 324, 327

U

Uptake efficiency, 10, 29, 63, 320, 330, 350, 351, 353, 385, 403
Urea, 4, 10, 13, 33, 53, 61, 63, 64, 73, 81, 85, 86, 87, 89, 91, 93, 102, 103, 110, 132, 136, 137, 143, 145, 169, 170, 172, 173, 174, 175, 176, 177, 178, 179, 180, 181, 183, 184, 185, 186, 187, 188, 191, 192, 201, 202, 204, 205, 206, 211, 215, 216, 217, 218, 219, 220, 221, 313, 314, 317, 320, 325, 327, 328, 344, 345, 346, 350, 351, 352, 353, 355, 360, 368, 369, 386, 390, 400, 401, 402, 403, 411, 412, 413, 414, 415, 418, 448, 449, 455, 457, 461, 462, 470
Urea production, 4, 170, 172, 173, 174, 179, 180, 183, 184, 185, 187, 192, 412, 414, 415
Urease inhibitors, 101, 102, 103, 320, 355, 366, 367, 368, 400, 401, 402, 469, 470

V

Vacuole, 20
Variable rate technology, 5, 346, 347, 349, 353, 393, 394, 395, 467, 468
Volatilisation, 5, 9, 10, 34, 47, 50, 62, 63, 76, 83, 86, 92, 96, 101, 102, 103, 104, 191, 268, 317, 319, 320, 321, 324, 333, 344, 348, 349, 350, 351, 352, 354, 355, 366, 368, 400, 401, 402, 403, 404, 411
VRT, 5, 346, 347, 348, 349, 350, 353, 393, 394, 395, 468

W

Waste management, 5, 132, 243, 357, 363, 364
Water management, 97, 354
Water pollution, 5, 10, 13, 31, 34, 64, 86, 357, 360
Water quality, 32, 95, 324, 361, 362, 363, 377, 380, 404, 406, 442, 469
Water retention, 38, 39, 40, 41, 45, 66, 67, 373
Water-use efficiency, 24, 53, 55
Weathering, 38, 48, 49, 65, 67, 68, 463
Wet-process phosphoric acid, 4, 129, 130, 223, 233, 234, 236, 237, 238, 239, 241, 244, 247, 248, 253, 262

X

Xylem, 18, 19, 20, 29

Z

Zinc, 27, 28, 30, 54, 55, 59, 60, 62, 69, 71, 72, 73, 83, 96, 105, 147, 153, 323, 345, 454